Wan Lei • Anthony C. K. Soong • Liu Jianghua
Wu Yong • Brian Classon • Weimin Xiao
David Mazzarese • Zhao Yang • Tony Saboorian

5G System Design

An End to End Perspective

 Springer

Wan Lei
Huawei Technologies
Beijing, China

Anthony C. K. Soong
Huawei Technologies, USA
Plano, TX, USA

Liu Jianghua
Huawei Technologies
Beijing, China

Wu Yong
Huawei Technologies
Beijing, China

Brian Classon
Huawei Technologies, USA
Rolling Meadows, IL, USA

Weimin Xiao
Huawei Technologies, USA
Rolling Meadows, IL, USA

David Mazzarese
Huawei Technologies
Beijing, China

Zhao Yang
Huawei Technologies
Shanghai, China

Tony Saboorian
Huawei Technologies, USA
Plano, TX, USA

ISBN 978-3-030-22238-3   ISBN 978-3-030-22236-9   (eBook)
https://doi.org/10.1007/978-3-030-22236-9

This Springer imprint is published by the registered company Springer Nature Switzerland AG
The registered company address is: Gewerbestrasse 11, 6330 Cham, Switzerland

In memory of Kim Chang, our colleague who passed away at a standards meeting in Japan.

Preface

It is hard to overstate the impact that HSPA and LTE technology have had on our global society. Mobile subscriptions for these technologies are now counted in the billions; touching lives and changing business everywhere on the planet. How did this come to pass? The dominance of these Third Generation Partnership (3GPP) technologies was not known a priori when their studies were first approved. There are many factors to explain the success, but surely these factors include the quality of the standards and technology as well as economies of scale from global deployment. For 5G, the fifth generation of telecommunication systems, the situation is different in that it is expected, even before the first equipment delivery, that 3GPP technology will dominate future global deployments. The vision for 5G also goes beyond traditional mobile broadband services. In 2015, ITU-R (International Telecommunication Union—Radio communications sector) established the 5G requirements for IMT-2020 (International Mobile Telecommunication system—2020), targeting diverse requirements from three key usage scenarios: enhanced mobile broadband (eMBB), massive machine type communication (mMTC), and ultra-reliable and low latency communication (URLLC).

Mobile broadband services like web browsing, social apps with text messaging, file sharing, music downloading, video streaming, and so on are already very popular and supported by 4G communication systems. In the 5G era, these and other applications such as ultra-high definition (UHD) video, 3D video, and augmented reality (AR), and virtual reality (VR) will be better served with data rates up to hundreds of megabits per second or even gigabits per second. In addition, the demand of uplink high data rate service is also emerging, for example with HD video sharing. These service requirements, together with the anywhere any-time experience requirements with high user density and user mobility, define new limits for the eMBB scenario.

In the future, any object that can benefit from being connected will be connected, either partially or dominantly, through wireless technologies. This trend poses a huge demand for connecting objects/machines/things in a wide range of applications. Driverless cars, enhanced mobile cloud services, real-time traffic control optimization, emergency and disaster response, smart grid, e-health and industrial communications, to name just a few, are all expected to be enabled or improved by wireless connectivity. By a closer observation of these applications, there are two major characteristics of these services: one is the number of desired connections; the other is the requested reliability within a given latency budget. These two significant characteristics drive the definitions of the mMTC and URLLC scenarios.

Accordingly ITU-R defines the IMT-2020 requirements for all the above potential use cases, including eight key capabilities: 20 Gbps peak rate, 100 Mbps user perceived data rate at cell edge, 3 times spectrum efficiency of IMT-Advanced, mobility up to 500 km/h, low latency with less than 1 ms air-interface round-trip time (RTT), connectivity density of 10M connections per square kilometer, 100 times energy efficiency of IMT-Advanced, and traffic density with 10 Mbps per square meter. It is anticipated that the area traffic capacity is predicted to be increased by at least 100 times in 5G, with available bandwidth increased by 10 times. Altogether, the 5G system can provide a basic information infrastructure to be used by both people and machines for all of these applications similar to the transportation system and electric power system infrastructures that we use now.

Each region in the world is planning their 5G spectrum. In Europe, multiple countries have allocated 5G spectrum mainly in C-band and released 5G operational licenses among mobile network operators. The UK has already auctioned 190 MHz sub-6 GHz spectrum for 5G deployment, with another 340 MHz low frequency spectrum auction ongoing. In Asia, China has newly allocated 390 MHz bandwidth in 2.6 GHz, 3.5 GHz, and 4.9 GHz frequency bands among the three major operators for 5G deployment by 2020; Japan has allocated totally 2.2 GHz 5G spectrum including 600 MHz in C-band and 1.6 GHz in 28 GHz mm-wave spectrum among four mobile carriers, with 5G investment around JPY 1.6 trillion ($14.4 billion) over the next 5 years; South Korea has auctioned 280 MHz bandwidth in 3.5 GHz and 2.4 GHz bandwidth in 28 GHz spectrum for 5G network among the three telco carriers, with SKT already released the first 5G commercial services since April 3rd, 2019. In the USA, 5G licenses are permitted in the existing 600 MHz, 2.6 GHz, and mm-wave frequency bands of 28 GHz and 38 GHz, with additional 3 mm-wave spectrum auctions in 2019 on 28 GHz, 24 GHz, as well as higher mm-wave spectrum at 37 GHz, 39 GHz, and 47 GHz. Europe, Asia, and North America have all announced early 5G network deployment by 2020.

This treatise elaborates on the 5G specifications of both the 5G new radio (5G-NR) and 5G new core (5G-NC) and provides a whole picture on 5G end-to-end system and key features. Additionally, this book provides the side-by-side comparison between 5G-NR and Long-Term Evolution (LTE, also called as 4G) to address the similarities and the differences, which benefits those readers who are familiar with LTE system. 3GPP Release 15, i.e., the first release of 5G standard has completed the standardization of both 5G non-standalone (NSA) and standalone (SA) architecture. 5G deployment will eventually go to SA deployment based on 5G-new carrier (NC) with advanced core network features as slicing, MEC, and so on. For some operators, however, due to their business case balance between the significant investments and quick deployment, may consider NSA deployment from the beginning, i.e., with a primary connection to LTE and a secondary connection to NR. In addition to the network architecture, NR has built-in provisions in configuration and operation for coexistence with LTE. These include same-band (and even same channel) deployment of NR and LTE in low-band spectrum. An especially important use case is a higher frequency NR TDD deployment that, for coverage reasons, includes a supplemental uplink (SUL) carrier placed in an existing LTE band.

The book is structured into six main chapters. The first chapter looks at the use cases, requirements, and standardization organization and activities for 5G. These are 5G requirements and not NR requirements, as any technology that meets the requirements may be submitted to the ITU as 5G technology including a set of Radio Access Technologies (RATs) consisting of NR and LTE; with each RAT meeting different aspects of the requirements. A second chapter describes, in detail, the air interface of NR and LTE side by side. The basic aspects of LTE that NR builds upon are first described, followed by sections on the NR specific technologies such as carrier/channel, spectrum/duplexing (including SUL), LTE/NR co-existence, and new physical layer technologies (including waveform, Polar/LDPC channel coding, MIMO, and URLLC/mMTC). In all cases, the enhancements made relative to LTE are made apparent. The third chapter contains description of NR procedures (IAM/Beam Management/Power control/HARQ), protocols (CP/UP/mobility, including grant-free), and RAN architecture. The fourth chapter has a detailed discussion related to end-to-end system architecture, and the 5G Core (5GC), network slicing, service continuity, relation to EPC, network virtualization, and edge computing. The fifth chapter describes the ITU submission and how NR and LTE meet the 5G requirements in significant detail, from the rapporteur responsible for leading the preparation and evaluation. Finally, the book concludes with a look at the 5G market and the future.

Beijing, China   Wan Lei
Plano, TX, USA   Anthony C. K. Soong
Beijing, China   Liu Jianghua
Beijing, China   Wu Yong
Rolling Meadows, IL, USA   Brian Classon
Rolling Meadows, IL, USA   Weimin Xiao
Beijing, China   David Mazzarese
Shanghai, China   Zhao Yang
Plano, TX, USA   Tony Saboorian

Acknowledgement

The authors would like to thank our colleagues from all over the world that participated in the different standardization and industrial fora for 5G development. It is through our collective hard labor in harmonization from which 5G was born. In addition, the efforts made by operators from all the three regions are appreciated very much, by identifying focused use cases, key features, and solutions, as well as prioritizing architecture option, spectrum bands, and terminal configurations, especially for the early deployment. Without those support, 3GPP could hardly complete the 5G standardization within such a short time. The first release of 5G standardization was done within 3 years, including study item and work item, which breaks the record in the history of 3GPP. Such an astonishing speed is a testimony to the joint cooperation and collaboration from all the members in the ecosystem; including operators, network vendors, devices, and chipset vendors. All from the industry are expecting 5G, and all contributed to 5G standardization.

We also thank the editorial and publication staff at Springer Natural for their support of this manuscript; chief among them our editor Susan Lagerstrom-Fife, editorial assistance Karin Pugliese and production project manager Mohanarangan Gomathi.

Most importantly, we thank the support of our family members for putting up with the many days that we were away from home at the various meetings of the standard bodies and industry fora.

Acronyms

3GPP	3rd Generation Partnership Project
5GAA	5G automotive association
5GACIA	5G alliance for connected industries and automation
5GC	5G core network
AMBR	Aggregate maximum bit rate
AMC	Adaptive modulation and coding
AMF	Access and mobility management function
ARP	Allocation and retention priority
AS	Application server
AUSF	Authentication server function
BLER	Block error rate
BWP	Bandwidth part
CA	Carrier aggregation
CBG	Code block group
CBGTI	CBG transmission information
CC	Component carrier
CCE	Control channel element
CDF	Cumulative distribution function
CM	Cubic metric
CMP	Cubic metric preserving
CORESET	Control resource set
CP	Cyclic prefix
CQI	Channel quality indication
CRB	Common resource block
CRI	CSI-RS resource indicator
CRS	Cell-specific reference signal
CSI	Channel State Information
CSI-IM	Channel state information-interference measurement
CSI-RS	CSI reference signal
D2D	Device to device
DC	Dual connectivity

DCI	Downlink control information
DCN	Dedicated core network
DM-RS	UE-specific reference signal (also known as "demodulation reference signal")
DN	Data network
DRS	Discovery reference signal
EC-GSM	Extended coverage for GSM
EDT	Early data transmission
eIMTA	Enhanced interference management traffic adaption
eMBB	Enhanced mobile broadband
eMTC	Enhanced machine type communication
EN-DC	E-UTRA-NR dual connectivity
EPDCCH	Enhanced physical downlink control channel
FDD	Frequency division duplex
FDMA	Frequency division multiple access
FR	Frequency range
FSTD	Frequency switching transmit diversity
GBR	Guaranteed bit rate
GoS	Grade of service
GSCN	Global synchronization channel number
GUTI	Globally unique temporary identifier
IAM	Initial access and mobility
IMT	International mobile technology
IoT	Internet of things
ITU	International telecommunication union
ITU-R	ITU-Radiocommunication sector
LAA	License assisted access
LBRM	Limited buffer rate-matching
LI	Layer indicator
LPWA	Low power wide area
LTE	Long-term evolution
MBMS	Multimedia broadcast multicast services
MBSFN	MBMS Single-Frequency Network
MCL	Maximum coupling loss
MCS	Modulation coding scheme
MIB	Master information block
MIMO	Multiple input multiple output
MME	Mobility management entity
mMTC	Massive machine type communication
MSD	Maximum sensitivity deduction
MTC	Machine type communication
NAS	Non-Access Stratum
NB-CIoT	NarrowBand cellular IoT
NB-IoT	Narrow Band-Internet of Things
NB-M2M	Narrow Band M2M

NE-DC	NR-E-UTRA dual Connectivity
NGEN-DC	NG-RAN E-UTRA-NR dual Connectivity
NGMN	Next generation mobile networks
NR	New radio
NSA	Non-standalone
NSSAI	Network slice selection assistance information
NZP	Non-zero power
OCC	Orthogonal cover code
OFDM	Orthogonal Frequency-Division Multiplexing
OFDMA	Orthogonal frequency division multiple access
PAPR	Peak to average power ratio
PBCH	Physical broadcast channel
PCC	Primary component carrier
PCF	Policy control function
PCFICH	Physical control format indicator channel
PDCCH	Physical downlink control channel
PDCP	Packet data convergence protocol
PDSCH	Physical downlink shared channel
PF	Paging frame
PGW	Packet gateway
PHICH	Physical hybrid ARQ indicator channel
PLMN	Public land mobile network
PMCH	Physical multicast channel
PMI	Precoding matrix indicator
PO	Paging occasion
PRACH	Physical random access channel
PRB	Physical resource block
PRG	Precoding resource block groups
PSM	Power saving mode
PSS	Primary synchronization signal
PT-RS	Phase tracking reference signal
PUCCH	Physical uplink control channel
PUSCH	Physical uplink shared channel
QCL	Quasi co-location
QoS	Quality of service
QRO	Quasi-Row Orthogonal
RAN	Radio access network
RAR	Random access response
RB	Resource block
REG	Resource element group
RI	Rank indication
RMSI	Remaining master system information
RRC	Radio resource control
RRM	Radio resource management
RS	Reference signal

RSFP	RAT frequency selection priority
RSRP	Reference signal received power
RSRQ	Reference signal received quality
RTT	Round trip time
SBA	Service-based architecture
SCC	Secondary component carrier
SC-FDMA	Single carrier-frequency division multiplexing access
SC-PTM	Single-cell point to multipoint transmission
SCS	Sub-carrier spacing
SD	Slice differentiator
SDAP	Service data adaptation protocol
SDL	Supplementary downlink
SEPP	Edge protection proxies
SFBC	Space frequency block coding
SFN	System Frame Number
SGW	Serving Gateway
SI	Study item
SIB	System information block
SINR	Signal to interference plus noise ratio
SLA	Service level agreement
SMF	Session management function
SMS	Short message service
SR	Scheduling request
SRI	SRS resource indicator
SRS	Sounding reference signal
SSB	Synchronization signal/PBCH block
SSBRI	SSB resource indicator
SSS	Secondary synchronization signal
SST	Slice/service type
SUL	Supplementary uplink
SUPI	Subscription permanent identifier
TBCC	Tail-biting convolutional code
TBS	Transport block size
TCI	Transmission configuration indicator
TDD	Time division duplex
TDMA	Time division multiple access
telco	Telephone company
TM	Transmission mode
TPMI	Transmit precoding matrix indicator
TRI	Transmit rank indicator
TRS	Tracking reference signal
TS	Time slot
TTI	Transmission time interval
UCI	Uplink control information
UDM	Unified data management

UPF	User plane function
URLLC	Ultra-reliable and low latency communication
URSP	UE route selection policy
V2X	Vehicle to Everything
WRC	World radiocommunication conference
WUS	Wake up signal
ZP	Zero power

Contents

Contributors

Wan Lei Huawei Technologies, Beijing, China

Anthony C. K. Soong Huawei Technologies, USA, Plano, TX, USA

Liu Jianghua Huawei Technologies, Beijing, China

Wu Yong Huawei Technologies, Beijing, China

Brian Classon Huawei Technologies, USA, Rolling Meadows, IL, USA

Weimin Xiao Huawei Technologies, USA, Rolling Meadows, IL, USA

David Mazzarese Huawei Technologies, Beijing, China

Zhao Yang Huawei Technologies, Shanghai, China

Tony Saboorian Huawei Technologies, USA, Plano, TX, USA

John Kaippallimalil Huawei Technologies, USA, Plano, TX, USA

Amanda Xiang Huawei Technologies, USA, Plano, TX, USA

Chapter 1
From 4G to 5G: Use Cases and Requirements

This chapter investigates the motivations and driving forces of 5G development as well as introduces the 5G use cases and technical requirements. 5G is the first generation that devotes itself to connecting both humans and machines. Accordingly, the service requirements and the technical performance requirements are extended from mobile broadband (MBB) to the new use cases. The diverse requirements pose significant challenges to system design.

This chapter also presents how 5G development is made based on industry collaboration, where ITU-R and 3GPP play the central role in this process. ITU-R procedure on IMT-2020 development is introduced, and 3GPP 5G standardization process is reviewed. With the guidance of ITU-R and the well-harmonized technical development in 3GPP, 5G technology is well developed, which is one of the major keys for 5G success.

1.1 Introduction

Mobile cellular network has been developing since the 1970s.[1] The first generation (1G) mobile network was based on frequency division multiple access (FDMA) (Fig. 1.1). It provided analog voice service to mobile users. After approximately 10 years, time division multiple access (TDMA) was developed in the second generation (2G) network which enabled the digital voice service and low data rate service. In mid-1990 to 2000s, coding division multiple access (CDMA) was employed to develop the third generation (3G) mobile network. The CDMA access enabled more efficient multiple user access through the specified bandwidth. By this means, the data rate can reach several kilo bits per second to several mega bits per second, which enables fast data transmission for multimedia.

---

[1] Illinois Bell Telephone Co. conducted a trial development cellular system in the Chicago area in 1979. Full commercial service began in Chicago in October of 1983.

© Springer Nature Switzerland AG 2020
Wan Lei et al., *5G System Design*, https://doi.org/10.1007/978-3-030-22236-9_1

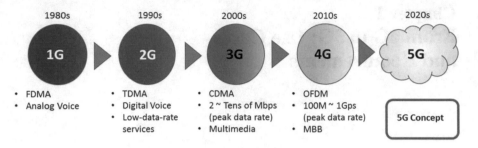

Fig. 1.1 A schematic view of the history of cellular communications

By the mid-2000s, ubiquitous data began to transform the meaning of mobile telephony service. The ever-increasing demand on data service stressed the capability of the 3G mobile network. Users are expecting to connect to the network anywhere, on the go or at home. More frequent data transmission happens, and faster data rates are becoming essential. Users start demanding broadband experience from their wireless mobile network service.

In 2005, the industry started the development of the fourth generation (4G) mobile network that aims to provide ubiquitous mobile broadband (MBB) service. 3GPP developed the Long Term Evolution (LTE) that employs Orthogonal Frequency Division Multiple Access (OFDMA) to offer a good compromise on multi-user data rates and complexity. LTE development received a wide range of industry support. The OFDMA concept is well incorporated with multiple-input multiple-output (MIMO) technology, and the complexity is significantly reduced.

International Telecommunication Union (ITU), especially its Radiocommunication Sector (ITU-R), has played an important role in mobile network development since 3G. Due to the great success of 1G and 2G mobile networks, the industry and research interests were increased exponentially for 3G development. Therefore, many stakeholders were involved in 3G development with respect to previous generation mobile networks. A global standardization was becoming necessary due to the involvement of a variety of network vendors, user terminal manufacturers, and chipset providers. Global spectrum harmonization also becomes a critical issue for the successful development and deployment of the mobile network. In order to harmonize the spectrum use in different regions for appropriate technologies for 3G mobile network, ITU-R established procedures to address the allocation of spectrum for 3G mobile network, and to identify the appropriate radio interface technology that could be deployed on those spectrums globally. In this context, International Mobile Technology-2000 (IMT-2000) was specified by the ITU, where a family of technologies were identified as radio interface technologies for 3G mobile network (IMT-2000 system). Such procedures provide fair opportunity for the proponents that are interested in mobile network development, as well as set the necessary performance requirements to guarantee that candidate technology can effectively meet the requirements. The procedure is further developed and applied to 4G and 5G development in ITU-R. This resulted in the IMT family specification: IMT-2000 for "3G," IMT-Advanced for "4G," and IMT-2020 for "5G."

While ITU-R plays the central role for defining appropriate technology for each generation of mobile network, the technology development is conducted in standard development organizations (SDOs). In 1998, the third generation partnership project (3GPP) was initiated by the key players of the mobile network development, and gains the support from six regional SDOs from Europe, China, Japan, Korea, and America. It lays the foundation of global development for mobile technologies, and attracts the participation of a variety of industry and academy players. 3GPP has grown to be the essential standard organization for technology development for mobile networks since 3G.

1.2   Global 5G Development

In July 2012, ITU-R started the development of the vision for IMT for 2020 and beyond, which is later known as "IMT-2020." Following the ITU-R activity, in 2013 to 2015, several regional promotion groups and research forums were established in China, Europe, Korea, Japan, and America for 5G development. The regional studies provided extensive investigations of 5G use cases and capability requirement, which formed a global foundation for 5G vision developments in ITU-R. In 2015, the 5G vision was set up in ITU-R based on the regional convergence.

Along with the gradual maturity of the 5G vision, in late 2014, 5G technology studies received increasing attention from industry and academy: with many new technologies and concepts proposed. In 2015, when ITU-R created the 5G vision, 3GPP as one of the most widely supported global standardization organizations started the technical requirement study and deployment scenarios investigation, targeting to fulfilling the 5G vision. In 2016, 3GPP initiated the 5G new radio (NR) technology study. Industry members, institutions, and universities are actively engaged in 3GPP study, which form the solid foundation for 5G radio interface technology that encompasses a wide range of usage scenarios and features.

In December 2017, 3GPP accomplished the first milestone of 5G specification. An initial characteristics description was submitted to ITU-R in February 2018. The ongoing global 5G development opens the gate to the fully connected world. The global harmonization and coordination are the essential keys throughout the 5G development.

1.2.1   ITU-R Development on 5G/IMT-2020

For different radio services to coexist, it must rely on the allocated spectrum resources to deliver their capability. ITU-R is responsible for coordinating the spectrum allocation and spectrum usage for many different radio services, including satellite, broadcast, scientific research, and mobile service. The radio spectrum resources are rather limited and, thus, efficient use of spectrum by each of the radio services should be guaranteed. This is also part of the responsibility of the ITU-R. It conducts technology studies to continuously improve the efficiency of spectrum usage, and compatibility studies are conducted among the radio services to make sure different radio services with appropriate spectrum allocation can coexist in an efficient manner. To this end, a range of study groups (SGs) and, under a specific SG, a number of working parties (WPs) are established in ITU-R. The technical and spectrum experts with rich expertise on related fields are gathered under WPs and SGs for the study of applying appropriate technologies for the specific radio services, and to investigate the spectrum allocation for specific radio service purposes.

WP 5D is the expert group under SG 5 for the study of International Mobile Technology (IMT) services in ITU-R. To guarantee that the IMT technologies can utilize the spectrum efficiently, and can well coexist with other radio services, the expert group devotes itself to develop a globally implemented radio interface standard that has high spectral efficiency and other key capabilities, which are widely supported by a variety of network vendors, user terminal manufacturers, and chipset providers.

The history of the development of IMT-2000 (3G), IMT-Advanced (4G), and IMT-2020 (5G) is depicted in Fig. 1.2 (see also [1]).

As seen from Fig. 1.2, ITU-R usually spent around 10 years for development of each generation of the mobile network (or IMT network under ITU-R context). From IMT-Advanced, a vision development was always made before the technical development. The vision study usually covers the investigation of high-level requirements, use cases, and key capabilities desired by the next generation mobile network. This is becoming an important step to converge the regional desires from different parts of the globe. It is observed that one ITU-R study period (3–4 years) is usually

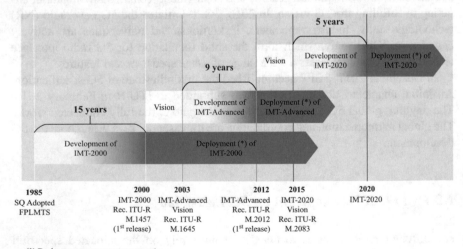

(*) Deployment may vary across countries

Fig. 1.2 Overview of timeline for IMT development and deployment

spent for this purpose. After that, another one to two (or even more) study periods are used to develop the technical aspects and to conduct the compatibility study if the new IMT spectrum that is requested are already used by other radio services. The time plan might be depending on the technical complexity of technology to achieve the vision, and the new spectrum it may utilize to reach the demanded capability.

For 5G development, the vision study was from 2012 to 2015. During this time period, regional 5G promotion groups and research forums were established to gather and converge the regional interest of 5G vision, and the regional views are contributed to ITU-R through the representative administration members and sector members.

From 2015, ITU-R started the technical development of 5G which aims to define an IMT-2020 (5G) global specification by October 2020. The technical development phase usually contains the definition of minimum technical requirements and evaluation guidelines, as well as an IMT-2020 submission and evaluation procedure. The minimum technical requirements guarantee that the candidate proposal can achieve the vision and can utilize the spectrum (including the potential new IMT spectrum) effectively. The submission and evaluation procedure defines the criteria of acceptance for IMT-2020 proposal, the procedure of inviting external organizations to submit IMT-2020 proposal, inviting the independent evaluation groups to evaluate the proposal, as well as ITU-R's procedure of approving the proposal. Following the ITU-R procedure, external organizations like 3GPP initiated their technical development of 5G from 2015 which will be reviewed in Sect. 1.2.3. More detailed discussion on the 5G timeline and the submission and evaluation procedure can be found in Sect. 1.4.

Concurrently ITU-R also started the compatibility study of potential new spectrum for 5G deployment. Such studies consist of two parts. One part is for the new IMT spectrums that were allocated in world radio communication conference (WRC) 2015. For example, the frequency range of 3.3–3.8 and 4.5–4.8 GHz (usually referred to as C-band) were identified as IMT spectrum globally or regionally. ITU-R continued to study the remaining issues on compatibility study on these bands in WP 5D. The other part is for potential new IMT spectrums that will be discussed at WRC-19. ITU-R has already begun its work on evaluating new spectrum by requesting technical development on feasible technologies to guarantee the efficient use of these potential new bands, and the compatibility studies to guarantee that they can efficiently coexist with other radio services.

The 5G spectrum will be discussed in more detail in Sect. 2.3.1.1.

1.2.2   Regional Development/Promotion on 5G

After ITU-R started 5G vision study in 2012, several regional promotion groups and research forums were founded in China, Europe, Korea, Japan, and America. These regional activities includes the study of 5G requirements, use cases, and deployment scenarios, as well as exploring the key technologies and the nature of 5G spectrum. These activities are briefly discussed in this section.

1.2.2.1   NGMN

Next Generation Mobile Networks (NGMN) is an alliance with worldwide leading operators and vendors that aims to expand the communications experience which will bring affordable mobile broadband services to the end user. It has a particular focus on 5G while accelerating the development of LTE-Advanced and its ecosystem.

NGMN published the "5G white paper" [2] in Feb 2015 which provided a list of requirements from the operator perspective to 5G networks. The requirements indicated a demand for capacity increase and the uniform user experience data rate across urban areas to rural areas. It also indicated that 5G networks should be capable of delivering diverse services including massive internet of things like sensor networks, extreme real-time communications like tactile internet, and ultra-reliable communications like e-Health services, to name a few. The services should be delivered in diverse scenarios, including high-speed trains, moving hotspots, and aircrafts. Such diversity fits in ITU's envisioned 5G use cases as described in Sect. 1.3.

1.2.2.2   IMT-2020 (5G) Promotion Group

In February 2013, the IMT-2020 (5G) promotion group was established in China by three ministries: the Ministry of Industry and Information Technology (MIIT), the National Development and Reform Commission, and the Ministry of Science and Technology. It is the major platform to promote the research and development of 5G in China. The promotion group consists of leading Chinese operators, network equipment vendors, research institutions, and universities.

The IMT-2020 (5G) promotion group published its 5G vision white paper [3] in May 2014. The white paper indicated two important 5G usage categories: mobile broadband and internet of things. The mobile broadband use cases will continue to be addressed by 5G network that provides 1 Gbps user experienced data rate in many deployment scenarios. The internet of things use cases on the other hand would be the other major driver to 5G network deployment, which requires a very capable 5G network to provide massive connections, very low latency, and high reliability. This forms a first overview of 5G use cases that were later developed into three 5G usage scenarios identified by ITU-R.

1.2.2.3   Europe: 5G PPP

The 5G Infrastructure Public Private Partnership (5G PPP) is a joint initiative between the European Commission and European ICT industry for 5G study in Europe. The first phase of 5G PPP started from July 2015, and continued to its second phase in 2017.

Before the first phase of 5G PPP initiative, an important 5G project was founded in Nov 2012, called Mobile and wireless communications Enablers for

Twenty-twenty (2020) Information Society (METIS). In April 2013, METIS published their study on 5G use cases, requirements, and scenarios (see [4]). In this study, METIS mentioned a number of new industrial and machine type communications for 5G besides mobile broadband applications. In April 2015, METIS summarized these use cases into three categories in [5], in accordance with ITU-R development on 5G use cases. The three use cases are extreme mobile broadband (xMBB), massive machine-type communications (mMTC), and ultra-reliable machine-type communications (uMTC), which is converged to ITU-R vision on 5G requirement and use cases as will be discussed in Sect. 1.3.1.

1.2.2.4   Korea: 5G Forum

The 5G Forum was founded by the Ministry of Science, ICT and Future Planning and mobile industries in Korea in May 2013. The members of 5G Forum consisted of mobile telecommunication operators, manufacturers, and academic professionals. The goal of the 5G Forum is to assist in the development of the standard and contribute to its globalization.

Five core 5G services were foreseen by the 5G forum, including social networking services; mobile 3D imaging; artificial intelligence; high-speed services; and ultra- and high-definition resolution capabilities; and holographic technologies. Such new services will be enabled by the 5G network with powerful capabilities that provide ultra-high capacity and data rates.

1.2.2.5   Japan: 5GMF

The Fifth Generation Mobile Communications Promotion Forum (5GMF) was founded in September 2014 in Japan. 5GMF conducts research and development related to 5G including the standardization, coordination with related organizations, and other promotion activities.

5GMF published the white paper "5G Mobile Communications Systems for 2020 and beyond" [6] in July 2016, which highlighted the 5G use cases of high data rate services, self-driving, location-based services, etc. It foresaw that 5G network needs to be extremely flexible to reach the requirements of these diverse requirements.

1.2.2.6   North and South America: 5G Americas

5G Americas is an industry trade organization composed of leading telecommunications service providers and manufacturers. It was continued in 2015 from the previously known entity: 4G Americas. The organization aims to advocate for and foster the advancement and full capabilities of LTE wireless technology and its evolution beyond to 5G, throughout the ecosystem's networks, services, applications, and wirelessly connected devices in the Americas. 5G Americas is invested in developing a connected wireless community while leading 5G development for the Americas.

5G Americas published its white paper on 5G Services and Use Cases [7] on November 2017. It provided an insightful report on 5G technology addressing new trends in a broad range of use cases and business models, with technical requirements and mappings to 5G capabilities. This is a continued study on 5G use cases for future uses.

1.2.2.7   Global 5G Event

The 5G regional developments call for a global coordination to form a unified 5G standard that is applicable worldwide. The regional 5G developers, including IMT-2020 (5G) PG, 5G PPP, 5G forum, 5GMF, 5G Americas, and 5G Brazil, have answered this call by setting up global 5G event to share the views and development status in each region. These events will be instrumental in building global consensus on 5G with the world's 5G promotion organizations. This series of events have been putting its efforts in promoting the usage of 5G for different vertical industries, 5G eco-systems and invite key industry players, administrations and regulators to participate in the discussion.

The first global 5G event was hosted by IMT-2020 PG in Beijing, May 2016, and the event is held twice a year in rotation. The next event will be at Valencia, Spain, in June 2019.

1.2.3   Standard Development

Along with ITU-R and regional activities towards 5G development, standardization organizations have also focused their attention on 5G since 2014. The global partnership project, 3GPP, has become the key standard organization of 5G development. It has become a global initiative for mobile cellular standard since the development of the 3G network. The current 4G, LTE mobile broadband standard is one of the most successful mobile standards and it is used worldwide.

3GPP is an organization that unites telecommunications standard development organizations (SDOs) all over the world. These SDOs are known as organization partners (OPs) in 3GPP, and currently there are seven OPs: ARIB and TTC from Japan, ATIS from America, CCSA from China, ETSI from Europe, TSDSI from India, and TTA from Korea. The seven OPs provide the members with stable environment to develop 3GPP technology. The OP members include key industry players, leading operators, vendors, user terminal manufacturers, and chipset developers. Research institutes with regional impacts, academic organizations and universities are also included. It covers almost all the key parties as far as a mobile cellular standard is concerned. The members of the OPs are heavily involved in the technical standard development, which ensure the 3GPP technology addresses the concerns and issues from different parties and different regions. Based on the consensus of the members from different OPs, the technical specifications defined by 3GPP are transposed by the OPs into their regional specifications. By this means, global mobile standards are developed. It can be argued that this is the essential key for the success of 3GPP standard development.

LTE is an early example under this consensus-based spirit of development. The global and wide range participation laid the foundation for the success of LTE development, standardization, and implementation. Due to the great success of LTE, 3GPP has become the essential standard development body for 5G. In late 2014, 3GPP initiated 5G studies and development along with the gradual maturing of the 5G vision.

During late 2015 to early 2017, when 3GPP was in its 14th release time frame (known as Release-14), 5G studies on technical requirements and deployment scenarios were conducted. These studies aimed to achieve the 5G vision as defined by ITU-R in June 2015. The study of a new radio (NR) interface was initiated following the requirement study. The key technical components were identified for the NR development, which form the basis of the specification work in the next release (i.e., Release-15) that last from early 2017 to June 2018. It is planned that a full capability 3GPP 5G technology, including NR and LTE, will be developed in Release-16 during the time frame of 2018 to the end of 2019. With this phased approach, 3GPP will bring its 5G solution to ITU-R as IMT-2020 in the year 2020.

1.3   Use Case Extensions and Requirements

Cellular communication system, from its first generation, is focused on connecting humans. The 1G and 2G communication system provided ubiquitous voice service between persons, which enables us to talk freely to our friends when we are at home, in office, or on the move. From 3G to 4G, multimedia and other mobile broadband applications are supported, and we are able to do much more, such as browsing the web, sharing nice pictures with our friends, and chatting on short videos, while on the move. For 5G, however, the study of the use cases reveals some new demands that are beyond the mobile broadband (MBB) which primarily aims to connect human beings and machines. This section provides a review of 5G use cases, requirements, and key capabilities.

1.3.1   5G Usage Cases and Service Requirement

5G communication is envisioned to enable a full connected world for 2020 and beyond. This full connectivity is targeted not only to communicating the people, but also enables the communication of machines and things that can bring added value to improving the operational efficiency of the society, and facilitate our everyday life. Under this vision, 5G use cases are extended from mobile broadband (MBB) to internet of things (IoT).

1.3.1.1 Extended Usage Scenarios: From eMBB to IoT (mMTC and URLLC)

ITU-R established the 5G vision in 2015 through its Recommendation ITU-R M.2083 [1], indicating that 5G will extend its usage scenario from enhanced mobile broadband (eMBB) to massive machine type communication (mMTC) and ultra-reliable and low latency communication (URLLC). The mMTC and URLLC services are a subset of internet of things (IoT) services. They are considered as the first step for 5G network stepping into the broad range of IoT services that are characterized by the key service requirements. On the other hand, 5G is as well the first generation communication that targets to extend wireless connection to other than human-to-human connections.

The extension of 5G from eMBB to mMTC and URLLC comes from the observation and the demand of the user and application trend. On one hand, high data rate video streams both down to the people (e.g., video streaming downloading) and up to the cloud server (e.g., user share their produced videos to their friends) are expected, and instantaneous and low latency connectivity becomes very vital to the user experiences for including augmented reality (AR) and virtual reality (VR). The user density with such high demand also increases, especially in urban areas; while in rural and/or high mobility cases, a satisfactory end-user experience is also desired by the users. Therefore, the challenging per-user high data rate request combined with high user density and user mobility becomes a driven force for 5G development, which requests significantly enhanced capability for mobile broadband services.

On the other hand, in the future, any object that can benefit from being connected will be connected, through, partially or dominantly, wireless technologies. This trend poses a huge amount of connectivity demand for connecting objects/machines/things in a wide range of applications. For example, driverless cars, enhanced mobile cloud services, real-time traffic control optimization, emergency and disaster response, smart grid, e-health, or efficient industrial communications are what are expected enabled or improved by wireless technologies/connections to just a few named in [1]. By a closer observation of these applications, one can find two major characteristics of these services: one is the number of desired connections; the other is the requested reliability within a given latency budget. These two significant characteristics provide the nature of mMTC and URLLC.

Therefore, 5G puts itself on the target to support diverse usage scenarios and applications including eMBB, mMTC, and URLLC. The following is a first glance on what the three usage scenarios indicates, which comes from [1],

- *Enhanced Mobile Broadband:* Mobile Broadband addresses the human-centric use cases for access to multi-media content, services, and data. The demand for mobile broadband will continue to increase, leading to enhanced Mobile Broadband. The enhanced Mobile Broadband usage scenario will come with new application areas and requirements in addition to existing Mobile Broadband applications for improved performance and an increasingly seamless user experience. This usage scenario covers a range of cases, including wide-area coverage and hotspot, which have different requirements. For the hotspot case, i.e., for an area with high user density, very high traffic capacity is needed, while the requirement for mobility is low and user data rate is higher than that of wide area coverage. For the wide area coverage case, seamless coverage and medium to high mobility are desired, with much improved user data rate compared to existing data rates. However, the data rate requirement may be relaxed compared to hotspot.
- *Ultra-reliable and low latency communications:* This use case has stringent requirements for capabilities such as throughput, latency, and availability. Some examples include wireless control of industrial manufacturing or production processes, remote medical surgery, distribution automation in a smart grid, transportation safety, etc.
- *Massive machine type communications:* This use case is characterized by a very large number of connected devices typically transmitting a relatively low volume of non-delay-sensitive data. Devices are required to be low cost, and have a very long battery life.

Additional use cases are expected to emerge, which are currently not foreseen. For future IMT, flexibility will be necessary to adapt to new use cases that come with a wide range of requirements.

In the following, we will first give a survey on the diverse services that are envisioned as 5G use cases, and then investigate the service requirements. Then technical performance requirements are extracted by grouping the similar service requirements to one characterized technical requirement. The importance of the technical requirement will be mapped to different usage scenarios based on the group of services under the specific usage scenario.

1.3.1.2   Survey of Diverse Services Across 5G Usage Scenarios and the Diverse Requirements

There are a broad range of emerging services that are expected to appear for the year 2020 and beyond that are under the scope of the 5G study. Generally, these 5G services are categorized into three groups according to the three usage scenarios elucidated in the last section.

1.3.1.2.1 eMBB Services

Mobile broadband services like web browsing, social apps with text messaging, file sharing, music downloading, and so on are already very popular and well supported by 4G communication systems. In the future, it is expected that higher data rate services like ultra-high definition (UHD) video, 3D video, and augmented reality and virtual reality will be dominating the human-to-human communication requirements. In addition to the abovementioned downlink high data rate services, the demand of uplink high data rate service is also emerging, e.g., the HD video sharing from the users. These service requirements define new horizon limits for eMBB development, together with the anywhere anytime experience requirements.

UHD/3D Video Streaming

The 4K/8K UHD video streaming requires up to 300 Mbit/s experienced data rate. From the study of technology enablers of enhanced Mobile Broadband conducted in 3GPP [8], the requested data rate of 4K and 8K UHD video are listed in Table 1.1:

Video Sharing

With the increased popularity of social applications, video sharing from the users to the cloud is becoming popular. The full HD (1080p) video can be assumed, and 4K/8K UHD video sharing can be expected in the 2020 and beyond. The requested data rate of 4K and 8K UHD video are the same as that listed in Table 1.1. The required data rate of 1080p and 720p video are given in Table 1.2 [9].

AR/VR Delivering to User

Augmented reality (AR) and virtual reality are the applications that bring fresh new experience to the eMBB users. Augmented reality provides the users with interactive experience of a real-world environment with "augmented" views generated by computer graphics. Virtual reality (VR) provides another interactive experience with an immersive environment created by computer graphics which can be

Table 1.1 Required data rate of UHD video

Video type	Video resolution	Frames rate (frame per second, FPS)	Encoding scheme	Quality requirement	Required data rate (Mbit/s)
4K UHD	3840 × 2160	50	HEVC	Medium quality	20–30
4K UHD	3840 × 2160	50	HEVC	High quality	~75
4K UHD	3840 × 2160	50	AVC	High quality	~150
8K UHD	7680 × 4320	50	HEVC	High quality	~300

Table 1.2 Required data rate of 1080p and 720p video

Video type	Video resolution	Frames rate (FPS)	Encoding scheme	Required data rate (Mbit/s)
720p	1280 × 720	60	H.264	3.8
1080p	1920 × 1080	40	H.264	4.5
1080p	1920 × 1080	60	H.264	6.8

fantastical or dramatic compared to physical reality. Both AR and VR require very high data rate and low latency to deliver the computer generated graphics and multimedia contents to the end users with guaranteed and smooth experience.

The required data rate and round trip delay of VR application is listed in Table 1.3 (see [10]).

In general, it is observed from the above examples that future eMBB services will require very high data rate compared with the requirement today. In addition, for AR/VR, the latency requirement is becoming more important. Therefore, the high data rate and low latency would become dominant for 5G eMBB services.

1.3.1.2.2   mMTC Services

Massive machine type communication (mMTC) refers to a group of emerging services that typically use massive number of sensors to report the sensing data to the cloud or a central data center in order to make smart decisions and/or reduce the human work load for collecting these data.

In the study of technology enablers of Massive Internet of Things conducted in 3GPP (see [11]), various services are investigated. Here we just mention some of them.

For example, the electric company deploys a large number of smart meters for each apartment within an apartment complex, and the smart meters report electricity usage to the company periodically.

Another example is that numbers of video recorders that are installed along or at the corner of the streets. The camera records continuous video, and stores the content for some period of time. The device periodically sends a status update to the traffic police indicating how the traffic is moving. When an accident occurs at the intersection, the device begins sending high quality video to the traffic police of the accident and ensuing traffic congestion.

Farm machinery is increasingly being automated. Farm machinery can report various sensor data, such as soil condition and crop growth, so that the farmer can remotely monitor the farm condition and control machinery.

There are other examples that has very similar paradigm to the abovementioned applications. The common service requirements can be summarized as follows:

- These services usually require large amount of sensors within a specific area. If we further consider the network needs to provide connection to multiple types of sensor or sensor applications, the amount will soon become incredibly large.

Table 1.3 Required data rate of VR

VR level		Video resolution	Single-eye resolution	Frames rate (FPS)	Encoding scheme	Color depth	Required instantaneous data rate		Round trip delay
							Field of view (FOV):	Full view:	
Entry-level VR	Weak interaction	Full-view 8K 2D/3D video (full frame resolution 7680 × 3840)	1920 × 1920 [with view angle of 110°]	30	H.265	8	40 Mbps (2D); 63 Mbps (3D)	75 Mbps (2D); 120 Mbps (3D)	30 ms (2D) 20 ms (3D)
	Strong interaction			90			120 bps (2D); 200 Mbps (3D)		10 ms
Advanced VR	Weak interaction	Full-view 12K 3D video (full frame resolution 11,520 × 5760)	3840 × 3840 [with view angle of 120°]	60	H.265	10	FOV: 340 Mbps	Full view: 630 Mbps	20 ms
	Strong interaction			120			1.4 Gbps		5 ms
Ultimate VR	Weak interaction	Full-view 24K 3D video (full frame resolution 23,040 × 11,520)	7680 × 7680 [with view angle of 120°]	120	H.266	12	FOV: 2.34 Gbps	Full view: 4.4 Gbps	10 ms
	Strong interaction			200			3.36 Gbps		5 ms

In this respect, it is not difficult to imagine that there might be one sensor per square meter for the year 2020 and beyond. For example, if an apartment is of the size of 80 m^2, and in one apartment building we have 10 floors, then if one apartment has 8 sensors (for example, various smart meters for electricity use, water use, indoor air quality monitoring, and temperature monitoring), we will have 80 sensors in the apartment building. It indicates one sensor per square meter.

– These services have a range of data rate request; however, the recent applications might be dominant by small data packet sizes. For example, the smart metering reports usually small amount of data.[2] The farm machinery also reports small amount of data at a time. In the video recorder application, although sometimes video needs to be transmitted when the accident occurs, the small data is dominant in daily case.
– Device battery life is critical to the economic success of these applications. Otherwise frequent re-installation of the sensors will be required. Considering the huge amount of sensor deployment, such re-installation can be very costly. In [12], it is required that the device battery life should be longer than 10 years.
– Coverage is very important in this use case. The connection service needs to be provided to the sensors deployed in deep coverage environment, e.g., in basement. Therefore, it is important to guarantee the network can reach such sensors.

In summary, the service requirements for mMTC applications are primarily on connection density (the number of devices per unity area) under a specific quality of service (e.g., the data rate), the battery life, and the coverage capability to the sensors. There might be different specific values for the above requirements. Nevertheless, high connection density, long battery life, and deep coverage capability are foreseen as the major service requirements for mMTC.

1.3.1.2.3   URLLC Services

Ultra-reliable and low latency communication (URLLC) is the group of emerging services that are very sensitive to latency and loss of data packets. In [1], some examples are given, which include wireless control of industrial manufacturing or production processes, remote medical surgery, distribution automation in a smart grid, transportation safety, etc. In the study of technology enablers of Massive Internet of Things conducted in 3GPP (see [13]), similar examples are mentioned, with more detailed descriptions on the service applications and envisaged application scenarios.

In [13], the service requirements for this type of applications are summarized. As indicated by its name "URLLC," the primary requirements are low latency and ultra-reliability. They need to be achieved at the same time. That is, for a transmitted data packet, the high reliability (very low loss of data) should be guaranteed within a given time duration.

---

[2] Small as compared to the data usage in eMBB.

1.3.1.3   Supporting Requirements and Operational Requirements to Enable 5G Service Deployment

The service requirement is from the perspective of an individual user or device, and it is independent to the deployment environment. However, from system design perspective, how the network can guarantee, as much as possible, a service is supported is needed, especially in a cost-effective manner. Therefore, besides the service requirement that is raised from a service perspective, it is also desired to define the related supporting and operational requirements to guarantee certain level of service support from system design and operating perspective.

1.3.1.3.1   eMBB

For eMBB services, it is needed to consider to what extent the high data rate can be supported. From a system perspective, there are multiple users in a given environment. If one assumes that data transmission of a specific service type simultaneously happens on these users, it is straightforward to consider the area traffic volume requirement, as well as to define a "cell edge" user experienced data rate requirement such that most of the users can achieve the required data rate for this service.

To this end, area traffic capacity and edge user experienced data rate requirement are defined.

Edge User Experienced Data Rate

Edge user (e.g., 5th percentile) experienced data rate is defined to guarantee that most of the users (e.g., larger than 95% of the users) can exceed a specific user data rate. In NGMN white paper [2], the edge user experienced data rate requirement is given in Table 1.4. By recalling the different service requirement in eMBB, it is seen

Table 1.4 Edge user experienced data rate requirement in NGMN white paper

Environment	Use case category	User experienced data rate	Mobility
Dense urban	Broadband access in dense areas	DL: 300 Mbps UL: 50 Mbps	On demand, 0–100 km/h
Indoor	Indoor ultra-high broadband access	DL: 1 Gbps, UL: 500 Mbps	Pedestrian
Crowded environment (e.g., Stadium)	Broadband access in a crowd	DL: 25 Mbps UL: 50 Mbps	Pedestrian
Rural	50+ Mbps everywhere	DL: 50 Mbps UL: 25 Mbps	0–120 km/h
High-speed vehicle	Mobile broadband in vehicles (cars, trains)	DL: 50 Mbps UL: 25 Mbps	On demand, up to 500 km/h

that with such requirement, UHD video is expected to be well supported in dense urban environment, and advanced or "ultimate" AR/VR are expected to be well supported in indoor. For crowded environment as well as in rural and high-speed vehicle environment, 1080p and 720p video is expected to be well supported.

It is also noted that mobility is defined to support the specific user experienced data rate requirement. This is from the motivation that specific services need to be well supported also on-the-go.

Area Traffic Capacity

Area traffic capacity is the amount of the total traffic in the given area. The total amount of traffic is related to the number of simultaneous users with specific service requirement. It provides the capability requirement on the number of supported users with a guaranteed average user data rate.

For example, for indoor environment, if we assume 20 users per 1000 m^2 are simultaneously transmitting data with service required data rate of 500 Mbps, then the area traffic capacity requirement would be 20×500 Mbps/1000 $m^2 = 10$ Mbps/m^2.

Another perspective is from the prediction of the amount of increase traffic. It is predicted that for the year of 2020 and beyond, the total traffic amount will be increased by 100 times to 1000 times compared to that in 2010s. In this case, by investigating the traffic amount in the 2010s, and by applying the increasing time prediction, one can derive the area traffic capacity requirement for 5G time frame. Besides the requirements that support successful deployment of eMBB services from system perspective, the operational requirements that aim to improve the resource utilization efficiency so as to reduce the cost are also defined.

Spectral Efficiency

Unlike wireline communication, radio spectrum is a scarce resource. Consequently, Spectral efficiency is very critical for eMBB services since the 3G era.

For 5G, it is anticipated that the requested data rate is significantly increased. The area traffic capacity is predicted to be increased by at least 100 times. This means in a given area, the total amount of data rate will be increased by at least 100 times. The total amount of data rate is given by averaged spectrum efficiency (bps/Hz per base station) multiplying with the available system bandwidth (Hz) and the number of base stations in the area. If we assume that the base station deployment within the given area can be increased by three times, and the available bandwidth can be increased by ten times, then the average spectrum efficiency should be increased by at least three times.

Besides the average spectral efficiency, there is a strong desire in the industry to improve the edge user spectral efficiency. Here we note that it is a *desideratum* and leave the quantitative requirement to be discussed in Chap. 5.

Energy Efficiency

Energy efficiency is motivated by recognizing that it is not affordable to provide 100 times network area traffic capacity with 100 times energy consumption increase. Indeed, the network energy consumption increase should be very limited, e.g., see analysis in [14].

Specifically, it is reported that currently the radio access network infrastructures consume about 0.5% of the global energy consumption [15], which is already a considerable portion. To avoid that the 5G network become a major consumer of the global energy production, it is reasonable to constrain the increase of energy consumption of 5G radio access network, such that its energy consumption ratio is below or at least at the same level compared to today. That is to say, the energy consumption increase of radio access needs to follow the pace of global energy consumption increase, or even slower.

Based on the data from [16] reproduced in Fig. 1.3, one can see that between 1990 and 2013, the CAGR of global energy consumption is about 1.91%, which indicates 1.2 times increase in one decade. Thus, one can forecast that in 2020 the global energy consumption would increase by no more than 20% compared to 2010. Therefore, the power consumption of 5G network in 2020s should be no more than 1.2 times as compared to 4G network in 2010s.

This implies that, if the area traffic capacity increases by 100 times, with the assumption of very limited, or no energy consumption increase, the network energy efficiency measured as bit per Joule (or bit/s per Watt) needs to be improved by approximately 100 times.

Energy efficiency is also important for mMTC and URLLC as operating requirements. However, eMBB may come as the first urgency in terms of this requirement.

1.3.1.3.2   mMTC

For mMTC, area traffic capacity can also be used as one of the requirements for system design. However, if the current focus is on small data packet transmission, the requirement would be very loose. However, for the services with larger

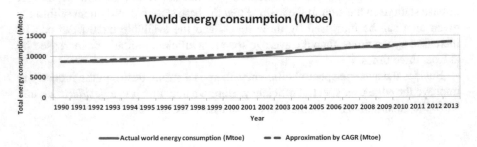

Fig. 1.3 World energy consumption trend in millions of tonnes of oil equivalent (Mtoe)

device-wise data rate requirement (e.g., video recorder applications), the area traffic capacity would become important in addition to the service requirement.

1.3.1.3.3   URLLC

Availability

To guarantee the users in different regions within the area achieving the service requirement (low latency and ultrahigh reliability), the availability is defined in [2] as follows: the network is available for the targeted communication in X% of the locations where the network is deployed and X% of the time. Within the X% of the locations the URLLC service requirement is achieved.

NGMN requires 5G to enable 99.999% network availability. This is very stringent and is a long-term target for system design for URLLC applications.

1.3.1.3.4   General

In addition to the above requirements, coverage capability of a network is the key for network operational cost.

Coverage

Coverage is defined as the maximum geographical range of a radio access point to provide a certain quality of service (QoS). For different QoS level, the coverage of a network will be different. For example, the uplink coverage will be more challenging for a service requirement of data rate of 1 Mbps than that of 100 kbps. Coverage is also highly related on site density. If the coverage for a certain QoS is small, operators need to deploy much dense network to provide the desired QoS. For example, in eMBB, if the coverage for providing UL data rate of 10 Mbps is 50 m, then the inter-site distance (ISD) could be hardly larger than 100 m. In this case, the number of sites required would be huge, leading to unacceptable cost. In this sense, coverage capability is fundamental for eMBB, URLLC, and mMTC usage scenarios.

1.3.2   *5G Key Capabilities and Technical Performance Requirements*

The 5G services form a solid basis for the definition of 5G key capability and the related technical requirements. In ITU-R, the key capabilities for 5G are defined together with 5G envisaged usage scenarios. Technical performance requirements are defined based on the 5G vision and key capabilities.

1.3.2.1   Key Capabilities for 5G

The key capabilities for 5G (also known as IMT-2020 in ITU-R context) are shown in Fig. 1.4 (see [1]). Figure 1.4a demonstrates the targets of 5G key capabilities as well as the enhancement compared to 4G (also known as IMT-Advanced). Figure 1.4b gives the importance of each key capability in the three usage scenarios. In the following, the key capabilities for each usage scenario are discussed.

1.3.2.1.1   eMBB

In ITU-R, the user experienced data rate and area traffic capacity are identified as part of the key capabilities most relevant for eMBB. As discussed in previous section, the two capabilities guarantee the system support of a specific eMBB service to be successfully delivered to most of the users, with a given number of users in the system.

User Experienced Data Rate

User experienced data rate is defined as the achievable data rate that is available ubiquitously across the coverage area to a mobile user/device (in Mbps). The term "ubiquitous" is related to the considered target coverage area and is not intended to relate to an entire region or country [1]. It can be seen that ten times user experienced data rate is expected compared to 4G. It can also be seen that 100 Mbps edge user experienced data rate is indicated for downlink in dense urban areas (Fig. 1.4). Although the target capability is a bit lower than the NGMN requirement, it still guarantees that most users experience good UHD video and have partial AR/VR capabilities. Undoubtedly this capability is well beyond the service requirement of 1080p and 720p video.

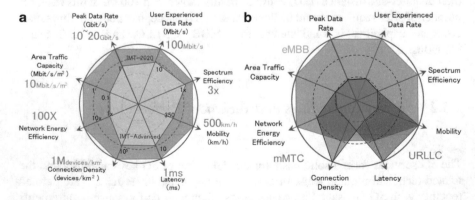

Fig. 1.4 Key capability of 5G. (**a**) Enhancement of key capability for 5G. (**b**) The importance of key capabilities in different usage scenarios

Area Traffic Capacity

Area traffic capacity is defined [1] as the total traffic throughput served per geographic area (in Mbps/m^2)

Area traffic capacity of 10 Mbps/m^2 is expected in high user density scenarios, e.g., indoor. This means that if we assume 20 users per 1000 m^2 simultaneously transmitting data, the service required data rate of 500 Mbps can be supported.

This capability demonstrates a 100 times improvement compared to 4G. This is the result of ten times user experienced data rate improvement together with ten times connection density capability.

Mobility

Mobility is an important capability that allows the support of high data rate transmission inside high-speed vehicles.

It is defined as the maximum speed at which a defined QoS and seamless transfer between radio nodes which may belong to different layers and/or radio access technologies (multi-layer/-RAT) can be achieved (in km/h) [1].

The capability of up to 500 km/h is to be supported. This mobility class is 1.4 times of 4G (where up to 350 km/h mobility class is defined.)

Peak Data Rate

The peak data rate is defined as the maximum achievable data rate under ideal conditions per user/device (in Gbps) [1]. It indicates the maximum capability of the data rate of a device. It is seen that 10–20 Gbps is expected. This means the device under the ideal condition can support AR/VR applications. Besides the above capabilities, the efficiency capabilities that are helpful to reduce the operational cost are identified as key capabilities for 5G.

Energy Efficiency

Network energy efficiency is identified as one of the key capabilities that enables the economics of 5G.

Energy efficiency has two aspects [1]:

– On the network side, energy efficiency refers to the quantity of information bits transmitted to/received from users, per unit of energy consumption of the radio access network (RAN) (in bit/Joule)
– On the device side, energy efficiency refers to quantity of information bits per unit of energy consumption of the communication module (in bit/Joule)

For network energy efficiency, it is required that the energy consumption should not be greater than the radio access networks deployed today, while delivering the enhanced capabilities. Therefore, if area traffic capacity is increased by 100 times, the network energy efficiency should be improved by a similar factor.

Spectral Efficiency

Spectral efficiency includes two aspects. The average spectrum efficiency is defined as the average data throughput per unit of spectrum resource and per transmit-receive point (bps/Hz/TRxP). The edge user spectral efficiency is defined as the 5th percentile user data throughput per unit of spectrum resource (bps/Hz).

As discussed in Sect. 1.3.1.3, the requirement on average spectral efficiency improvement is related to the required amount of increase in area traffic capacity. ITU-R defines 100 times improvement capability for area traffic capacity. In this case, the total amount of data rate will be increased by at least 100 times. The total amount of data rate is given by averaged spectrum efficiency (bps/Hz/TRxP) multiplying with the available system bandwidth (Hz) and the number of transmit-receive point (TRxP) in the area. If we assume that the TRxP deployment within the given area can be increased by 3 times, and the available bandwidth can be increased by 10 times, then the average spectrum efficiency should be increased by at least 30 times.

For edge user spectral efficiency, the requirement of improvement is related to the supported edge user experienced data rate, the number of users under one TRxP, and the available bandwidth. If we assume the connection density within the area is increased by ten times (see below on "connection density" capability) while the number of TRxPs within the area can be increased by three times, then the number of users in one TRxP will be increased by three times. On the other hand, one may assume the available bandwidth in one TRxP is increased by ten times. This means for an individual user, the available bandwidth is increased by 10/3 = 3.3 times. In this case, to consider the support of ten times edge user experienced data rate (see above on "user experienced data rate" capability), the edge user spectral efficiency should be improved by three times.

1.3.2.1.2   mMTC

For mMTC usage scenario, connection density and energy efficiency are identified as the two most relevant key capabilities. Besides key capabilities, operational lifetime is also identified as desired capabilities for 5G network for mMTC. These capabilities will now be discussed *in ordinem.*

Connection Density

Connection density is the total number of connected and/or accessible devices per unit area (per km^2) [1]. For mMTC usage scenario, connection density is expected to reach 1,000,000 devices per km^2 due to the demand of connecting vast number of devices in the time frame of 2020 and beyond. This is a ten times improvement compared to 4G (IMT-Advanced).

Network Energy Efficiency

Network energy efficiency is also identified as one of the important key capabilities for mMTC. This is because the large coverage provided for mMTC devices should not be at the cost of significantly increased energy consumption.

Operational Lifetime

Operational lifetime refers to operation time per stored energy capacity. This is particularly important for machine-type devices requiring a very long battery life (e.g., more than 10 years) whose regular maintenance is difficult due to physical or economic reasons [1].

1.3.2.1.3   URLLC

As discussed in the previous sections, latency, mobility, and reliability are identified as the two most relevant key capabilities of URLLC.

Latency

In [1], user plane latency is highlighted. It is defined as the contribution by the radio network to the time from when the source sends a packet to when the destination receives it (in ms).

The 1 ms latency is desired. Reliability is further defined in technical performance requirements and will be discussed in the following sections.

Mobility

Mobility is relevant to URLLC usage scenario because the transportation safety applications, etc. are usually under high-speed mobility.

Besides the above key capabilities, reliability and resilience are also identified as desired capabilities for 5G network for URLLC.

Reliability

Reliability relates to the capability to provide a given service with a very high level of availability [1].

Resilience

Resilience is the ability of the network to continue operating correctly during and after a natural or man-made disturbance, such as the loss of mains power [1].

1.3.2.1.4 Other Capabilities

Other capabilities for 5G are also defined in [1].

Spectrum and Bandwidth Flexibility

Spectrum and bandwidth flexibility refer to the flexibility of the system design to handle different scenarios, and in particular to the capability to operate at different frequency ranges, including higher frequencies and wider channel bandwidths than today.

Security and Privacy

Security and privacy refer to several areas such as encryption and integrity protection of user data and signaling, as well as end user privacy preventing unauthorized user tracking, and protection of network against hacking, fraud, denial of service, man in the middle attacks, etc.

   The above capabilities indicate that 5G spectrum and bandwidth flexibility, as well as security and privacy will be further enhanced.

1.3.2.2 Technical Performance Requirements for 5G

Based on the key capabilities and IMT-2020 vision defined in [1], technical performance requirements are defined in Report ITU-R M.2410 (see [17]).

   The technical performance requirements are summarized in Tables 1.5, 1.6, and 1.7. The detailed definition on the technical performance requirements can be found in [17].

   To reach the 5G vision defined by ITU-R, 3GPP further studied the deployment scenarios and the related requirements associated with the three usage scenarios as documented in 3GPP TR 38.913 (see [12]). These requirements are usually higher than ITU's technical performance requirement, showing 3GPP's ambition of providing higher capability than ITU required. A detailed description of the 3GPP requirements is beyond the scope of this book. This interested reader is encouraged to consult [12].

1.3.3 Summary on 5G Requirements

From the analyses in the previous sections, it can be seen that 5G has diverse requirements.

   For the eMBB usage scenario, the edge user experienced data rate should be enhanced to deliver high quality video at anywhere and anytime to the end users, and high data rate transmission should also be possible under high mobility class.

Table 1.5 eMBB technical performance requirements

Technical performance requirement	DL	UL	Comparison to IMT-Advanced requirement
Peak data rate	20 Gbit/s	10 Gbit/s	*~6× LTE-A (Release-10)*
Peak spectral efficiency	30 bit/s/Hz	15 bit/s/Hz	*2× IMT-Advanced*
User experienced data rate (5th percentile user data rate)	100 Mbit/s	50 Mbit/s	–
5th percentile user spectral efficiency	~3× IMT-Advanced	~3× IMT-Advanced	*~3× IMT-Advanced*
Average spectral efficiency	~3× IMT-Advanced	~3× IMT-Advanced	*~3× IMT-Advanced*
Area traffic capacity	10 Mbit/s/m^2	–	–
Energy efficiency	High sleep ratio and long sleep duration under low load		–
Mobility class with traffic channel link data rates	–	Up to 500 km/h, with 0.45 bit/s/Hz	*1.4× mobility class; 1.8× mobility link data rate*
User plane latency	4 ms	4 ms	*>2× reduction compared to IMT-Advanced*
Control plane latency	20 ms	20 ms	*>5× reduction compared to IMT-Advanced*
Mobility interruption time	0	0	*Much reduced*

Table 1.6 URLLC technical performance requirements

Technical performance requirement	DL	UL	Comparison to IMT-Advanced requirement
User plane latency	1 ms	1 ms	*>10× reduction compared to IMT-Advanced*
Control plane latency	20 ms	20 ms	*>5× reduction compared to IMT-Advanced*
Mobility interruption time	0	0	*Much reduced*
Reliability	99.999% within 1 ms	99.999% within 1 ms	–

Table 1.7 mMTC technical performance requirements

Technical performance requirement	DL	UL	Comparison to IMT-Advanced requirement
Connection density	–	1,000,000 devices/km^2	–

The area traffic capacity is required to be improved by 100 times or more to enable more users to enjoy high data rate service. Latency is also found to be of increasing importance for eMBB: for example, in AR/VR applications. The above performance requirements need to be provided in affordable manner, which in turn requires spectral efficiency to be improved at least by three times, and energy efficiency to be significantly enhanced. Note that not all of these requirements are unique to the eMBB service, for example, energy efficiency is also desired for network providing mMTC and URLLC services.

For the mMTC usage scenario, connection density of the order of 1,000,000 devices per km^2 should be supported to enable mMTC services that are delivered to huge number of sensors, with variable data rates. Long battery life and deep coverage are desired.

For the URLLC usage scenario, low latency together with high reliability is required. High availability should also be targeted from a long-term perspective which further ensures that most of the locations within the network coverage are able to achieve the URLLC service requirement.

Coverage is a fundamental requirement when providing eMBB, URLLC, and mMTC services, because otherwise the site deployment would become very dense. A balance will need to be strike between density and network cost to make the system design economically feasible.

Other requirements including privacy and security enhancement, which makes 5G a safer network, are also necessary. This is of vital importance especially given that 5G targets to connect everything.

1.4   Standard Organization and 5G Activities

5G development is a vast process that needs wide support from industries, SDOs, and administrations. As is seen from Sect. 1.2, different regions show their visions for 5G development and applications. Consequently, for efficient 5G development, global efforts are needed to harmonize the regional concepts, from both technology perspective and spectrum perspective, so as to develop a unified 5G technology that is applicable on 5G potential bands.

To achieve this goal, the following organizations play vital role in 5G standardization and development. ITU-R, as the leading organization of identifying qualified 5G technology (known as IMT-2020 in ITU-R context) for efficient spectrum use, develops strict procedures for IMT-2020 submission and evaluation. These procedures are open to all technical standard development organizations, and they guarantee that the proposed technology could fulfill a specific set of requirements that are suitable for 5G deployment and to reach the vision of the envisaged network system for 2020 and beyond. On the other hand, 3GPP, as a very active technical partnership project, becomes the essential standard development organization for 5G development for the SDOs. This section discusses the ITU procedures for IMT-2020 submission and 3GPP developments towards ITU-R submission.

1.4.1 ITU-R Procedure/Process of IMT-2020 Submission

ITU-R is the leading organization for IMT-2020 development. ITU-R's role is to identify the IMT-2020 (5G) vision which will serve as the overall target for 5G development for 2020 and beyond, and to invite the technical standard development organizations to submit their candidate technologies that are capable to achieve the IMT-2020 vision. The qualified technology proposals are allowed to be deployed on IMT bands that are licensed to operators.

ITU-R Working Party 5D (WP 5D), which is the responsible working party in ITU-R for IMT system development, made an overall time plan for the above process in Fig. 1.5.

There are, generally, three stages to develop IMT-2020:

1.4.1.1 Stage 1: IMT-2020 Vision Development (2012–2015)

From July 2012 to June 2015, ITU-R WP 5D developed the IMT-2020 vision (Recommendation ITU-R M.2083) to define the framework of IMT-2020. At the same time, the study on technology trends and IMT feasibility for above 6 GHz were conducted in the preparation for 5G development.

1.4.1.2 Stage 2: IMT-2020 Technical Performance and Evaluation Criteria Development (2015–2017)

After the IMT-2020 vision is developed, WP 5D started the definition of IMT-2020 requirements in 2016. The requirements include three aspects: technical performance requirements, service requirements, and spectrum requirements. These

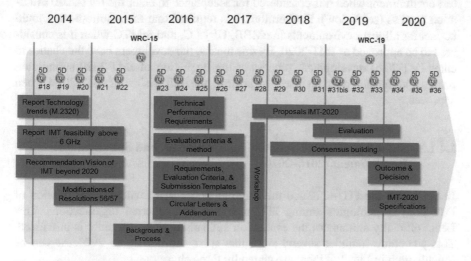

Fig. 1.5 IMT-2020 submission and evaluation procedure [18]

requirements are used to evaluate whether a candidate proposal is a qualified technical proposal that has the potential to be included as IMT-2020 standard for global
deployment. At the same time, the evaluation criteria and methods are developed by
ITU-R to evaluate whether a specific requirement is achieved by the candidate proposal. Submission templates are also developed such that a unified submission format can be used among the proponents, which is helpful to ease the work of ITU-R
and independent evaluation groups when evaluating the received candidate technologies. In 2017, ITU-R published three reports related to the above aspects:

- Report ITU-R M.2411: *Requirements, evaluation criteria and submission templates for the development of IMT-2020,* which defined IMT-2020 requirements,
  evaluation criteria and submission templates. Technical performance requirements
  are detailed in Report ITU-R M.2410. And evaluation criteria and method are
  detailed in Report ITU-R M.2412.
- Report ITU-R M.2410: *Minimum requirements related to technical performance
  for IMT-2020 radio interface(s),* which defines the 13 technical performance
  requirements for IMT-2020 development.
- Report ITU-R M.2412: *Guidelines for evaluation of radio interface technologies
  for IMT-2020, which defines the test environment,* evaluation method, and
  detailed evaluation parameters for the technical performance requirement.
  Evaluation criteria on how to conduct these test environment, method and parameters, and how to determine a requirement is fulfilled is defined in sufficient
  details.

Besides the above three important reports, Document IMT-2020/02 (Rev. 1) [18]
further defines the criteria for accepting a proposal to enter the IMT-2020 evaluation
procedure and the criteria for accepting a proposal to be approved as IMT-2020. It
is, currently, required that the proposed radio interface technology should meet the
requirement of at least two eMBB test environments and one URLLC or one mMTC
test environment when it is considered for acceptance to enter the IMT-2020 evaluation process (see below). Furthermore, it is required that the requirements should
be met for all test environments in eMBB, URLLC, and mMTC when it is considered to be approved as IMT-2020. By this means, the developers have the chance to
enter IMT-2020 development process at the beginning of IMT-2020 submission
phase, and further develop itself to achieve the full vision of IMT-2020 when
approved.

1.4.1.3   Stage 3: IMT-2020 Submission, Evaluation, and Specification
Development (2016–2020)

In Feb 2016, the ITU-R issued the Circular Letter that invites the submission of
IMT-2020 technologies among ITU members and external organizations. This
Letter officially announces the evaluation and submission procedure is initialized.
This procedure includes several steps that spans from 2017 to 2020. As can be
roughly seen in Fig. 1.5, there are generally three phases:

– "Submission phase" (see "Proposals IMT-2020" in Fig. 1.5). In this phase, the proponents can propose IMT-2020 candidate technologies. Self-evaluation will need to be provided by the proponent to demonstrate that their proposed technology can pass the least number of requirements requested by ITU-R.
– "Evaluation phase." In this phase, independent evaluation groups are invited to evaluate the qualified candidate technologies. Evaluation reports from the evaluation groups will be reviewed by WP 5D to determine which candidates pass the requested number of requirements based on a predefined criteria. These technologies will go into specification phase.
– "Specification phase." In this phase, the qualified technologies will be specified and captured in a recommendation of ITU-R for global IMT-2020 standard. The technologies captured in IMT-2020 standard will be able to be deployed on IMT bands licensed to operators.
– Consensus building is brought across the procedure to achieve a unified IMT-2020 technology.

The detailed procedure is given in Document IMT-2020/02 (Rev. 1) [18].

Based on the guidance of ITU-R procedure for IMT-2020 submission, the technical standard development organizations can submit their candidate technologies to ITU-R, and ITU-R will invite independent evaluation groups to make an independent evaluation to all the received candidates that declare themselves capable of meeting IMT-2020 requirements. 3GPP as one of the technical standard partnership projects has made such efforts for IMT-2020 submission.

1.4.2   3GPP Development Towards ITU-R Submission

3GPP is a partnership project formed during the 3G development era that targets to develop a global harmonized technical specification for wireless cellular systems. It unites seven telecommunications standard development organizations from different regions worldwide. The seven organizations, also known as "Organizational Partners" (OPs) of 3GPP, are ARIB and TTC from Japan, ATIS from North America, CCSA from China, ETSI from Europe, TSDSI from India, and TTA from Korea. The seven OPs provide their members with a stable environment to produce the Reports and Specifications that define 3GPP technologies.

It should be observed that the seven OPs have good representation and worldwide coverage for the global development on wireless telecommunications. Through the seven OPs, the member companies, universities, and research institutes can join the technical discussions of 3GPP and contribute to the technology development. This is a very essential base that 3GPP can gather a wide range of support across industry, academy, and other stakeholders for wireless system development. Therefore, 3GPP continues to be the essential technology standard partnership project for 5G.

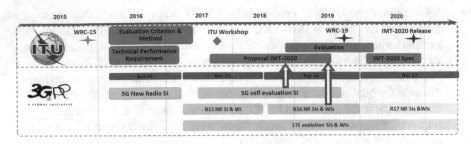

Fig. 1.6 3GPP timeline for 5G development and IMT-2020 submission

In 2015, after ITU-R published the IMT-2020 vision (Recommendation ITU-R
M.2083), 3GPP started the new radio (NR) study in its 14th release (Release-14)
timeframe, which targets to fulfill IMT-2020 requirements (Fig. 1.6). This study
item (SI) included new numerology and frame structure design, coding scheme
development, flexible duplexing study, and multiple access study. At the first stage,
NR study focuses on eMBB and URLLC usage scenarios as defined in IMT-2020
vision. NR mMTC may be developed in a later stage, since the early-stage mMTC
applications are expected to be supported by NB-IoT or eMTC developed earlier in
LTE.

In 2017, 3GPP started the normative work in Release-15 timeframe for NR. The
NR work item (WI) was set up to specify the technologies identified useful for 5G
development in Release-14 NR SI. The NR numerology, frame structure, duplex
and multiple access were specified. NR MIMO were discussed and developed. The
specification considers the IMT-2020 requirement and the potential spectrum
deployment for 5G, including the high data rate and high spectral efficiency requirement, low latency requirement and high reliability requirement, etc., together with
wide area coverage. Specifically, the C-band characteristics played a key role in 5G
development. To address the high data rate and low latency target simultaneously
with large coverage issue, 3GPP developed several key technical components. They
include numerologies and frame structures that are suitable for large bandwidth,
massive MIMO technology that are suitable for C-band deployment, and downlink
uplink decoupling that is suitable for achieving high data rate (especially for UL) on
C-band with existing inter-site distance.

In the Release-16 timeframe, 3GPP will continue to enhance its NR capability to
address higher spectral efficiency for eMBB, and to enhance URLLC in multi-user
scenarios. 3GPP also targets to extend NR usage to more vertical scenarios, including vehicular communications.

5G self-evaluation study was set up in 2017 to conduct the self-evaluation of
3GPP technologies developed in Release-15 and Release-16 for IMT-2020 submission. The preliminary self-evaluation was submitted to ITU-R in Oct 2018, and the
final evaluation results are planned to be submitted in June 2019.

1.4.3   Independent Evaluation Groups to Assist ITU-R Endorse IMT-2020 Specification

To assist ITU-R on evaluating whether a candidate technology can fulfill the IMT-2020 requirements, ITU-R invites independent evaluation groups to conduct evaluation on the received candidates and reports to WP 5D on their conclusions. As of Dec. 2018, there are 13 registered independent evaluation groups:

- 5G Infrastructure Association: https://5g-ppp.eu/
- ATIS WTSC IMT-2020 Evaluation Group: http://www.atis.org/01_committ_forums/WTSC/
- ChEG Chinese Evaluation Group: http://www.imt-2020.org.cn/en
- Canadian Evaluation Group: http://www.imt-ceg.ca/
- Wireless World Research Forum: http://www.wwrf.ch/
- Telecom Centres of Excellence, India: http://www.tcoe.in/
- The Fifth Generation Mobile Communications Promotion Forum, Japan: http://5gmf.jp/en/
- TTA 5G Technology Evaluation Special Project Group: https://www.tta.or.kr/eng/index.jsp
- Trans-Pacific Evaluation Group: http://tpceg.org/
- ETSI Evaluation Group: http://www.etsi.org/
- Egyptian Evaluation Group: N/A
- 5G India Forum: https://www.coai.com/5g_india_forum

1.5   Summary

In this section, the motivation and driving forces of 5G development are investigated, and the 5G use cases and technical requirements are introduced. Cellular communication system, from its first generation, is focused on connecting humans. From 5G, however, the use cases are extended beyond MBB for connecting human beings, to mMTC and URLLC for connecting different machines. 5G will therefore be the first generation that devotes itself to connecting both humans and machines.

Accordingly, the service requirements and the technical performance requirements are extended from eMBB to the new use cases (or usage scenarios). For eMBB, it requires significant increase of data rate to adapt to the emerging high definition video and AR/VR transmission, at the cost of affordable energy consumption and spectrum occupation. This implies that both energy efficiency and spectral efficiency should be considerably increased. Furthermore, the cell edge data rate should be significantly increased to several, or tens of megabits per second, with the similar coverage of existing network deployment, which poses big challenge for eMBB 5G design. For URLLC, very low latency is required and, at the same time high reliability should be guaranteed. On the other hand, such capability should be provided under various conditions including mobility. For mMTC, the large connection density is desired to provide massive connection in an efficient manner. Besides, in both use cases, coverage is needed to enable an affordable network deployment. Finally, the above requirements should be achieved by a unified air interface framework from cost efficiency perspective. In summary, 5G requirements are diverse, but at the same time the cost of energy, spectrum, and deployment should be restricted to affordable level. All these issues pose significant challenges to system design.

Like previous generations, 5G development is based on industry collaboration, where ITU and 3GPP play the central role in this process. ITU-R defined the 5G vision, technical performance requirements, and the evaluation criteria for the candidate 5G technologies, from the perspective of efficient use of spectrum and other resources, e.g., energy and site deployment, to address the diverse requirements from 5G use cases. ITU-R has a well-established procedure of the IMT-2020 development, which guarantees that only the technologies that pass the above-mentioned requirements could be approved as IMT-2020 (the name of 5G in ITU-R), and only such technologies can be allowed to use IMT spectrums. By this means, the technologies that is able to be deployed on IMT spectrum is guaranteed to be of high performance and high efficiency when using the scarce spectrum and other resources. On the other hand, 3GPP is the major organization for 5G technical development which unites the telecommunications standard development organizations (SDOs) all over the world. By the support of the seven SDOs (also known as organization partners, OPs): ARIB and TTC from Japan, ATIS from America, CCSA from China, ETSI from Europe, TSDSI from India, and TTA from Korea, 3GPP provide the members with stable environment for 5G technology development. The OP members cover almost all the key parties, which ensure the 3GPP technology addresses the concerns and issues from different parties and different regions. With the guidance of ITU-R and the well-harmonized technical development in 3GPP, 5G technology and 5G standardization is well developed. This is the key for 5G success.

Now that we have an understanding of the requirements of 5G and the process by which it was developed, the next chapter will begin our study of the enablers of the 5G wireless system. The discussion will begin with a brief technical overview of LTE. This will naturally lead us to understand the 5G air interface by comparing and contrasting with the LTE design.

References

1. Recommendation ITU-R M.2083, IMT vision—Framework and overall objectives of the future development of IMT for 2020 and beyond, Sept 2015
2. NGMN, NGMN 5G white paper, Feb 2015, https://www.ngmn.org/fileadmin/user_upload/NGMN_5G_White_Paper_V1_0_01.pdf
3. IMT-2020 (5G) PG, 5G vision and requirement white paper, May 2014, http://www.imt-2020.cn/zh/documents/download/1
4. METIS, D1.1 scenarios, requirements and KPIs for 5G mobile and wireless system, Apr 2013

5. METIS, D6.6 final report on the METIS 5G system concept and technology roadmap, Apr 2015
6. 5GMF, 5G Mobile Communications Systems for 2020 and beyond, July 2016, http://5gmf.jp/en/whitepaper/5gmf-white-paper-1-01/
7. 5G Americas, 5G services and use cases, Nov 2017, http://www.5gamericas.org/files/9615/1217/2471/5G_Service_and_Use_Cases__FINAL.pdf
8. 3GPP TR22.863, Feasibility study on New Services and Markets Technology Enablers—Enhanced Mobile Broadband, Sept 2016
9. https://en.wikipedia.org/wiki/Bit_rate
10. Cloud VR Bearer Networks, Huawei, 2017, https://www-file.huawei.com/-/media/CORPORATE/PDF/ilab/cloud_vr_oriented_bearer_network_white_paper_en_v2.pdf?source=corp_comm
11. 3GPP TR22.861, Feasibility Study on New Services and Markets Technology Enablers for Massive Internet of Things, Sept 2016
12. 3GPP TR38.913, Study on Scenarios and Requirements for Next Generation Access Technologies, June 2017
13. 3GPP TR22.862, Feasibility Study on New Services and Markets Technology Enablers for Critical Communications, Sept 2018
14. Document 5D/757, Discussion on energy efficiency and the requirement for future IMT, China (People's Republic of), Sept 2014
15. The Climate Group, Smart 2020: enabling the low carbon economy in the information age, 2008
16. Enerdata, Global energy statistical yearbook 2014, http://yearbook.enerdata.net/
17. Report ITU-R M.2410, Minimum requirements related to technical performance for IMT-2020 radio interface(s), Nov 2017
18. Document IMT-2020/02 (Rev. 1), Submission, evaluation process and consensus building for IMT-2020, Feb 2017

Chapter 2
5G Fundamental Air Interface Design

This chapter describes, in detail, the air interface of NR and LTE side by side. The basic aspects of LTE that NR builds upon are first described, followed by sections on the NR-specific designs for the carrier, frame structure, physical channels, and reference signals. Then the global candidate spectrum for 5G NR is introduced which demonstrates the new characteristic with wider frequency range and bandwidth impacting on NR design. With respect to C-band as NR deployment typical spectrum, the coverage issue is identified and the mechanism of UL/DL decoupling scheme (i.e., LTE/NR spectrum sharing) to solve the coverage issue is mainly introduced. Afterwards, NR physical layer technologies including waveform, Polar/ LDPC codes, MIMO, and mMTC are described, and in all cases the enhancements of NR made relative to LTE are made apparent.

2.1 LTE Air Interface Overview

As the NR air-interface borrows some design from LTE, one straightforward way to understand NR for those who are at least passingly familiar with LTE is to look at the differences between LTE and NR. This section will review the significant parts of LTE that are critical to the understanding of NR. The interested reader on details of LTE is referred to some literatures such as [1] and [2].

LTE standardization starts from Release-8 in 3GPP, and then many features were standardized in the following releases from Release-8 to Release-14. The evolution in standard from Release-15 which parallels NR is still going on to support different new features from industry interests. The designed LTE network can support diverse services including eMBB, Voice, Multimedia Broadcast Multicast Services (MBMS), D2D, IoT, Vehicle, etc. These services can be transmitted in the form of unicast transmission, or Single-Cell point to multipoint transmission (SC-PTM), or MBSFN (MBMS Single-Frequency Network) transmission. The transmission is mainly operated in the licensed spectrum, but eMBB can also be transmitted in the

© Springer Nature Switzerland AG 2020
Wan Lei et al., *5G System Design*, https://doi.org/10.1007/978-3-030-22236-9_2

unlicensed spectrum. For conciseness of description, the overview of LTE air interface in the following sections will focus on unicast transmission for both downlink and uplink in the licensed spectrum.

2.1.1 LTE Frame Structure

The frame structure is dependent on the duplex scheme applied in the spectrum. Two duplex schemes FDD and TDD are supported. FDD is operated in the paired spectrum (see Sect. 2.3.1), where the downlink and uplink transmission are performed in different carrier frequency. For FDD, whether downlink and uplink transmission can occur simultaneously depends on UE capability. Half duplex FDD is for the case when the UE cannot transmit and receive at the same time while there are no such restriction for full duplex FDD. TDD is operated in the unpaired spectrum (see Sect. 2.3.1), where the downlink and uplink transmission occur in different time instance of the same carrier frequency. For TDD, switching time is needed to guarantee the switching from downlink to uplink transmission.

Two different frame structures, denoted as Type 1 and Type 2, are supported, which are applicable to FDD and TDD operated in licensed spectrum, respectively.

For frame structure type 1, one radio frame consists of ten subframes and each subframe is 1 ms consisting of two slots. There are ten subframes available for downlink transmission and ten subframes available for uplink transmission in each 10 ms interval as shown in Fig. 2.1.

For frame structure type 2, the length of one radio frame is 10 ms and each radio frame consists of ten subframes with 1 ms each. Each subframe consists of two slots of length 0.5 ms. The subframes in one radio frame are reserved for downlink transmissions, switching from downlink to uplink transmission and uplink transmissions separately. The subframe(s) reserved for switching transmission direction is called special subframe. There are three fields in the special subframe which are DwPTS, GP, and UpPTS, where DwPTS and UpPTS are used for downlink and uplink

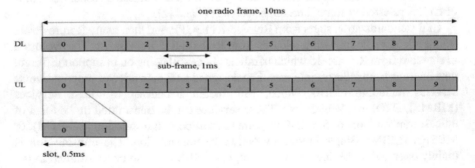

Fig. 2.1 Frame structure type 1

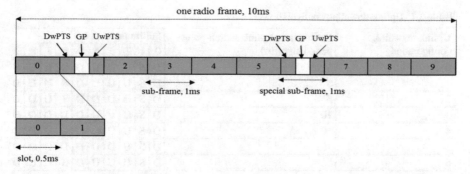

Fig. 2.2 Frame structure type 2

transmission separately as shown in Fig. 2.2. GP is the time duration without both downlink and uplink transmission, which is to mainly avoid the overlap of uplink reception and downlink transmission at the base station side. The length of GP is related to the cell size as determined by the round trip time of propagation. As there are different deployment scenarios in the practical network, several GP configurations are supported in [3]. The supported downlink-to-uplink switching periodicity are 5 and 10 ms.

The uplink-downlink subframe configuration in one radio frame depends on the UL/DL traffic ratio. Normally, the DL/UL traffic is asymmetric and DL has more traffic than UL statistically [4], and therefore more DL subframes are needed in one radio frame. There are seven different uplink-downlink subframe configurations defined in [3] to match different DL/UL traffic pattern, which is provided in Table 2.1 [3]. In the existing TDD commercial networks over the world, uplink-downlink configuration 2 is a typical configuration.

As downlink and uplink transmission are operated in the same carrier frequency for TDD, serious cross-link interference may occur without network synchronization, e.g., uplink transmission suffers from downlink transmission of neighboring cell as illustrated in Fig. 2.3. In other words, the same uplink-downlink configuration needs to be configured among the different cells of the same operator or between different operators. Due to the requirement of such strict configuration, the uplink-downlink configuration is semi-static and cannot be changed dynamically; otherwise it will cause severe cross-link interference.

2.1.2 Physical Layer Channels

The properties of the physical layer channels are described in this section starting with the multiple access schemes.

Table 2.1 Uplink-downlink configurations

Uplink-downlink configuration	Downlink-to-uplink switch-point periodicity (ms)	Subframe number									
		0	1	2	3	4	5	6	7	8	9
0	5	D	S	U	U	U	D	S	U	U	U
1	5	D	S	U	U	D	D	S	U	U	D
2	5	D	S	U	D	D	D	S	U	D	D
3	10	D	S	U	U	U	D	D	D	D	D
4	10	D	S	U	U	D	D	D	D	D	D
5	10	D	S	U	D	D	D	D	D	D	D
6	5	D	S	U	U	U	D	S	U	U	D

Note: "D", "S", and "U" represent the subframe is reserved for downlink, downlink to uplink switching, and uplink transmission, respectively

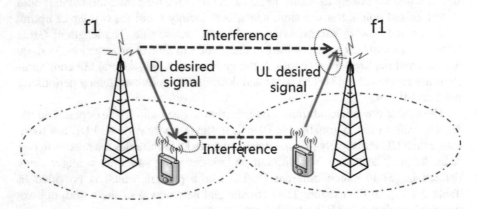

Fig. 2.3 Cross-link interference

2.1.2.1 Multiple-Access Scheme

Orthogonal Frequency-Division Multiplexing (OFDM)-based scheme is the basic transmission scheme for LTE downlink and uplink transmission. The basic principle of OFDM can be found in some textbooks [1, 5, 6]. OFDM is a multi-carrier technology which generates a number of orthogonal subcarriers with small bandwidth. Each subcarrier with small bandwidth faces a relatively flat fading in the propagation, which can effectively remove the impact of frequency selective fading and simplify the equalizer at the receiver especially for wideband transmission. OFDM can be implemented by IFFT which is illustrated in Fig. 2.4, and DFT is used for OFDM reception.

Some of the main aspects that led to the selection of OFDM for the downlink transmission include low cost and complexity implementation for wide (20 MHz) bandwidth, inherent and simple multipath protection through the use of a cyclic prefix, and a naturally scalable implementation where multiple system bandwidths can be supported.

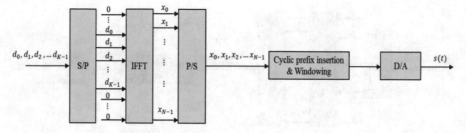

Fig. 2.4 OFDM implementation

Due to the multi-carrier transmission, the instantaneous OFDM signal has a large variation and results in high peak to average power ratio (PAPR). For the LTE uplink, PAPR was a hot discussion, soon to be replaced with the more accurate but not-as-straightforward to calculate cubic metric (CM) [7]. The power amplifier is required to have a large linear dynamic range, which will cause high cost and low efficiency of power amplifier. Alternatively, the nonlinearity of the device results in the distortion of the signal with large variation distortion, and power back-off is needed which reduces the cell coverage. Hence OFDM is not a good choice for UE in the uplink transmission. The selection of a single carrier waveform allowed for a substantially lower CM for the uplink, thus facilitating lower-cost lower-power consumption UE. The SC-FDMA waveform used is "DFT"-spread OFDM (DFT-S-OFDM), using with an extra DFT block making the net signal single carrier, which is illustrated in Fig. 2.5. SC-FDMA with cyclic prefix in the form of DFT-S-OFDM is to achieve uplink inter-user orthogonality and to enable efficient frequency-domain equalization at the receiver side. This allows for a relatively high degree of commonality with the downlink OFDM scheme, and the same parameters, e.g., clock frequency, can be reused [8].

Based on the downlink and uplink transmission scheme described above, UE multiplexing in the downlink and uplink transmission is operated by OFDMA and SC-FDMA as the multiple access scheme, respectively. As DFT-S-OFDM is based on OFDM, different UEs in the downlink or uplink transmission can be multiplexed over different subcarriers within a certain time interval in terms of several OFDM symbols. An exemplary downlink and uplink UE multiplexing is illustrated in Fig. 2.6.

The downlink supports the transmission to UE in either localized or distributed manner. For localized transmission, a set of resources being consecutive in the frequency domain is allocated to achieve scheduling gain. In this case, it implies that the base station knows the part of the wideband channel that is of good quality for a certain UE, and then schedules the corresponding part for transmitting data to the UE. The distributed transmission is that a set of resources allocated for a certain UE is distributed in the frequency domain to achieve frequency diversity gain. Normally, due to the high mobility or the worse channel condition, the UE cannot accurately track or timely report the channel state information to the base station. In this case, since scheduling gain cannot be achieved, it is better to schedule for frequency diversity gain. It can be seen that knowledge of the channel information is key to making good scheduling decision and, thus, the UE is required to report the channel state information to the base station.

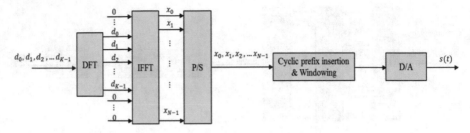

Fig. 2.5 DFT-S-OFDM operation

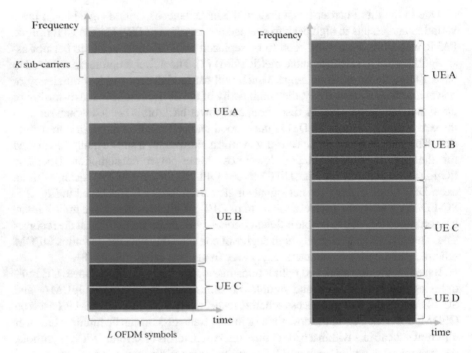

Fig. 2.6 Illustration of DL multiplexing and UL multiplexing

In the uplink, in order to keep the single carrier property with low CM, the UE multiplexing is different from downlink. All the subcarriers allocated for one UE have to be localized.

2.1.2.2 System Bandwidth

LTE targets for operating below 6 GHz and most of the existing LTE commercial networks are deployed in sub-3 GHz licensed spectrum. As the spectrum has multiple channel bandwidths existing in sub-3 GHz, LTE is designed to support scalable channel bandwidth in order to facilitate the deployment. The supported channel bandwidth is shown in Table 2.2 [9].

Table 2.2 Channel bandwidth and the transmission bandwidth configuration

Channel bandwidth [MHz]	1.4	3	5	10	15	20
Transmission bandwidth configuration N_{RB}	6	15	25	50	75	100

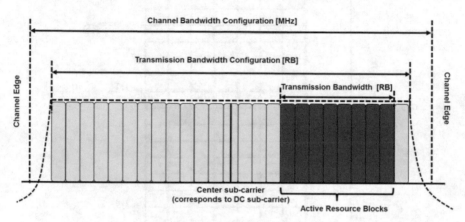

Fig. 2.7 Definition of channel bandwidth and transmission bandwidth configuration for one LTE carrier

If all the subcarriers corresponding to a certain channel bandwidth are utilized, there is the problem of out-of-band emissions because the added window for the OFDM symbol cannot completely remove the side lobes in the frequency domain. The out-of-band emissions will interfere with the adjacent channel. To avoid the serious impact of out-of-band emissions, a number of subcarriers at the edge of channel bandwidth need to be used as guard band without data transmission. In LTE, at least 10% of channel bandwidth is required for the guard band. The transmission bandwidth is the actual bandwidth available for data transmission, which is smaller than the channel bandwidth (see Fig. 2.7) [9]. The transmission bandwidth is expressed in terms of the number of resource blocks (RB), and the bandwidth of one resource block is 180 kHz. The spectrum utilization is defined as the ratio of transmission bandwidth over channel bandwidth. It can be observed there is 90% spectrum utilization for LTE except for the 1.4 MHz bandwidth.

2.1.2.3 Numerology

The subcarrier spacing of OFDM in LTE is selected as 15 kHz for both downlink and uplink unicast transmission. It is noted that LTE also supports 7.5 kHz subcarrier spacing for dedicated MBMS network and 3.75 kHz subcarrier spacing for NB-IoT uplink. The selection of subcarrier spacing takes into account the impact of Doppler spread which is related to the candidate carrier frequency of LTE network and the supported velocity of up to 350 km/h [10]. With 15 kHz subcarrier spacing, the number of subcarriers for a channel bandwidth can be determined. A cyclic prefix is used to mitigate the impact of delay spread at the cost of overhead. Two cyclic prefix length including normal cyclic prefix and extended cyclic prefix are supported.

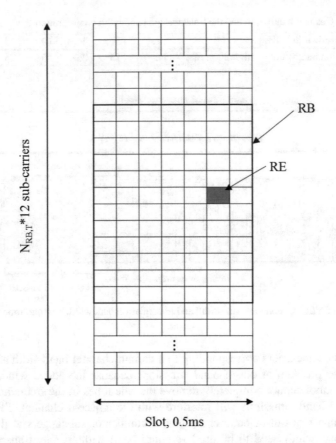

Fig. 2.8 Resource grid of normal cyclic prefix

The extended cyclic prefix is applied for the very large cell, e.g., 100 km cell radius and MBSFN transmission by using large number of cells simultaneously. The selection of cyclic prefix is a balance of overhead and mitigating delay spread impact.

In case of normal cyclic prefix, there are seven OFDM symbols in one slot with 0.5 ms (Fig. 2.8). For the extended cyclic prefix, the number of OFDM symbols per slot is 6. A resource block (RB) is defined as seven consecutive OFDM symbols with normal cyclic prefix in the time domain and 12 consecutive subcarriers in the frequency domain. Resource element (RE) is defined as one subcarrier over one OFDM symbol, which is the smallest resource unit. There are 84 (7∗12) REs in one resource block in the case of normal cyclic prefix while there are 72 REs in the case of extended cyclic prefix. The selection of RB size is a trade-off between resource allocation size and padding for small packets.

A resource grid consists of a number of $N_{RB,T}$ RBs, where $6 \leq N_{RB,T} \leq 110$. The functionality defined in the specification [3] supports the downlink and uplink transmission from 6 to 110 RBs; however, the actual number of used RBs is restricted by the transmission bandwidth defined in Table 2.2.

The scheduling unit in the time domain is one subframe. A physical resource block (PRB) pair is defined as the two physical resource blocks in one subframe having the same frequency position. For the downlink and uplink localized transmission, the resource allocation unit is physical resource block pair, i.e., the same resource blocks in the two slots within one subframe are allocated. However, for the downlink distributed transmission or uplink transmission with intra-subframe frequency hopping, the allocated resource blocks for a certain UE in the two slots of one subframe are different.

2.1.2.4 Physical Channel Definition

A downlink physical channel is defined as a set of resource elements carrying information originating from higher layers [3]. The following downlink physical channels defined:

- Physical Downlink Shared Channel, PDSCH
- Physical Broadcast Channel, PBCH
- Physical Multicast Channel, PMCH
- Physical Control Format Indicator Channel, PCFICH
- Physical Downlink Control Channel, PDCCH
- Physical Hybrid ARQ Indicator Channel, PHICH

The information from higher layer is carried on the channels defined in higher layer including logical channels and downlink transport channels. The mapping from higher layer channels to physical channels are given in Fig. 2.9 [11].

The following uplink physical channels are defined [3, 11] (Fig. 2.10):

- Physical Uplink Shared Channel, PUSCH
- Physical Uplink Control Channel, PUCCH
- Physical Random Access Channel, PRACH

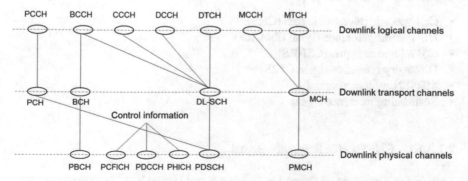

Fig. 2.9 Downlink channel mapping

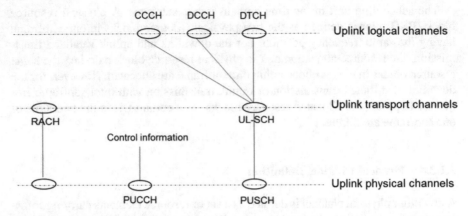

Fig. 2.10 Uplink channel mapping

2.1.3   Reference Signal

Reference signal refers to the so-called "pilot signal" used for channel estimation by the receiver. Specifically in LTE, a reference signal is a predefined signal transmitted over a set of predefined resource elements in a resource grid. The concrete purpose of the estimated channel depends on the definition of the corresponding reference signal. From the receiver perspective, the reference signal is utilized to obtain estimates of the channel of the resource grid conveying it, and therefore the channel of symbols conveyed on the same resource grid can be inferred from the derived channel. Each reference signal is associated with one antenna port.

2.1.3.1   Downlink Reference Signals

Downlink reference signals are used for downlink channel measurement and/or coherent demodulation of downlink transmission. There are the following reference signals defined in downlink [3]:

- Cell-specific reference signal (CRS)
- UE-specific reference signal (DM-RS)
- CSI reference signal (CSI-RS)
- Discovery reference signal (DRS)
- MBSFN reference signal
- Positioning reference signal

2.1.3.1.1   Cell-Specific Reference Signal

The CRS is a basic signal defined in Release-8 and is used by UEs accessing to LTE network. The demodulation of PBCH, PDCCH/PCFICH/PHICH, and PDSCH in transmission mode (TM) 1–6 [12] uses the channel estimation based on CRS. When precoding is used for PDSCH transmission in case of multiple CRS, the used pre-coding matrices for PDSCH is signaled in downlink control information (DCI). The UE utilizes the estimated channel from CRS and the signaled precoding matrices to generate the equivalent channel for coherent demodulation of PDSCH. CRS is also used to derive the channel state information (CSI) in the form of CQI/RI/PMI for PDSCH scheduling in TM 1–8. In addition, CRS can be used for both synchroniza-tion tracking after initial synchronization by PSS/SSS and RRM measurement for cell selection.

CRS is transmitted in all downlink subframes for frame structure Type 1, all downlink subframes and DwPTS for frame structure Type 2. In the normal sub-frame, CRS is distributed over the two slots within one subframe. CRS is only transmitted in the first two OFDM symbols of MBSFN subframe. CRS is defined for 15 kHz subcarrier spacing only. The number of CRS in one cell can be one, two, and four, which corresponds to one, two, and four antenna ports. The CRS antenna ports are named as antenna port 0, 1, 2, and 3. The structure of CRS in one resource block pair in case of normal cyclic prefix is illustrated in Fig. 2.11 [3]. CRS spans over the whole system bandwidth in its located OFDM symbol. It is noted that the frequency position of different CRS in Fig. 2.11 is relative.

For each CRS port, the distance between two adjacent reference signals in one OFDM symbol is 6 REs, and the starting position of CRS has six frequency shifts. The frequency shift is cell-specific, which is obtained by performing $N_{cell\ ID}$ mod 6. $N_{cell\ ID}$ is the cell ID index selected from [0, 503], where there are 504 cell IDs defined. The frequency shift is to enable the adjacent cells to avoid the CRS inter-cell interfer-ence. The CRS for antenna port 0/1 is located at the first and fifth OFDM symbol of one slot in a staggered manner; however, the CRS for antenna port 2/3 is only located in the second OFDM symbol of each slot in order to reduce the overhead of reference signal. The CRS overhead in case of two antenna ports is 16/168 = 9.52%, and it would be 19.05% if assuming the same density applied for antenna port 2 and 3 in case of four antenna ports which is too high. The lower density of antenna port 2/3 may degrade the channel estimation performance at very high velocity.

The reference signal of antenna port 0 and 1 is separated by FDM. The resource elements on antenna port 0 are not used for transmission when these resource ele-ments correspond to the positions of the reference signal of antenna port 1. The power of the resource elements not used for transmission can be transferred to the resource elements of reference signal within the same OFDM symbol, which can improve the channel estimation performance. This is applied for antenna port 2 and 3 as well.

In addition to using CRS for CSI measurement and coherent demodulation of physical channels, CRS is also used for measurement for cell selection/reselection and cell handover. CRS port 0 is used for determining reference signal received power (RSRP), and CRS port 1 can be optionally used depending on whether the UE can reliably detect CRS port 1 is available. RSRP is defined as the linear average over the power contributions (in [W]) of the resource elements that carry cell-specific reference signals within the considered measurement frequency bandwidth [13]. The network makes the decision whether to perform cell selection/reselection or cell handover based on the reported RSRP and another parameter, the reference signal received quality (RSRQ) [13].

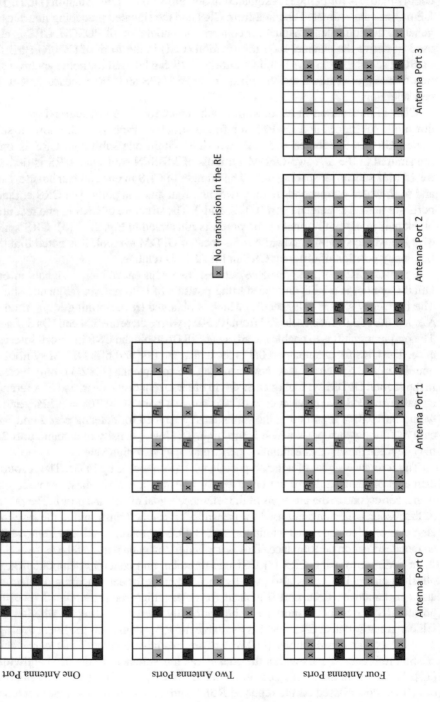

Fig. 2.11 Structure of CRS in one subframe with normal cyclic prefix

2.1.3.1.2   UE-Specific Reference Signal

UE-specific reference signal[1] is used for demodulation of PDSCH in TM 7–10 [12]. The UE-specific reference signal is only present in the scheduled resource blocks of the PDSCH transmission. For the UE-specific reference signal-based PDSCH transmission, the used precoding matrix for PDSCH is transparent to UE and, hence, the signaling in DCI is not needed. As the UE does not have the information how the precoding matrix is used over different resource blocks, it cannot perform channel interpolation among different scheduled resource blocks. In order to balance the precoding gain and channel estimation performance loss, PRB bundling is supported in TM9 and it is assumed that the same precoding matrix is applied in a precoding resource block groups (PRGs). The PRG size depends on the system bandwidth (see Table 2.3) [12].

UE-specific reference signal-based PDSCH transmission TM7 was first introduced in Release-8, which only supports single layer beamforming. The defined UE-specific reference signal corresponds to antenna port 5 shown in Fig. 2.12 [12].

In Release-9, TM8 was introduced to support dual-layer beamforming and two new reference signals were defined corresponding to antenna 7 and 8 in Fig. 2.13. Due to the support of backward compatibility, the introduced UE-specific reference signal shall avoid the collision with cell-specific reference signal, PDCCH and PBCH/PSS/SSS. The position is located at the last two OFDM symbols of each slot within one subframe. There are 12 REs for each UE-specific reference signal in one resource block pair. The two UE-specific reference signals occupy the same resource elements but separated by CDM. The two codes [+1   +1   +1   +1] and [+1   −1   +1   −1] spanning in the time domain are used for antenna port 7 and 8, respectively. The designed UE-specific reference signal in TM8 supports the dynamic switching between SU-MIMO and MU-MIMO. The MU-MIMO transmissions are transparent to UE, i.e., SU-MIMO operation is always assumed from the UE perspective. There are two candidate predefined sequences for the reference signal in TM8, and the two sequences are quasi-orthogonal. For each of the predefined sequence, the generated UE-specific reference signals corresponding to antenna port 7 and 8 are orthogonal to each other; however, the UE-specific reference signals generated from the two predefined sequences are quasi-orthogonal. Which sequence is used for the UE-specific reference signal of a certain UE is indicated by DCI. Hence, it can support up to four UEs for MU-MIMO in TM8.

In Release-10, TM9 was introduced to support up to eight layers transmission based on UE-specific reference signal, and thus up to eight UE-specific reference signals need to be defined. The UE-specific reference signals are extended from antenna port 7 and 8 by extending the number of orthogonal cover code (OCC) from 2 to 4 and the number of resource elements from 12 to 24. The corresponding antenna ports are 7, 8, 9, 10, 11, 12, 13, and 14. For each group of 12 resource elements, there are four antenna ports which are separated by four codes. The mapping between the antenna port and the OCC code is shown in Table 2.4. The mapping of

---

[1] UE-specific reference signal are also known as demodulation reference signal (DM-RS)

Table 2.3 PRG size

System bandwidth (number of PRBs)	PRG size (number of PRBs)
≤10	1
11–26	2
27–63	3
64–110	2

Fig. 2.12 Structure of antenna port 5 with normal cyclic prefix

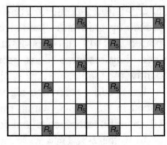

Antenna Port 5

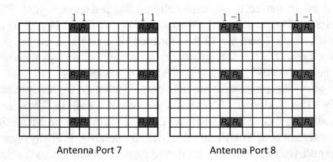

Antenna Port 7   Antenna Port 8

Fig. 2.13 UE-specific reference signals for antenna port 7 and 8 (normal cyclic prefix)

Table 2.4 The mapping between antenna port and OCC code

Antenna port	OCC code
7	$[+1 \ +1 \ +1 \ +1]$
8	$[+1 \ -1 \ +1 \ -1]$
9	$[+1 \ +1 \ +1 \ +1]$
10	$[+1 \ -1 \ +1 \ -1]$
11	$[+1 \ -1 \ +1 \ -1]$
12	$[-1 \ -1 \ +1 \ +1]$
13	$[+1 \ -1 \ -1 \ +1]$
14	$[-1 \ +1 \ +1 \ -1]$

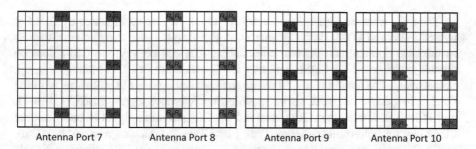

Fig. 2.14 UE-specific reference signals for antenna port 7, 8, 9, 10 (normal cyclic prefix)

UE-specific reference signal for antenna port 7, 8, 9, and 10 in case of normal cyclic prefix is illustrated in Fig. 2.14.

2.1.3.1.3 CSI Reference Signal

CSI reference signal is used for measuring the channel state information, which was introduced in LTE Release-10 and applicable for both TM9 and TM10. To increase the peak data rate, eight layers transmission support was a requirement for Release-10 development. Extending the methodology of CRS antenna ports 0/2/4 design for eight antenna ports is inefficient and complicated. First, the resource overhead of eight CRS ports would not be acceptable. As described in Sect. 2.1.3.1, the overhead of four CRS antenna ports is already 14.28% and the overhead of eight antenna ports may be about 30%. Furthermore, in a practical network, the number of UEs operating larger than rank 4 transmission is small. Second, due to the requirement of backward compatibility, the legacy UEs operating in LTE Release-8 are required to work in LTE Release-10 and the legacy CRS antenna ports have to be presented in Release-10 network. It would be complicated to coexist the newly defined eight antenna ports and the legacy CRS antenna ports. These considerations necessitated a new methodology for reference signal design that are different from CRS design. The functionalities of reference signal including channel measurement and demodulation are performed by CSI reference signal and UE-specific reference signal separately. The UE-specific reference signal is described in Sect. 2.1.3.1.2.

Although the CSI reference signal is introduced because of supporting eight layers transmission, the CSI reference signal is also defined for one, two, and four antenna ports in Release-10. The CSI reference signal are transmitted on one, two, four, or eight antenna ports, and the corresponding antenna ports are 15, 15/16, 15/16/17/18, and 15/16/17/18/19/20/21/22, respectively. As the CSI reference signal is only for channel measurement, low density CSI reference signal is desired to reduce the overhead. Optimizing the trade-off between CSI reference signal overhead and channel measurement performance, the transmission of each CSI reference signal is wideband and the average number of RE per port in one resource block pair is 1 in the frequency domain. The periodicity in the time domain are configurable with values of 5, 10, 20, 40, and 80 subframes.

The CSI reference signals in one resource block pair are multiplexed by CDM or hybrid FDM and TDM. Each CSI reference signal spans over two consecutive REs in the time domain, and the two consecutive REs can convey two CSI reference signals multiplexed by CDM. The two OCC codes are [+1   +1] and [+1   −1], respectively. The locations of CSI reference signal should avoid collisions with the CRS, UE-specific reference signal, positioning reference signal, PDCCH, PBCH, and PSS/SSS. There are 40 REs reserved for CSI reference signal in one resource block pair for both frame structure Type 1 and Type 2, which correspond to 20, 10, and 5 CSI reference signal configurations for the number of CSI reference signals being 1 or 2, 4, and 8, respectively. The CSI reference signal configurations for different number of CSI reference signal applicable for both frame structure type 1 and type 2 in case of normal cyclic prefix are illustrated in Fig. 2.15, where the resource elements marked by the same color correspond to one CSI reference signal configuration and the number represents the index of CSI reference signal configuration; the resource elements marked with gray and "x" represent the positions of CRS or UE-specific reference signal. In addition, there are additional 12, 6, and 3 CSI reference signal configurations corresponding to a number of 1 or 2, 4, and 8 CSI reference signals defined for frame structure type 2 only [3].

The number of CSI reference signal and the corresponding CSI reference signal configuration are configured by higher layer in a UE-specific manner. The CSI reference signals within a CSI reference signal configuration are multiplexed by CDM and FDM, where the four pair of antenna ports 15/16, 17/18, 19/20, and 21/22 are multiplexed by FDM and the two antenna ports of each pair are code multiplexing. Figure 2.16 takes CSI reference signal configuration 0 as an example to illustrate the CSI reference signal of different antenna port.

As the CSI reference signal is a wideband transmission, it may collide with the resource blocks scheduled for Release-8 legacy UEs. In this case, data puncturing is performed for Release-8 legacy UEs. Since the base station knows the CSI reference signal configuration and Release-8 legacy UEs' scheduling, the impact of data puncturing on the legacy UEs can be mitigated by the scheduling and link adaption. For the Release-10 and later release UEs, rate matching is operated around the CSI reference signal. Due to the UE-specific CSI reference signal configuration, a UE does not know the configured CSI reference signal of other UEs, and therefore cannot do rate matching around the configured CSI reference signal. To enable the rate

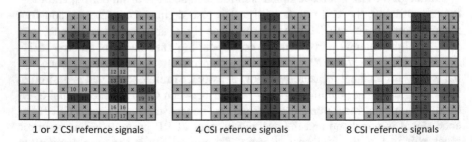

1 or 2 CSI refernce signals   4 CSI refernce signals   8 CSI refernce signals

Fig. 2.15 CSI reference signal configurations

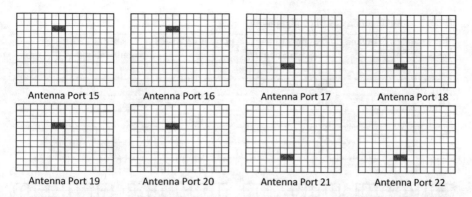

Fig. 2.16 Antenna ports of CSI reference signal configuration 0 (normal cyclic prefix)

matching, a zero power CSI reference signal configuration is signaled to a UE and the UE assumes zero power transmission in these resource elements. The zero power CSI reference signal configuration reuse the CSI reference signal configuration of four CSI reference signals. The CSI reference signal configurations for which the UE shall assume zero transmission power in a subframe are given by a 16-bit bitmap. For each bit set to one in the 16-bit bitmap, the UE shall assume zero transmission power for the resource elements corresponding to the CSI reference signal configuration.

The number of antenna ports was extended to 12, 16, 20, 24, 28, and 32 in Release-13 and -14 in the context of supporting full-dimension MIMO (FD-MIMO) [14, 15]. The indices for 12, 16, 20, 24, 28, and 32 antenna ports are labeled as $p = 15,..., 26, p = 15,..., 30, p = 15,..., 34, p = 15,..., 38, p = 15,..., 44, p = 15,...,$ 38, and $p = 15,..., 46$, respectively. In case the number of antenna ports is larger than 8, the corresponding CSI reference signals are generated by aggregating $N_{res}^{CSI} > 1$ CSI reference signal configurations within one subframe. The number of antenna ports N_{CSIres}^{ports} per CSI reference signal configuration is 4 or 8. Although two antenna ports per CSI reference signal configuration is the most flexible configuration for aggregation, it will result in significant number of configurations to increase signaling overhead and configuration complexity. The supported configurations of aggregated CSI-RS configurations are shown in Table 2.5.

The resource elements in one resource block pair corresponding to each aggregated CSI reference signal configuration with four or eight antenna ports is the same as shown in Fig. 2.15. In the case of CDM where OCC = 2 is used, each reference signal spans over two resource elements and the two reference signals multiplexed on the two same resource elements are separated by [+1   +1] and [+1   −1], which is shown in Fig. 2.16. When CDM OCC = 4 is applied, each reference signal spans over four resource elements within one resource block pair and separated by [+1   +1   +1   +1],   [+1   −1   +1   −1],   [+1   +1   −1   −1],   and [+1   −1   −1   +1]. An example of CSI reference signal configuration 0 with four and eight antenna ports in the case of CDM OCC = 4 is illustrated in Figs. 2.17 and 2.18, respectively.

Table 2.5 Aggregation of CSI configurations

Total number of antenna ports $N_{res}^{CSI} N_{CSI\,res}^{ports}$	Number of antenna ports per CSI-RS configuration $N_{CSI\,res}^{ports}$	Number of CSI-RS configurations N_{res}^{CSI}
12	4	3
16	8	2
20	4	5
24	8	3
28	4	7
32	8	4

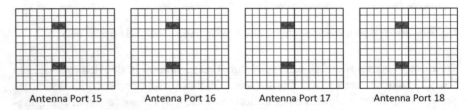

Fig. 2.17 CSI reference signal configuration 0 with 4 antenna ports (CDM OCC = 4, normal cyclic prefix)

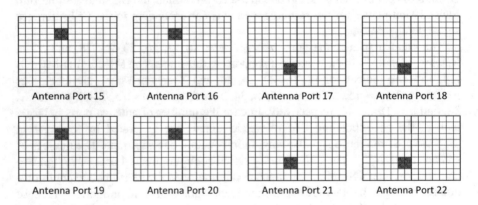

Fig. 2.18 CSI reference signal configuration 0 with 8 antenna ports (CDM OCC = 4, normal cyclic prefix)

The mapping between the antenna ports of one CSI reference signal configuration and the code of OCC = 4 is provided in Tables 2.6 and 2.7.

When CDM OCC = 8 is applied, each reference signal spans over eight resource elements within one resource block pair and are separated by the code with length equals to 8. It should be noted that CDM OCC = 8 is only applicable for 32 and 24 antenna ports. For the case of CDM OCC = 8 and the number of antenna ports equals to 32, the aggregated 4 CSI reference signal configurations with eight antenna ports each are restricted to one of {0, 1, 2, 3}, or {0, 2, 3, 4} or {1, 2, 3, 4}; for the number of antenna ports equals to 24, the aggregated 3 CSI reference signal

Table 2.6 OCC code assignment for one CSI reference signal configuration with four antenna ports

REs	[w(0) w(1) w(2) w(3)]			
Antenna port				
15	+1	+1	+1	+1
16	+1	−1	+1	−1
17	+1	+1	−1	−1
18	+1	−1	−1	+1

Table 2.7 OCC code assignment for one CSI reference signal configuration with eight antenna ports

REs	[w(0) w(1) w(2) w(3)]				[w(0) w(1) w(2) w(3)]			
AP								
15	+1	+1	+1	+1				
16	+1	−1	+1	−1				
17					+1	+1	+1	+1
18					+1	−1	+1	−1
19	+1	+1	−1	−1				
20	+1	−1	−1	+1				
21					+1	+1	−1	−1
22					+1	−1	−1	+1

configurations with eight antenna ports each are restricted to $\{1, 2, 3\}$. The aggregated CSI reference signal configurations can be found in Fig. 2.15. The OCC code assignment for different antenna ports in case of 32 and 24 antenna ports are illustrated in Tables 2.8 and 2.9, respectively.

For the number of antenna ports larger than 8, the total number of antenna ports equals $N_{\text{res}}^{\text{CSI}} N_{\text{CSI res}}^{\text{ports}}$ which are labeled as $p = 15, 16, \ldots 15 + N_{\text{res}}^{\text{CSI}} N_{\text{CSI res}}^{\text{ports}} - 1$. The number of antenna ports per CSI reference signal configuration is $N_{\text{CSI res}}^{\text{ports}}$, and the antenna ports are indexed as $p' = 15, 16, \ldots 15 + N_{\text{CSI res}}^{\text{ports}} - 1$. The relation between p and p' is [3].

if CDM OCC = 2

$$
p = \begin{cases}
p' + \dfrac{N_{\text{CSI res}}^{\text{ports}}}{2} i & p' \in \left\{ 15, 16, \ldots, 15 + \dfrac{N_{\text{CSI res}}^{\text{ports}}}{2} - 1 \right\} \\[3mm]
p' + \dfrac{N_{\text{CSI res}}^{\text{ports}}}{2} \left(i + N_{\text{res}}^{\text{CSI}} - 1 \right) & p' \in \left\{ 15 + \dfrac{N_{\text{CSI res}}^{\text{ports}}}{2}, \ldots, 15 + N_{\text{CSI res}}^{\text{ports}} - 1 \right\}
\end{cases}
$$

Table 2.8 CDM OCC = 8 for each antenna port with 32 antenna ports

REs of each antenna port in the aggregated CSI reference signal configurations				[w(0) w(1) w(2) w(3) w(4) w(5) w(6) w(7)]							
				First aggregated CSI reference signal configuration		Second aggregated CSI reference signal configuration		Third aggregated CSI reference signal configuration		Fourth aggregated CSI reference signal configuration	
15	17	19	21	+1	+1	+1	+1	+1	+1	+1	+1
16	18	20	22	+1	−1	+1	−1	+1	−1	+1	−1
23	25	27	29	+1	+1	−1	−1	+1	+1	−1	−1
24	26	28	30	+1	−1	−1	+1	+1	−1	−1	+1
31	33	35	37	+1	+1	+1	+1	−1	−1	−1	−1
32	34	36	38	+1	−1	+1	−1	−1	+1	−1	+1
39	41	43	45	+1	+1	−1	−1	−1	−1	+1	+1
40	42	44	46	+1	−1	−1	+1	−1	+1	+1	−1

Table 2.9 CDM OCC = 8 for each antenna port with 24 antenna ports

REs of each antenna port in the aggregated CSI reference signal configurations			[w(0) w(1) w(2) w(3) w(4) w(5) w(6) w(7)]							
CSI-RS Config. 1	CSI-RS Config. 2	CSI-RS Config. 3								
15	31	25	+1	+1	+1	+1	+1	+1	+1	+1
16	32	26	+1	−1	+1	−1	+1	−1	+1	−1
19	35	29	+1	+1	−1	−1	+1	+1	−1	−1
20	36	30	+1	−1	−1	+1	+1	−1	−1	+1
23	17	33	+1	+1	+1	+1	−1	−1	−1	−1
24	18	34	+1	−1	+1	−1	−1	+1	−1	+1
27	21	37	+1	+1	−1	−1	−1	−1	+1	+1
28	22	38	+1	−1	−1	+1	−1	+1	+1	−1

$$i \in \left\{0,1,\ldots N_{res}^{CSI} -1\right\}$$

else

$$p = p' + N_{CSI\,res}^{ports} i \quad p' \in \left\{15,16,\ldots,15 + N_{CSI\,res}^{ports} -1\right\}$$

$$i \in \left\{0,1,\ldots N_{res}^{CSI} -1\right\}$$

end

For each CSI reference signal configuration using CDM OCC = 2, the set of antenna ports $p' = 15, 16, \ldots, \dfrac{N_{\text{CSI res}}^{\text{ports}}}{2} - 1$ and $p' = \dfrac{N_{\text{CSI res}}^{\text{ports}}}{2}, \ldots N_{\text{CSI res}}^{\text{ports}} - 1$ represent the antenna ports corresponding to two different polarizations in case of cross-polarization antenna configuration. Hence, the mapping between p and p' can enable the overall antenna ports $p = 15, 16, \ldots, \dfrac{N_{\text{CSI res}}^{\text{ports}} N_{\text{res}}^{\text{CSI}}}{2} - 1$ and $p = \dfrac{N_{\text{CSI res}}^{\text{ports}} N_{\text{res}}^{\text{CSI}}}{2}, \ldots, N_{\text{CSI res}}^{\text{ports}} N_{\text{res}}^{\text{CSI}} - 1$ represent two different polarizations.

2.1.3.1.4 Discovery Signal

Discovery signal was introduced in Release-12 for small cell deployment scenario. In order to mitigate the interference between small cells and save energy, adaptive cell on/off that is adaptive with traffic load was defined [16]. Discovery signal is used to facilitate the fast and efficient discovery of small cells, which can reduce the transition time.

The discovery signal is transmitted in the downlink subframe(s) or DwPTS region of special subframe. A discovery signal consists of primary synchronization signal (PSS)/secondary synchronization signal (SSS), CRS port 0, and the configurable CSI-RS. The period of a discovery signal occasion for a cell can be configured. The period can be one to five consecutive subframes for frame structure type 1 and 2 to five consecutive subframes for frame structure type 2. The periodicity of the discovery signal occasion is configured by higher layer parameter.

During the period of a discovery signal occasion, PSS is in the first subframe of the period for frame structure type 1 or the second subframe of the period for frame structure 2, SSS is in the first subframe of the period, CRS port 0 is in all downlink subframes and in DwPTS of all special subframes in the period, and the CSI-RS is configured in some subframes within the period if there is.

2.1.3.1.5 Other Downlink Reference Signals

In addition to the reference signals described above, there are some other downlink reference signals in LTE including MBSFN reference signal which only applied for PMCH transmission, positioning reference signal for positioning, demodulation reference signals associated with EPDCCH, etc. Due to space limitations, these reference signals will not be discussed here and the details can be found in [3].

2.1.3.2 Uplink Reference Signals

Similar to the downlink reference signal, uplink reference signals are used for coherent demodulation and uplink channel measurement. In the downlink, the reference signal can be common for all the UEs in a broadcast manner, e.g., CRS;

however, for uplink the reference signals are UE-specific. Two types of uplink reference signals are supported:

• Uplink demodulation reference signal (DM-RS)
• Sounding reference signal (SRS)

2.1.3.2.1 Uplink Demodulation Reference Signal

Uplink DM-RS is used for channel estimation by the base station to perform coherent demodulation of PUSCH and PUCCH. There are two types of DM-RS, one type is associated with the PUSCH and the other with the PUCCH transmission. For DM-RS, only single antenna port is supported for PUSCH and PUCCH transmission in Release-8. In Release-10, it is extended to support two and four antenna ports for PUSCH transmission and two antenna ports for PUCCH transmission. The maximum number of DM-RS ports for PUSCH or PUCCH transmission is configured by higher layer in a UE-specific manner. The DM-RS antenna port labeling for PUSCH and PUCCH transmission is summarized in Table 2.10 [3].

DM-RS is described starting from single antenna port transmission for simplicity and then extended to the case with multiple antenna port transmission. As it is required to have low cubic metric transmission in uplink, DM-RS and data are time multiplexed in the assigned resource block(s) within one slot to preserve the single carrier property. DM-RS spans the same bandwidth as data in the frequency domain to estimate the channel of data transmission. For PUSCH, DM-RS is transmitted in the fourth OFDM symbol of the scheduled resource blocks within one slot and the remaining six OFDM symbols are used for data. The PUCCH transmission is constrained to one resource block. The multiplexing of PUCCH DM-RS and data in the resource block depends on PUCCH format, e.g., for PUCCH format 1/1a/1b, the third, fourth, and fifth OFDM symbols in one resource block are used for DM-RS transmission. The multiplexing of DM-RS and data is illustrated in Fig. 2.19.

Different from downlink reference signal, the uplink DM-RS sequence shall have constant amplitude in both frequency and time domain. The constant amplitude of DM-RS sequence in the frequency domain is for obtaining the evenly estimated channel performance over different subcarriers. The constant amplitude in the time domain is to keep low cubic metric. However, due to oversampling in IFFT

Table 2.10 Antenna ports for PUSCH and PUCCH

	Maximum number of antenna ports configured		
	1	2	4
PUSCH	10	20	40
	–	21	41
	–	–	42
	–	–	43
PUCCH	100	200	–
		201	–

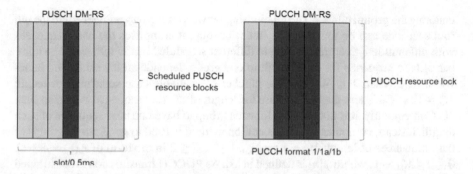

Fig. 2.19 TDM of DM-RS and data for PUSCH and PUCCH

operation, there would be some amplitude variation in the time domain but still with low cubic metric. In addition, it is also desirable that the DM-RS sequence have zero auto-correlation for optimizing the channel estimation and low cross-correlation property to mitigate inter-cell interference.

Zadoff-Chu (ZC) sequence was, therefore, selected as the uplink DM-RS sequence because of its good auto- and cross-correlational properties [17]. ZC sequence is defined by

$$x_q(m) = e^{-j\frac{\pi q m(m+1)}{N_{ZC}^{RS}}} \quad 0 \le m \le N_{ZC}^{RS} - 1$$

where N_{ZC}^{RS} is the length of ZC sequence and q represents the qth root ZC sequence. For a certain ZC sequence length N_{ZC}^{RS}, the number of available root sequence equals to the number of integers being relative prime to N_{ZC}^{RS}. That is to say the number of root sequence if the ZC sequence length is a prime number is maximized. More root sequences for a certain length means that different root sequences can be allocated to more different cells to mitigate inter-cell interference. For this reason, ZC sequence length is selected to be prime number in LTE.

DM-RS sequence spans the same bandwidth as PUSCH. The bandwidth of PUSCH depends on the number of assigned consecutive resource blocks, i.e., $M_{sc} = 12 * N_{RB}$, where N_{RB} is the number of scheduled resource blocks. The length of DM-RS sequence is equal to M_{sc}. As the ZC sequence length is a prime, a ZC sequence with length close to M_{sc} can be extended or truncated to generate the DM-RS sequence. The method of ZC sequence cyclic extension is adopted because of better cubic metric property [18]. The length N_{ZC}^{RS} of the ZC sequence is determined as the largest prime number being smaller than M_{sc}. Then the ZC sequence is extended to generate the DM-RS base sequence equal to M_{sc} by

$$r_{u,v}(n) = x_q\left(n \bmod N_{ZC}^{RS}\right), \quad 0 \le n \le M_{SC} - 1$$

There are multiple base sequences corresponding to each possible scheduled bandwidth. The base sequences are divided into 30 groups with $u \in \{0, 1, \ldots, 29\}$

denoting the group number. The relationship between the group number u and the qth root sequence can be found in [3]. In each group, it comprises the base sequences with different length corresponding to different scheduled bandwidth N_{RB}. The number of base sequence for each length in one group depends on the scheduled bandwidth N_{RB}. When $1 \leq N_{RB} \leq 2$, the number of available base sequences of length $M_{sc} = 12 * N_{RB}$ is less than 30 because the length of ZC root sequence is smaller than 30. Consequently, it is not possible for each group to have one base sequence of such length. Instead, computer search generation method is used to generate 30 groups of base sequences of length $M_{sc} = 12 * N_{RB}$, $1 \leq N_{RB} \leq 2$ in the form of $r_u(n) = e^{j\varphi(n)\pi/4}$, $0 \leq n \leq M_{SC} - 1$, where $\varphi(n)$ is defined in [3]. As PUCCH transmission is constrained within one resource block, DM-RS sequence for PUCCH will use the computer search sequences. The length N_{ZC}^{RS} of ZC sequence is larger than 30 when $3 \leq N_{RB}$ which allows for the generation of more than 30 base sequences for each length of M_{sc}. In this case, each group contains one base sequence ($v = 0$) of each length $M_{sc} = 12 * N_{RB}$, $3 \leq N_{RB} \leq 5$ and two base sequences ($v = 0, 1$) of each length $M_{sc} = 12 * N_{RB}$, $6 \leq N_{RB}$.

The DM-RS is located at the same position among different cells, which may cause serious inter-cell interference if the DM-RS base sequences are not assigned well. The base sequences are grouped to mitigate the inter-cell interference. The base sequences of each length are assigned to the different group, and one group consisted of the base sequences with different length. Each cell is associated with one group. In this case, the DM-RS interference between the cells using different group can be mitigated because of low cross-correlation property of ZC sequence. As there are only 30 base sequence groups, it can only enable 30 cells to associate with different base sequence group. In the case of 504 cells with different cell ID, each base sequence group will be reused by 17 cells. Clearly two cells associated with the same base sequence group will result in very serious interference. This situation is solved by cell planning, i.e., the adjacent cells are planned to associate different base sequence group. This method can avoid the adjacent cells using the same base sequence group, but it is inflexible and complicated because all the cell IDs need to be updated once a new cell is added in the network. Another method is to introduce base sequence group hopping over different time instance, i.e., the base sequence group of one cell varies over different slot to randomize the inter-cell interference.

There are 17 different hopping patterns and 30 different sequence shift patterns. The base sequence group number u is determined by

$$u = \left(f_{gh}\left(n_s\right) + f_{ss}\right) \bmod 30$$

The sequence group hopping can be enabled or disabled by the cell-specific higher layer parameter. The sequence group hopping pattern is generated by a pseudo-random sequence and the pseudo-random sequence generator is initialized with $\dfrac{N_{cell\,ID}}{30}$. There will be 30 cells associated with each sequence group hopping pattern. Then 30 sequence shift patterns associated with the 30 cells are used to generate different base sequence group number. The sequence shift pattern f_{ss} definition differs between PUSCH and PUCCH. The sequence shift pattern for PUSCH and PUCCH are provided by $f_{ss} = (N_{cell\ ID} + \Delta_{ss})$ mod 30, where $\Delta_{ss} \in \{0, 1, ..., 29\}$ and $f_{ss} = N_{cell\ ID}$ mod 30, respectively. Here TM10 uplink CoMP is not assumed; otherwise the physical cell ID $N_{cell\ ID}$ can be replaced by a higher configured virtual Cell ID. The detail can be found in [3].

In one cell, when the UEs are scheduled in different resource blocks for PUSCH transmission, the corresponding DM-RS base sequences from the same group are transmitted in FDM which are orthogonal to each other. However, when uplink multi-user MIMO or virtual MIMO is applied, multiple users are scheduled on the same resource blocks to transmit simultaneously. For PUCCH, multiple users are multiplexed by CDM in one resource block. In this case, the DM-RS sequences of different UEs need to be orthogonal. Since the DM-RS base sequences have the zero auto-correlation property, the cyclic shift of the base sequence in the time domain can be assigned to different UEs. The length of cyclic shift shall be larger than the delay spread in order to differentiate different UEs. With the cyclic prefix length selected to be larger than the delay spread, the cyclic shift length can be determined to be larger than the cyclic prefix. Based on such consideration, it can support a maximum of 12 cyclic shifts. The cyclic shift in the time domain is equal to the phase shift in the frequency domain. The phase shift of base sequence is given by

$$r_{u,v}^{\alpha}(n) = e^{j\alpha n} r_{u,v}(n), 0 \leq n \leq M_{sc} - 1$$

where $\alpha = \dfrac{2\pi n_{cs}}{12}, n_{cs} \in \{0,1,...11\}$. The value of n_{cs} is signaled by DCI for PUSCH and higher layer configured for PUCCH. The detail of signaling can be found in [19].

In Release-10, uplink multiple antennas transmission is introduced to support up to four layers PUSCH transmission. It is needed to define up to 4 DM-RS and each DM-RS is associated with one layer. Similar to uplink virtual MIMO, the cyclic shifts of the base sequence can be used for different layers. In addition, orthogonal cover code (OCC) in time domain is also used to generate two dimension DM-RS. The OCC code [+1   +1] and [−1   −1] is used for the DM-RS in the two slots of one subframe as shown in Fig. 2.20. As there are 12 cyclic shifts and 2 OCC code in time domain, in principle there are up to 24 DM-RS available. The cyclic shift is sensitive to frequency selectivity and the OCC code is sensitive to mobility. To achieve the better orthogonality of the DM-RS corresponding to different layer, there are eight combinations of cyclic shift and OCC code are defined for up to four layers [3].

2.1.3.2.2 Uplink Sounding Reference Signal (SRS)

Uplink sounding reference signal is not associated with PUSCH transmission, i.e., there is no need to transmit SRS and PUSCH or PUCCH together. The SRS is used by base station for uplink channel measurement. The channel measurement is for

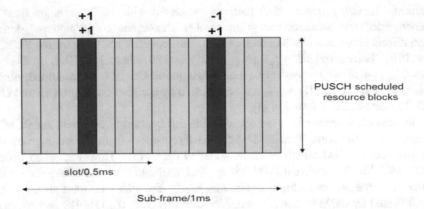

Fig. 2.20 PUSCH DM-RS for multiple antennas

Table 2.11 SRS antenna port number

	Number of configured SRS antenna ports		
	1	2	4
SRS antenna port number	10	20	40
	–	21	41
	–	–	42
	–	–	43

uplink link adaption, uplink scheduling, power control, etc. In addition, the SRS can also be used for downlink beamforming due to channel reciprocity with TDD system.

For SRS, single antenna port transmission was supported in Release-8, and then up to four SRS antenna ports transmission was introduced in Release-10. The number of SRS antenna ports is configured by higher layer, which can be configured as 1, 2, and 4. The SRS antenna port number and the number of configured SRS antenna ports are given in Table 2.11.

SRS Transmission in Time Domain

The subframes used for SRS transmission are configured by a cell-specific higher layer parameter, which enables all the UEs in the cell to know the position of the SRS. There are 16 SRS subframe configurations defined for both frame structure type 1 and 2 [3]. An extreme case is that all the uplink subframes can be configured for SRS transmission in the context of high load in the cell. For all the subframes other than special subframes configured for SRS, SRS is transmitted in the last symbol of the subframe. In the special subframes, it supports up to two symbols for SRS transmission before Release-13, and then was extended to support up to six symbols for SRS in Release-13 because of the requirement of SRS capacity increase. For each UE, a subset of subframes

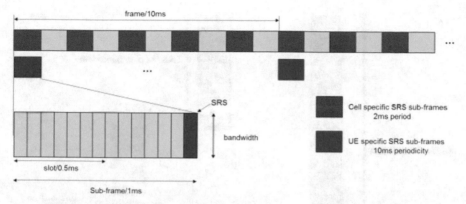

Fig. 2.21 SRS transmission in time domain

within the configured cell-specific SRS subframes are configured for its own SRS transmission. The starting subframe and periodicity of SRS transmission for each UE is configured by a UE-specific parameter. The location of SRS in time domain is illustrated in Fig. 2.21. In the subframes configured for SRS transmission, PUSCH and PUCCH will not be transmitted in the last symbol to avoid the collision.

SRS Transmission in Frequency Domain

There is a cell-specific SRS bandwidth configured by a cell-specific parameter which is common for all the UEs in the cell. There are eight cell-specific SRS bandwidth configurations for different system bandwidth. For each UE, there is one UE-specific SRS bandwidth configuration. The UE-specific SRS bandwidth is in multiples of four resource blocks. There are two methods to transmit SRS which are wideband transmission and frequency hopping. The wideband transmission is that the UE transmits the SRS spanning the whole UE-specific SRS bandwidth in one shot. It can obtain the whole channel measurement result in one subframe. However, the power spectrum density is lower which will degrade the channel estimation performance especially for power limited UE. Hence, the wideband SRS transmission is useful in the scenarios with good channel condition and the UE power is not limited. In the UE power limited scenario, SRS frequency hopping can be used, i.e., UE transmits SRS in a subset of SRS bandwidth in one shot and spans multiple time instance to obtain the whole channel measurement result. It can improve the channel estimation performance in each shot, but there is latency to obtain the whole channel information (Fig. 2.22).

SRS is transmitted in every second subcarrier within the SRS transmission bandwidth in the form of comb transmission. The SRS on the two combs are multiplexed by FDM and the SRS on the same comb are multiplexed by CDM as shown in Fig. 2.23.

If the SRS bandwidth is M_{SRS} resource blocks, then the length of SRS sequence is $M_{SC} = M_{SRS} * 12/2$. The SRS sequence is defined by

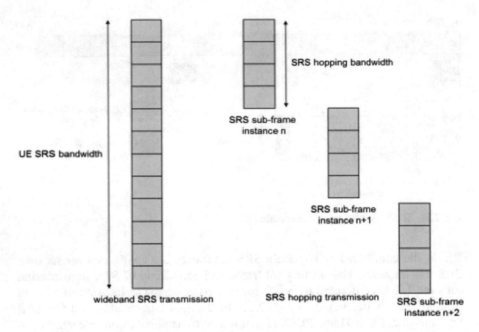

Fig. 2.22 Wideband SRS transmission and SRS hopping transmission

Fig. 2.23 Multiplexing of SRS

$$r_{\mathrm{SRS}}^{p}(n) = e^{j\alpha_{p}n} r_{u,v}(n), 0 \leq n \leq M_{\mathrm{sc}} - 1$$

where u is the base sequence group number and v is the sequence number. The determination process of u is the same as that with DM-RS, but the difference is that the sequence shift pattern is $f_{\mathrm{ss}} = N_{\mathrm{cell\ ID}}$ mod 30. The cyclic shift of SRS is given as

$$\alpha_p = 2\pi \frac{n_{CS}^p}{8}, \quad n_{CS}^p \in \{0,1,\dots 7\}$$

$$n_{CS}^p = \left(n_{SRS}^{CS} + \frac{8p}{N_{ap}} \right) \bmod 8$$

$$p \in \{0,1,\dots,N_{ap}-1\}$$

where N_{ap} is the number of configured SRS antenna ports, and $n_{SRS}^{CS} \in \{0,1,2,3,4,5,6,7\}$ is the cyclic shift signaled by the higher layer parameter. For the comb-like transmission, the signal is a repetition in time domain which reduces the number of available cyclic shifts.

Aperiodic SRS

In addition to periodic SRS transmission, aperiodic SRS transmission was introduced in Release-10. Since uplink MIMO is supported in Release-10, more SRS resources were needed. Given the existing SRS capacity, the periodicity of SRS needs to be increased in order to support multiple antennas. In this case, the channel measurement will not track the channel variation timely due to the longer SRS periodicity especially for high mobility. To obtain accurate channel information, aperiodic SRS is supported. Different from periodic SRS transmission, aperiodic SRS is triggered by uplink grant and not with a higher layer parameter. The configurations of aperiodic SRS transmission are still higher layer configured, e.g., the comb, SRS bandwidth, and cyclic shift. Once it is triggered, aperiodic SRS is transmitted in the configured SRS subframe immediately after the PDCCH UL grant. With proper aperiodic trigger, the channel variations can be properly tracked.

2.1.4 Downlink Transmission

Now that the role of the reference signals are understood, the different physical channels as introduced that are essential to grasp the operation of LTE in Sect. 2.1.2 can be described in detail. In this section, the transmission of physical downlink broadcast channel, control channels and data channel are introduced. There are also some other physical channels including PMCH, EPDCCH, MPDCCH, etc., which will not be discussed here for simplicity.

2.1.4.1 PBCH

Before the UE sets up the connection with a cell, the UE first acquires the cell synchronization (including time/frequency/frame synchronization and cell ID) based on Primary Synchronization Signal (PSS) and Secondary Synchronization Signal (SSS).

Then the UE needs to determine the system information, such as system bandwidth and number of antenna ports. The system information is divided into the *MasterInformationBlock* (MIB) and a number of *SystemInformationBlocks* (SIBs) [20]. The MIB is transmitted on the PBCH and includes a limited number of the most essential and most frequently transmitted parameters that are needed to acquire the other information from the cell.

The information in the MIB includes [11]:

- Downlink bandwidth: indicates the system (for both downlink and uplink) bandwidth in terms of the number of resource blocks from {6, 15, 25, 50, 75, 100, }, 3 bits
- PHICH Configuration: indicates the PHICH resource used to derive the resources for control channel and data channel, 3 bits
- System Frame Number (SFN): Eight most significant bits of SFN are included in the MIB, which indicates the number of the starting frame with PBCH transmission.
- Spare bits: 10 spare bits are reserved

Since the UE does not know the system information before PBCH detection, the PBCH is transmitted in a predefined position so that the UE can detect it after cell synchronization. The PBCH is located at the central 72 subcarriers (i.e., central six resource blocks) of the carrier which is the same as that of PSS/SSS, and the first four consecutive OFDM symbols in slot 1 in subframe 0 of each radio frame. The transmission time interval (TTI) of the PBCH is 40 ms (four consecutive radio frames). Consequently, it is transmitted on each radio frame fulfilling $n_f \bmod 4 = 0$. Recalled from the description of the CRS, that some of OFDM symbols within PBCH are for the transmission of the CRS, and therefore rate matching around the CRS is needed when performing resource mapping for PBCH. However, the number of CRS antenna ports is unknown before PBCH detection. To simplify the UE receiver design, it is assumed that there are always four CRS antenna ports for PBCH resource mapping regardless of the actual number of CRS antenna ports. Then the number of available resource elements for PBCH within one TTI are $240*4 = 960$. The resource mapping of PBCH is illustrated in Fig. 2.24.

The system designer will need to guarantee that the performance of PBCH transmission is such that all the UEs in the cell can detect the PBCH. As the payload size of PBCH is small with only 40 bits (24 information bits+16 CRC bits), convolutional coding and QPSK modulation is used because of their better performance. The rate matching is done by using a 1/3 mother coding rate with repetition to obtain a very low coding rate ($40/1920 \cong 0.02$) for the PBCH, which enables the PBCH in each radio frame within one TTI to be decodable. In addition, transmit diversity scheme space-frequency block coding (SFBC) and SFBC+frequency shift transmit diversity (FSTD) are used in the case of two and four CRS antenna ports, respectively. The information on the number of CRS antenna ports is embedded in the CRC mask of the PBCH, i.e., there are three different CRC masks corresponding to one, two, and four CRS antenna ports [19]. The number of CRS antenna port is blindly detected by the UE.

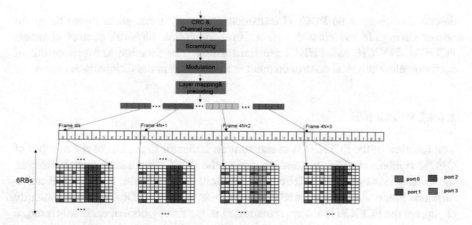

Fig. 2.24 Processing and resource mapping of PBCH

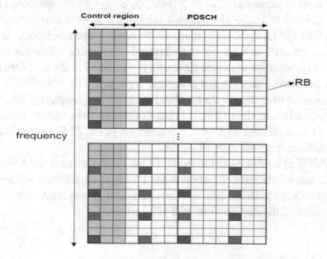

Fig. 2.25 Multiplexing of control region and PDSCH (e.g., three OFDM symbols)

2.1.4.2   Control Channel

The downlink physical control channel and data channel are time multiplexed as shown in Fig. 2.25. Normally, in one subframe, the control region is located within the first $n = 1/2/3$ OFDM symbols or $n = 2/3/4$ for bandwidth smaller than 10 RBs. The maximum number of OFDM symbols in the control region is dependent on subframe type [3]. The number of OFDM symbols in the control region is dynamically changing. Time division multiplexing (TDM) of the control region with the data region has the benefit of low latency and power saving for the UE. For example, with TDM, once the UE completes the PDCCH detection, it can immediately start decoding the scheduled data and reduce the waiting time. Furthermore, if the UE detects that there is no PDCCH transmission for it, it can go to micro sleep for power saving. In the control region, three downlink physical control channels PCFICH, PDCCH, and PHICH are transmitted. The function and processing of each downlink physical control channel are described in the following sections.

2.1.4.2.1   PCFICH

The function of the PCFICH is to transmit an indicator to the UE of the number of OFDM symbols used for the transmission of the PDCCHs in a subframe. Moreover, the starting position of PDSCH can be implicitly derived from the PCFICH transmission. Since the number of OFDM symbols used for PDCCH is dynamically changing, the PCFICH is always transmitted in the first symbol of each-subframe so that the UE knows where to detect it.

To guarantee the reliability of PCFICH detection, three codewords with length 32 bits are used to represent 1, 2, or 3 (or 2, 3, or 4) OFDM symbols [19]. The 32 bits are modulated to 16 QPSK symbols which are divided into four groups. Each group with four QPSK symbols is mapped onto one resource element group (REG) which is the resource unit for control channel to resource elements mapping [3]. There are six consecutive resource elements in one REG in the first OFDM symbol, where two resource elements are reserved for CRS. The four REGs are almost evenly distributed over the whole bandwidth to achieve frequency diversity shown in Fig. 2.26. In addition, the PCFICH is transmitted on the same set of antenna ports as PBCH. In case of two or four CRS antenna ports, SFBC or SFBC+FSTD is applied to achieve transmit diversity gain.

As the PCFICH is located as the first OFDM symbol of each subframe for all the cells, it may cause interference between the PCFICH of different adjacent cells. To randomize the interference, cell-specific scrambling sequence and cell-specific starting position of the first REG of PCFICH are used.

2.1.4.2.2   PDCCH

The PDCCH transmits downlink control information (DCI) including downlink scheduling assignments, uplink scheduling assignments (i.e., UL grant), and power control information to the UE. A number of DCI formats are defined such as format 0/1/1A/1B/1C/1D/2/2A/2B/2C/3/3A/4, which depends on downlink or uplink scheduling assignment, the carried control information (e.g., power control, MCC change), transmission scheme, and the payload size. For simplicity, the DCI formats in Release-10 are summarized in Table 2.12. It should be noted that some new DCI formats are introduced after Release-10. These new DCI formats and the detail of the information field in each DCI format can be found in [19].

A PDCCH is transmitted on an aggregation of one or several consecutive control channel elements (CCE), where one CCE corresponds to nine REGs and there are

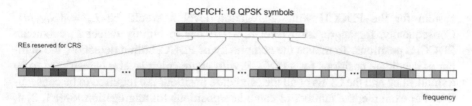

Fig. 2.26 Resource mapping of PCFICH

Table 2.12 DCI formats

DCI formats	Control information	Usage of DCI
0	Uplink grant	PUSCH scheduling with one antenna port
1	Downlink assignment	PDSCH scheduling with one codeword
1A	Downlink assignment	Compact scheduling of one PDSCH codeword
1B	Downlink assignment	CRS-based PDSCH with rank 1 precoding
1C	Downlink assignment	Very compact scheduling of one PDSCH codeword
1D	Downlink assignment	CRS-based MU-MIMO
2	Downlink assignment	Closed loop spatial multiplexing using CRS
2A	Downlink assignment	Open loop spatial multiplexing using CRS
2B	Downlink assignment	Dual layer beamforming with DM-RS
2C	Downlink assignment	Up to 8 layers transmission with DM-RS
3	Power control	TPC commands for PUCCH/PUSCH with 2-bit power adjustments
3A	Power control	TPC commands for PUCCH/ PUSCH with single bit power adjustments
4	Uplink grant	Uplink spatial multiplexing

four available REs per REG. The supported CCE aggregation levels are 1, 2, 4, and 8. The number of CCEs assigned for one PDCCH depends on the channel quality and the payload size of PDCCH. The channel quality is based on the reported CQI from the scheduled UE, and the payload size depends on the DCI format.

The number of available CCEs in the system is $N_{CCE} = \lfloor N_{REG}/9 \rfloor$ and are enumeratedfrom 0 to $N_{CCE} - 1$, where N_{REG} is the number of REGs not assigned to PCFICH and PHICH (see Sect. 2.1.4.2.1). When assigning the CCEs for PDCCH, a PDCCH consisting of n consecutive CCEs can only start on a CCE fulfilling $i \bmod n = 0$, where i is the CCE number. In principle, the number of candidate positions in the system for the PDCCH with aggregation level n would be $L_n = \lfloor N_{CCE}/n \rfloor$. Consequently, for aggregation level n, a UE has to blindly detect L_n candidate PDCCHs positions. To reduce the complexity of PDCCH blind detection, the number of candidate positions for a PDCCH with aggregation level n is restricted to be a small value and the CCEs of all the candidate positions are required to be consecutive. For example, the number of candidate positions for aggregation level 1, 2, 4, and 8 are 6, 6, 2, and 2, respectively. How to place the PDCCHs on the CCEs to minimize the number of PDCCH blind detection is part of the search space design by the system designer [12].

Given the PDCCHs to be transmitted in a subframe are placed on the CCEs as aforementioned, the block of information bits on each PDCCH are multiplexed. If there is no PDCCH on some CCEs, <NIL> elements are inserted during the multiplexing to enable the PDCCHs to start at the required CCE position. The block of multiplexed bits are scrambled with a cell-specific scrambling sequence and then modulated to QPSK symbols. The QPSK symbols are grouped into a sequence of symbol quadruplets, where each symbol quadruplet consists of four consecutive QPSK symbols. Then the sequence of symbol quadruplets is interleaved by using a sub-block interleaver [19]. The block of symbol quadruplets output from the interleaver are cyclically shifted, where the cyclic shift size is determined by the physical cell identity. The interleaving can enable CCEs of one PDCCH to be distribute in the frequency domain to take advantage of frequency diversity gain. The cell-specific scrambling sequence and cell-specific shift size are provided to the system designer to randomize the inter-cell interference to the PDCCH.

An illustration of REG is given in Fig. 2.27. The cyclically shifted symbol quadruplets are sequentially mapped onto the REGs not assigned to either the PCFICH or the PHICH. The mapping is time first and then going to the frequency domain. The size of REG depends on which OFDM symbol located at in the control region and the number of CRS antenna ports. In the first OFDM symbol, one REG consists of six consecutive REs in one resource block and there are two REGs per resource block, where two REs in each REG are reserved for CRS regardless of one or two CRS antenna ports. In case of four CRS antenna ports, the REG definition in the second OFDM symbol within control region is the same as that of first OFDM symbol. With two CRS antenna ports, each REG in the second OFDM symbol consists of four consecutive REs in one resource block and there are three REGs per resource

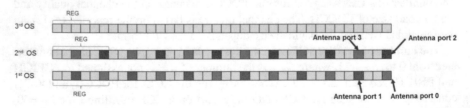

Fig. 2.27 REG definition assuming four CRS antenna ports

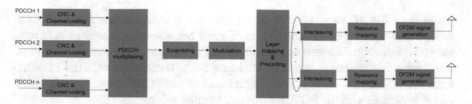

Fig. 2.28 Illustration of PDCCH processing

block. For the third OFDM symbol, the REG is defined as that of the second OFDM symbol with two CRS antenna ports.

The same set of antenna ports as the PBCH are used for PDCCH transmission. In case of two or four CRS antenna ports, SFBC or SFBC+FSTD is applied to achieve transmit diversity gain. The processing of PDCCH is illustrated in Fig. 2.28.

2.1.4.2.3 PHICH

The PHICH carries the hybrid-ARQ (HARQ) ACK/NACK in response to uplink PUSCH transmission. For proper uplink HARQ system operation, PHICH performance should be maintained with very low error probability to reduce the number of retransmission and avoid the fail reception of a transport block in MAC layer[2]. The target performance is up to the system designer but it is nominally acknowledged that the probability of ACK erroneously detected as NACK is in the order of 10^{-2} and the error rate of NACK detected as ACK shall be at least lower than 10^{-3}.

In the case of PUSCH transmission with one or two transport blocks, there are the corresponding one or two bits HARQ ACK/NACK. Multiple PHICHs are mapped to the same set of resource elements and code multiplexed by using different orthogonal sequences n_{PHICH}^{seq}, which is known as a PHICH group n_{PHICH}^{group}. A PHICH resource is identified by the index pair $\left(n_{PHICH}^{group}, n_{PHICH}^{seq}\right)$.

The number of PHICH groups supported in the system are configurable and is transmitted to the UE in the MIB of the PBCH. In the PBCH, PHICH configuration includes 1 bit PHICH-duration and 2 bits-PHICH resource. The PHICH-duration indicates that PHICH spans 1 or 3 OFDM symbols in the control region. If the PHICH is only located at the first OFDM symbol of the control region, the room for power boosting of PCFICH would be limited because of the power use to transmit the PCFICH and PDCCH if they also exist on the first OFDM symbol and therefore the coverage of PCFICH would be restricted. LTE, thus, provide the PHICH resource to indicate the number of PHICH groups expressed as a fraction of the downlink system bandwidth in terms of resource blocks so that the system designer can spread the PHICH over multiple OFDM symbols. For frame structure type 1, the number of PHICH groups is constant in all subframes according to the configu-

---

[2] RLC retransmission with long latency is needed to recover from the fail reception of a transport block in the MAC layer.

ration from PBCH. However, for frame structure type 2, the number of PHICH groups may vary between downlink subframes because each downlink subframe may associate with different number of uplink subframes. For example, in uplink-downlink configuration 0, the PHICHs in subframe 0 associate with the PUSCHs from two uplink subframes [12] for other cases, the PHICHs in one downlink subframe associates with the PUSCHs from one uplink subframe.

In a PHICH group, there are up to eight orthogonal sequences in the case of normal cyclic prefix and four orthogonal sequences in the case of extended cyclic prefix. The orthogonal sequences used in one PHICH group are shown in Table 2.13.

For one UE, each of HARQ ACK-NACK bits transmitted on one PHICH in one subframe is encoded with three repetitions resulting in a block of coded bits. The block of coded bits is modulated using BPSK to generate a block of modulation symbols. There are three modulation symbols corresponding to one HARQ ACK/NACK bit and each of the modulation symbols is spread with the assigned orthogonal sequence. The spread modulation symbols from different UE are multiplexed and mapped onto REGs as shown in Fig. 2.29. The distance of two adjacent REGs of one PHICH group in the frequency domain is almost 1/3 the system bandwidth to achieve the frequency diversity gain.

In the case of two CRS antenna ports, SFBC is used for PHICH transmission. However, in case of four CRS antenna ports, SFBC+FSTD cannot be directly used. To keep the orthogonality of PHICH sequences in one REG, the four modulation symbols mapped onto the same REG have to be transmitted on the same antenna port. Hence it only performs SFBC on two antenna ports for the modulation symbols within one REG of PHICH, which causes the power imbalance among different antenna ports. To achieve more transmit diversity gain and enable the power balance

Table 2.13 Orthogonal sequences for PHICH

Sequence index	Orthogonal sequences	
	Normal cyclic prefix	Extended cyclic prefix
0	$[+1 \quad +1 \quad +1 \quad +1]$	$[+1 \quad +1]$
1	$[+1 \quad -1 \quad +1 \quad -1]$	$[+1 \quad -1]$
2	$[+1 \quad +1 \quad -1 \quad -1]$	$[+j \quad +j]$
3	$[+1 \quad -1 \quad -1 \quad +1]$	$[+j \quad -j]$
4	$[+j \quad +j \quad +j \quad +j]$	–
5	$[+j \quad -j \quad +j \quad -j]$	–
6	$[+j \quad +j \quad -j \quad -j]$	–
7	$[+j \quad -j \quad -j \quad +j]$	–

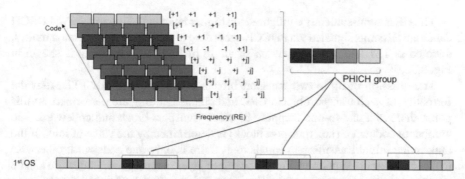

Fig. 2.29 Multiplexing of PHICH in one PHICH group (e.g., PHICH duration equals to 1 OFDM symbol)

over different antenna ports, SFBC operation on (port 0, port 2) or (port 1, port 3) is switched over the first, second, and third REG of one PHICH group.

2.1.4.3   PDSCH

PDSCH carries the data to be transmitted in the downlink with the following transmission schemes [12]:

- **Single antenna port scheme**
  PDSCH transmission is on one antenna port which can be on port 0 (Transmission mode (TM) 1), port 5 (TM7), port 7 or port 8 (TM8).
- **Transmit diversity scheme**
  It is an open loop rank 1 transmission scheme based on CRS antenna ports. In case of two and four CRS antenna ports, the transmit diversity schemes are SFBC and SFBC+FSTD, respectively. Transmit diversity scheme is supported in TM 2. It is also the fallback transmission scheme in TM3–10.
- **Large Delay CDD scheme**
  It is CRS-based open loop spatial multiplexing with rank = 2/3/4 transmission, which is supported in TM3.
- **Closed loop spatial multiplexing scheme**
  CRS-based closed loop spatial multiplexing with up to four layers, which is supported in TM4/6.
- **Multi-user MIMO scheme**
  Up to two users are multiplexed together for MU-MIMO, which is CRS-based transmit scheme and supported in TM5.
- **Dual layer scheme**
  It was introduced in Release-9 for dual-layer beamforming. PDSCH is performed with two transmission layers on antenna ports 7 and 8 in TM8.
- **Up to eight layers transmission scheme**
  It is introduced in Release-10 and DM-RS-based transmission scheme. PDSCH would be performed with up to eight transmission layers on antenna ports 7–14. It is supported in TM 9 and 10.

The UE is semi-statically configured via higher layer signaling to receive PDSCH data transmissions signaled via PDCCH according to one of ten transmission modes, denoted as TM 1 to 10. The general structure of PDSCH processing is shown in Fig. 2.30.

Transmission of up to two transport blocks is supported in one TTI. After the formation of each transport block, CRC and channel coding are performed. At this point, there is a one-to-one mapping between transport block and codeword. The number of codeword (i.e., transport block) is determined by the value of rank. If the rank in the initial transmission equals to 1, there is only one codeword; otherwise two codewords are supported. For codeword to layer mapping, one codeword can be mapped onto more than one layer when rank is large than 2. The selection of up to two codewords transmission is a trade-off of performance and signaling overhead. If each codeword is mapped onto one layer, it can enable UE to perform SIC operation for performance improvement; however, the overhead of required DL/UL signaling, e.g., MCS/CQI would be proportional to the number of codewords. The detailed mapping of the codeword to layers is rather complicated and is beyond the scope of this treatise. The interested reader should consult the standard [3] .

The precoding operation is different among the aforementioned transmission schemes. For the transmission schemes in TM 1–6, the antenna ports after the precoding in Fig. 2.30 refer to CRS antenna port which can be configured as 1/2/4. In this case, the precoding matrix is predefined or signaled to UE by DCI, i.e., UE knows what precoding matrix is used at the eNB side. The precoding matrices are defined in [3]. However, for the transmission schemes in TM 7–10, the antenna ports are DM-RS ports and the number of DM-RS ports is equal to the number of layers. The precoding is transparent to UE and the signaling of the precoding matrices is not needed any more. The precoding information is embedded in the DM-RS. The precoding matrix used for PDSCH is up to eNB implementation, which gives the freedom for eNB operation. Such operation is friendly to TDD because of utilizing channel reciprocity and without the need of precoding matrix indicator (PMI) feedback. For FDD, precoding matrices are defined only for CSI reporting. The UE performs channel measurement based on CSI-RS and reports the selected precoding matrices to the eNB for reference. Regarding how to use the reported precoding matrices, this is left for eNB implementation.

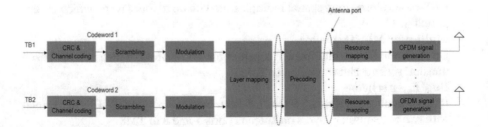

Fig. 2.30 General structure of PDSCH processing

The resource mapping is the same for all the transmission schemes. The pre-coded data is mapped onto the resource elements of each antenna port. The resource mapping is performed from frequency first and then time within the assigned PDSCH resource blocks. The mapping skips the resource elements occupied by different kinds of reference signal, e.g., CRS, DM-RS, and CSI-RS.

2.1.4.4   Modulation Coding Scheme (MCS)

Scheduling based on the reported channel state information (CSI), such as rank indication (RI), precoding matrix indicator (PMI), and channel quality indication (CQI), is an important characteristics of LTE system. CQI reflects the channel quality and is reported via the modulation scheme and coding rate. A number of 15 CQI values are defined, and the step size in terms of SINR is about 2 dB. When UE reports a CQI index based on the channel measurement, there is the assumption that "a single PDSCH transport block with a combination of modulation scheme and transport block size corresponding to the reported CQI index, and occupying a group of downlink physical resource blocks termed the CQI reference resource, could be received with a transport block error probability not exceeding 0.1" [12]. Then eNB performs PDSCH scheduling based on the reported CSI.

In order for the UE to receive the PDSCH transmission, it needs to determine the modulation order, transport block sizes (TBS), and the assigned resource blocks. The combination of modulation order and TBS index is indicated by a 5-bits MCS field in the DCI. The assigned resource blocks are also signaled in the DCI according to the resource allocation types. The 5-bits MCS table includes 29 combinations of modulation and TBS index and three states reserved for implicit modulation order and TBS signaling for retransmissions. The 29 MCS correspond to 29 spectral efficiency with finer granularity compared to the CQI values, where the spectral efficiency has an overlap between QPSK and 16QAM and also for 16QAM and 64QAM. In total there are 27 different spectral efficiencies because some values of the 5-bits MCS table have the same spectral efficiency. Although the spectral efficiency of two modulation schemes is same, different modulation order shows better performance in different fading channels. It is up to system designer to select the modulation order given the same spectral efficiency based on the fading channels.

There is a basic TBS table defined and the dimension of which is 27×110 for the case one transport block not mapping to more than one layer. From the signaled MCS in the DCI, the modulation order and TBS index can be determined. The TBS of PDSCH is determined by looking up the basic TBS table according to the TBS index and the number of assigned resource blocks. If the transport block is mapped to more than one layer, the determined TBS from the basic TBS table is then translated to a final TBS corresponding to the number of layers that the transport block mapped to. The three entries for retransmission in the 5-bits MCS table represents that the TBS is assumed to be as determined from DCI transported in the latest PDCCH for the same transport block.

In principle, the TBS can be directly calculated according to the MCS and the number of available resource elements, which results in an arbitrary TBS. The arbitrary

TBS would not match the size of Turbo QPP interleaver (the padding or depadding is needed) and the MAC payload size normally calculated with bytes [21, 22]. In addition, the TBS may vary during the initial transmission and retransmission because the number of available resource elements could be different, e.g., the control region is changed. Hence, it is desirable that TBS corresponding to different number of resource blocks is constant across different subframes, and the TBS be aligned with QPP sizes to remove the need for padding/depadding. Calculating the TBS from MCS requires a resource assumption: three OFDM symbols for control region, two CRS antenna ports, no PSS/SSS/PBCH and normal CP [21]. The generated TBS tables can be found in [12]. It is noted that the actual resource allocations may not be the same as the resource assumptions for generating TBS tables, e.g., the number of OFDM symbols in the control region is not 3, the existence of CSI-RS or DM-RS, etc. The resulted coding rate may not be the exact coding rate of MCS, but is close to the MCS.

2.1.5 Uplink Transmission

2.1.5.1 PUCCH

Uplink control information (UCI) including scheduling request (SR), HARQ-ACK, and periodic CSI is transmitted on PUCCH when there is no PUSCH transmission in the same subframe. In the case of concurrent transmission of PUCCH and PUSCH, UCI is piggybacked on PUSCH if the simultaneous transmission of PUCCH and PUSCH is not configured. The PUCCH is located at the resource blocks at the edge of the transmission bandwidth within one slot and frequency hopping across the two slots within one subframe is performed as shown in Fig. 2.31. The placement of PUCCH at the edge is to facilitate the continuous resource blocks allocation for PUSCH and for achieving a frequency diversity gain. The resource blocks allocated for PUCCH are defined by the scheduler and are transmitted via higher layer signaling.

Several PUCCH formats are defined according to the conveyed UCI on PUCCH. Some PUCCH format (e.g., format 3) is extended to accommodate new UCI as the releases evolve. For this treatise, only the original function of such kind of PUCCH format is introduced for simplicity. The extended function can be found in [12]. The conveyed UCI of different PUCCH formats is summarized in Table 2.14.

2.1.5.1.1   PUCCH Formats 1/1a/1b

For PUCCH format 1, scheduling request (SR) is carried by the presence or absence of transmission of PUCCH from the UE. The periodic resource for SR is UE-specifically configured by higher layer signaling. If there is the scheduling request in the SR time instance, UE transmits the bit "1" in the form of BPSK on

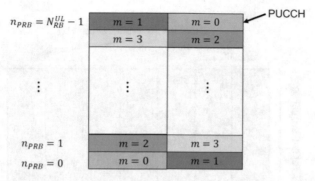

Fig. 2.31 Mapping of PUCCH onto physical resource blocks

Table 2.14 Conveyed UCI on PUCCH formats

PUCCH format	Conveyed UCI	Originally introduced release
1	SR	Release-8
1a/1b	HARQ-ACK	Release-8
2	periodic CSI	Release-8
2a/2b	CSI+HARQ-ACK	Release-8
3	HARQ-ACK	Release-10
4	HARQ-ACK or periodic CSI	Release-13
5	HARQ-ACK or periodic CSI	Release-13

PUCCH; otherwise, UE does not transmit anything. PUCCH format 1a/1b carries 1 bit and 2 bits HARQ-ACK, and the corresponding modulation schemes are BPSK and QPSK, respectively.

The modulation symbol is multiplied with a cyclically shifted sequence with length 12, where the sequence is the same as the uplink DM-RS sequence defined for one resource block. The multiplied sequence is block-wise spread with an orthogonal sequence with length 4 or 3 depending on the number of OFDM symbols for PUCCH 1/1a/1b within one slot. Then the two-dimensional spread symbol is mapped onto one resource block. The three OFDM symbols in the middle of the resource block are reserved for PUCCH DM-RS for coherent detection, and the remaining OFDM symbols are for PUCCH data as shown in Fig. 2.32. The PUCCH DM-RS is also a two-dimensional sequence, which is a cyclic shift (CS) of one sequence with length 12 is block-wise spread with an orthogonal sequence with length 3.

The PUCCH format 1/1a/1b from different UEs are multiplexed in one resource block by using different two-dimensional sequences. The PUCCH resource is indicated by a combination of CS index and orthogonal code covering (OCC) index. The sequence in the frequency domain is a CS of one sequence. The number of available CS depends on the CS distance with three values, 1, 2, or 3, which is configured by higher layer and dependent on the delay spread. The configured CS

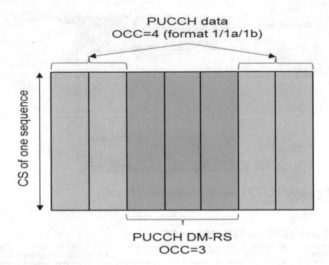

Fig. 2.32 PUCCH format 1/1a/1b

Table 2.15 Orthogonal sequences with OCC for PUCCH format 1/1a/1b

Sequence index	OCC = 4	OCC = 3
0	$\begin{bmatrix} +1 & +1 & +1 & +1 \end{bmatrix}$	$\begin{bmatrix} +1 & +1 & +1 \end{bmatrix}$
1	$\begin{bmatrix} +1 & -1 & +1 & -1 \end{bmatrix}$	$\begin{bmatrix} +1 & e^{j2\pi/3} & e^{j4\pi/3} \end{bmatrix}$
2	$\begin{bmatrix} +1 & -1 & -1 & +1 \end{bmatrix}$	$\begin{bmatrix} +1 & e^{j4\pi/3} & e^{j2\pi/3} \end{bmatrix}$

distance is applied for both PUCCH data and PUCCH DM-RS. The CS varies with the symbol number and slot number in a cell-specific manner to randomize the inter-cell interference. Considering the additional 4 OCC in time domain, the maximum number of available two-dimensional sequence for PUCCH data is 48, 24, 16 in principle. The OCC length for PUCCH DM-RS is 3, and therefore the maximum number of available DM-RS is 36, 18, 12,respectively. Due to the requirement of coherent detection and the restriction of the number of PUCCH DM-RS, only three OCCs are needed for PUCCH data. The sequences of OCC = 4 and 3 are summarized in Table 2.15. For PUCCH data, OCC = 4 is applied for both slots within one subframe or OCC = 4 is applied for the first slot and OCC = 3 is applied for the second slot in case the last OFDM symbol is reserved for SRS.

2.1.5.1.2   PUCCH Format 2/2a/2b

PUCCH format 2 is to convey periodic CSI including RI, PMI, and CQI. The Reed Muller (RM) code is used to encode the periodic CSI with up to 11 bits and the output of encoding is 20 bits [19]. QPSK modulation is performed to generate 10 QPSK symbols. Each modulation symbol is multiplied with a cyclically shifted

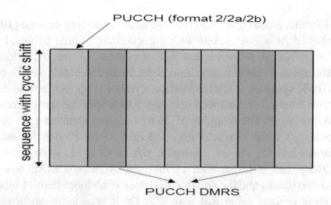

Fig. 2.33 Transmission structure of PUCCH format 2/2a/2b in case of normal CP

sequence with length 12 and it is mapped onto one symbol within one resource block. For the case of normal CP, there are two symbols reserved for DM-RS and the remaining five symbols for PUCCH data in one resource block as shown in Fig. 2.33. In the case of extended CP, there is only one symbol for DM-RS located at the fourth symbol of the resource block.

The total 10 QPSK symbols are mapped onto two resource blocks from two different slots within one subframe. Similar to PUCCH format 1/1a/1b, the CS of one base sequence is assigned to different UE for multiplexing and the number of available CS depends on the configured CS distance. Typically, it can support up to six UEs multiplexed together. The resource of PUCCH format 2 is semi-statically configured via higher layer signaling.

PUCCH format 2a/2b is used when periodic CSI and ACK/NACK feedback from a certain UE occurs at the same subframe, and it is only applicable in the case with normal CP. The ACK/NACK modulation symbol is multiplied by the second DM-RS sequence within one resource block, i.e., the ACK/NACK information is conveyed by the relative phase between the two DM-RS symbols. PUCCH format 2a/2b corresponds to 1 bit or 2bits ACK/NACK, respectively. In the case of extended CP, the periodic CSI and 1 or 2 bits ACK/NACK are concatenated and then encoded by the RM coding.

2.1.5.1.3   PUCCH Format 3

In Release-10, carrier aggregation with up to five component carriers was supported. As there is an independent HARQ entity per carrier, the number of reported ACK/NACK bits increased. For example, there are up to 10 bits ACK/NACK for FDD and even more for TDD depending on the DL-UL subframe configuration. PUCCH 1a/1b/2 is not able to convey so many ACK/NACK bits. PUCCH format 3 was introduced to support up to 11 bits for FDD corresponding to up to 10 bits ACK/NACK and 1 bit positive/negative SR, and up to 21 bits for TDD corresponding to up to 20 bits ACK/NACK and 1 bit positive/negative SR.

PUCCH format 3 uses RM $(\mathbf{32, O})$ as the channel coding scheme [19], where $\mathbf{O}$ is the number of ACK/NACK bits with the maximum value being 11. When the number of ACK/NACK/SR bits is less than or equal to 11 bits, the output of encoding is 32 bits, and then circular repetition is performed to obtain 48 bits corresponding to 24 QPSK symbols. Each modulation symbol is spread in time domain by a sequence with length 5 and mapped onto the 5 REs with the same subcarrier index in one resource block. The mapping of 24 modulation symbols span two resource blocks of two slots within one subframe. The DM-RS location is the same as that of PUCCH format 2/2a/2b. As the maximum size of $\mathbf{O}$ is 11, in case the number of ACK/NACK/SR bits is larger than 11 bits, the ACK/NACK/SR bits are evenly divided into two blocks and the size of each block is no more than 11 bits. The same channel coding scheme using RM with $(\mathbf{32, O})$ is separately performed for each block, i.e., dual-RM code [23] and the output is restricted to be 24 bits corresponding to 12 QPSK symbols. To achieve the frequency diversity gain, the modulations symbols from these two blocks are alternately mapped onto the REs in the frequency domain. This enables the total 12 QPSK symbols of each block to be distributed over the two PUCCH resource blocks. The transmission structure of PUCCH format 3 is demonstrated in Fig. 2.34.

2.1.5.1.4 PUCCH Format 4/Format 5

In Release-13, massive carrier aggregation with up to 32 component carriers was introduced. Normally, it is not possible for one operator to have so many licensed component carriers for aggregation. However, there is larger unlicensed spectrum bandwidth available for use below 6 GHz. Due to the larger number of carriers, the payload size of periodic CSI and/or ACK/NACK is further significantly increased, and therefore PUCCH format 4/5 were introduced as the container.

The transmission of the PUCCH format 4 is similar to the PUSCH. Up to eight resource blocks can be configured for PUCCH format 4, and frequency hopping

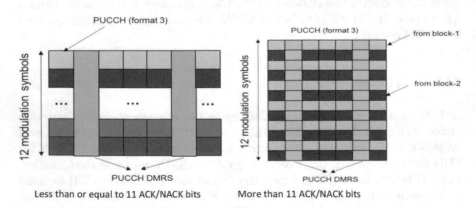

Fig. 2.34 Transmission structure of PUCCH format 3

across two slots of one subframe is applied to reap the frequency diversity gain. Only QPSK modulation is supported for the robust transmission of the PUCCH. Similar to the PUSCH, DFT-based transform precoding is applied for the modulation symbols. The DM-RS transmission is also the same as that of PUSCH. Normal PUCCH format and shorten PUCCH format are defined corresponding to whether SRS is present at the last symbol of the second slot. For normal PUCCH format, there are 12 (normal CP) and 10 (extended CP) symbols for the PUCCH data transmission. For the shorten PUCCH format, there are 11 (normal CP) and 9 (extended CP) symbols for PUCCH data. The number of coded bits conveyed by PUCCH format 4 is $N_{\mathrm{RB}}^{\mathrm{PUCCH4}} * 12 * N_{\mathrm{symbol}}^{\mathrm{PUCCH4}} * 2$, where $N_{\mathrm{RB}}^{\mathrm{PUCCH4}}$ is the number of configured resource blocks and $N_{\mathrm{symbol}}^{\mathrm{PUCCH4}}$ is the number of symbols. Tail biting convolutional code (TBCC) with CRC is use as the channel code for PUCCH format 4.

PUCCH format 5 transmission is restricted to one resource block and with frequency hopping across the two slots within the one subframe. The operation of channel coding, modulation, and DFT-based transform precoding is similar to that of PUCCH format 4. The difference is that there is a block-wise spread operation before DFT-based transform precoding. A block of six modulation symbols is spread using [+1 + 1] or [+1 − 1] to obtain 12 modulation symbols, and then the 12 modulation symbols are operated on by DFT-based transform precoding. The number of coded bits conveyed by PUCCH format 5 is $6 * N_{\mathrm{symbol}}^{\mathrm{PUCCH5}} * 2$, where $N_{\mathrm{RB}}^{\mathrm{PUCCH5}}$ is the number of symbols depending on normal or shorten PUCCH format. PUCCH format 5 is applicable for the medium payload size. Since two spread sequences are used, it can support up to two UEs multiplexed.

2.1.5.2   PUSCH

The two transmission schemes for the PUSCH are:

– **Single antenna port scheme**

  It is the transmission scheme in uplink TM1 and fallback transmission scheme in TM2

– **Closed loop spatial multiplexing**

  Up to four layers transmission is supported and it is the transmission scheme in TM2

The general structure of PUSCH processing is shown in Fig. 2.35. For the single antenna port scheme, there is only one transport block. In case of closed loop spatial multiplexing, similar to the downlink, up to two transport blocks are supported. The number of layers (i.e., rank) and the precoding matrix used for the PUSCH transmission are determined by the eNB and then signaled to UE via the UL grant. The maximum number of layers supported is 4. The codeword to layer mapping is the same as that of the PDSCH. The cubic metric preserving (CMP)

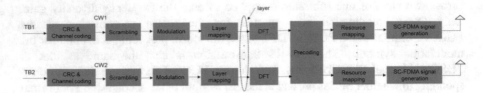

Fig. 2.35 General structure of PUSCH processing

codebook was introduced in order to preserve the single carrier property of PUSCH transmission on each antenna port. There is only one non-zero element in each row of the precoding matrix, which avoids the mix of the two signals on the same antenna port. In addition, antenna selection precoding matrices for two and four antenna ports are supported. It may occur that part of transmit antennas are blocked due to hand grabbing for mobile phone, which cause significant loss in the radio signal. In this case, antenna selection is used to save transmission power. The detail of CMP codebook for different rank are beyond the scope of this treatise and can be found in [3].

PUSCH transmission is based on DM-RS ports, and the number of DM-RS ports is equal to the rank signaled in the UL grant. It is different from downlink that the precoding of the PUSCH transmission is not transparent. In uplink, the precoding matrix and the corresponding MCS signaled to the scheduled UE are determined by the eNB. The UE is mandated in the standards and is tested via test specified by RAN 4 to use the signaled precoding matrix. MU-MIMO is also supported for the PUSCH transmission. That is, multiple UEs transmitting PUSCH data on the same time and frequency resources can be separated by eNB because of its multiple receive antennas capability. Uplink MU-MIMO is also known as virtual MIMO because the multiple transmit antennas are from different UEs. From the UE perspective, it does not know whether there are some other UEs multiplexed with it together. That is MU-MIMO is transparent to UE. To separate the PUSCH from different UEs, uplink DM-RS is assigned to different UEs by using different cyclic shifts of one DM-RS sequence. As there are eight cyclic shifts available, it can, in principle, support up to eight UEs for MU-MIMO in principle. The assignment of cyclic shifts for different UEs is applicable for the case of aligned resource allocation among the MU-MIMO UEs. In Release-10, OCC over time domain for uplink DM-RS was introduced (see Sect. 2.1.3.2), which supports two UEs with unaligned resource allocation for MU-MIMO. In a practical system, especially for TDD systems, there are a large number of receive antennas, e.g., 8 or 16, and the uplink resource is limited, consequently there is the desire to support more UEs with unaligned resource allocation for MU-MIMO. In Release-14, uplink DM-RS in the form of comb as SRS is introduced to support up to four UEs with unaligned resource allocation.

2.1.5.3 Modulation

In order for the eNB to receive the PUSCH transmission, it needs to determine the modulation order and transport block sizes (TBS). The procedure of determining MCS and TBS for PUSCH is the same as PDSCH. The combination of modulation order and TBS index is also indicated by a 5-bits MCS field in the uplink grant. The 5-bits MCS table for PUSCH is originated from that of PDSCH with some changes. The change is that for the first four MCS indices corresponding to 64QAM the modulation scheme is treated as 16QAM. The reason is that many UE categories do not support 64QAM or eNB does not support 64QAM transmission for PUSCH. In this case, the highest MCS corresponding to 16QAM is 2.41 bits/symbol which restricts the peak data rate, and therefore the related changes are done and the MCS can reach to 3.01 bits/symbol [24].

2.1.6 HARQ Timing

HARQ uses the stop-and-wait protocol for transmission. After a transport block transmission, the transmitter stops to wait until receiving the confirmation from the receiver. The receiver reports an indicator (i.e., ACK or NACK) to the transmitter based on the transport block detection result. If ACK is received, the transmitter will transmit a new transport block; otherwise, the erroneously detected transport block is retransmitted. Each stop-and-wait protocol transmission forms one HARQ process. The downlink HARQ operation is illustrated in Fig. 2.36.

Multiple parallel HARQ processes are supported to improve the overall system performance. The maximum number of HARQ process depends on the round trip time (RTT) as shown in Fig. 2.36, where RTT is restricted by the eNB and UE processing time [1]. In LTE, it is assumed that the eNB processing time is 3 ms and UE processing time is $3 - T_A$ ms in case of 1 ms TTI for PDSCH [25]. Based on that assumption, the maximum of HARQ process for FDD DL/UL was set to be 8. The maximum number of HARQ processes for TDD depends on DL/UL configuration and the eNB/UE processing capability [12, 26], which are summarized in Table 2.16.

There are two types of HARQ mechanisms defined as synchronous HARQ and asynchronous HARQ [8]

– Synchronous HARQ: the (re)transmissions for a certain HARQ process are restricted to occur at known time instants.
– Asynchronous HARQ: the (re)transmissions for a certain HARQ process may occur at any time, but the duration between the transmission and retransmission is larger than the RTT.

For synchronous HARQ, the time instants for the (re)transmission for a certain HARQ process are predefined, and therefore the HARQ process number can be derived from the time instant directly[3]. In this case, the retransmission can reuse the

---

[3] That is subframe number and the related explicit signaling are not required

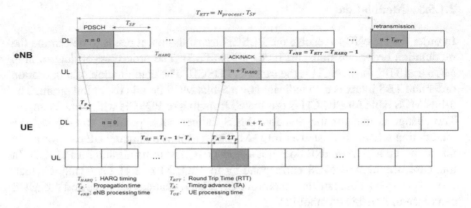

Fig. 2.36 Downlink HARQ operation (eNB as transmitter, UE as receiver)

Table 2.16 Maximum number of HARQ processes for TDD

TDD UL/DL configuration	Maximum number of HARQ processes	
	DL	UL
0	4	7
1	7	4
2	10	2
3	9	6
4	12	3
5	15	2
6	6	1

same transmission format as the initial transmission including the resource allocations and MCS without control signaling, which is known as non-adaptive
HARQ. This has the benefit of control signaling overhead reduction and the simple
scheduling. On the contrary, asynchronous HARQ requires the explicit signaling of
HARQ process number, but has the advantage of scheduling flexibility for the
retransmission(s). This is suitable for downlink transmission because it can flexibly
schedule the retransmission to avoid the MBSFN subframes or in the subframes
with paging or system information. In LTE, synchronous HARQ is used for uplink
and asynchronous HARQ is used for downlink.

Soft combing schemes for the HARQ retransmission include chase combining
and incremental redundancy (IR). Chase combing implies that the retransmission
bits are exactly identical to the initially transmitted information bits. For IR, however, it is not necessary that the retransmission bits must be the same as that of the
initial transmission. Instead, compared to the coded bits transmitted in the previous
transmission (initial or retransmission), there are incremental coded bits transmitted
in the next retransmission. In both cases soft combining after the retransmission will
result in a lower code rate to increase the detection probability. The incremental
coded bits in the retransmission are named as redundancy version (RV) which is

implicitly indicated by the MCS index in the DCI. In LTE, HARQ is based on IR. It is noted that CC is a special case of IR[4]. .

Regardless of synchronous or asynchronous HARQ, the HARQ timing in LTE is predefined, i.e., upon the detection of the transport block at subframe n, the corresponding ACK/NACK is transmitted in subframe $n + T_{HARQ}$ as illustrated in Fig. 2.36. For FDD, the HARQ timing T_{HARQ} for both DL and UL is 4, which is shown in Fig. 2.37. For TDD, the HARQ timing depends on the DL/UL configuration and the location of the PDSCH/PUSCH. It can be seen that the HARQ ACK/NACK corresponding to the PDSCHs transmitted in several downlink subframes are transmitted in one uplink subframe, but it requires that $T_{HARQ} \geq 4$, e.g., the HARQ timing for DL/UL configuration 2 in Fig. 2.37.

2.1.7   Carrier Aggregation (CA) and Band Combinations

For LTE Release-8, the maximum channel bandwidth of single carrier is 20 MHz and the peak data rate can reach to ~300 Mbps assuming 64 QAM and four layers [28]. In order to fulfill the requirement for peak data rate of 1 Gbps [29], CA was introduced in Release-10 to support up to five component carriers aggregation and

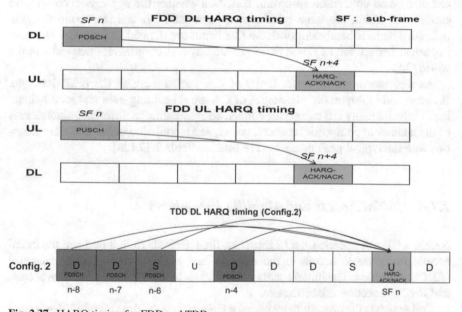

Fig. 2.37 HARQ timing for FDD and TDD

---

[4]It should be clear that Chase combining is just a special case of IR. In this case, the additional redundancy bits can be considered as based on repetitive coding. The details of IR and Chase combining are well known and can be found in [27].

Table 2.17 Several typical peak data rates UE categories

DL/UL	UE category	Peak data rate	Number of component carriers	Number of layers	Modulation
DL	4	150 Mbps	1	2	64QAM
DL	5	300 Mbps	1	4	64QAM
DL	8	3 Gbps	5	8	64QAM
DL	14	3.9 Gbps	5	8	256QAM
DL	17	25 Gbps	32	8	256QAM
UL	5	75 Mbps	1	1	64QAM
UL	13	1.5 Gbps	5	4	64QAM
UL	19	13.56 Gbps	32	4	256QAM

the maximum channel bandwidth of each component carrier is 20 MHz. In Release-12, CA was enhanced to support the aggregation of FDD and TDD carrier, which is a kind of FDD and TDD convergence because the merits of both FDD and TDD can be jointly utilized. In Release-13, CA was further enhanced to support up to 32 component carriers aggregation, and in principle it can support up to $32 \times 20 = 640$ MHz. However, it is difficult for one operator to have 640 MHz licensed spectrum in the practical system, and the aggregated component carriers can also be on unlicensed spectrum. Based on whether the aggregated component carriers are within the same band or not, intra-band CA and inter-band CA was defined. The LTE bands and different CA bands are defined in [28]. CA is a very important feature for LTE and there are already 241 commercial networks in the world [30].

As mentioned, an important target of CA was to increase the peak data rate. However, other important advantages of CA are scheduling gain and load balancing. There are many UE categories defined corresponding to different combinations of the number of component carriers, layers, and modulation order. The UE categories with the typical peak data rates are listed in Table 2.17 [28].

2.1.8 Initial Access and Mobility Procedures

Before a UE can operate on a LTE carrier, the UE needs to first perform the initial access procedure to access to the network and establish radio resource control (RRC) connection. The initial access procedure includes cell search, random access, and RRC connection establishment.

Cell search is the procedure by which a UE acquires the time (symbol/slot/frame) and frequency synchronization with a cell as well as physical layer cell identity (Cell ID). There are 504 physical layer cell IDs supported and many layer 1 transmission parameters such as reference signal sequence generation and scrambling sequence are dependent on the physical layer cell ID. The 504 physical layer cell IDs are grouped into 168 unique physical layer cell ID groups and each group contains three IDs. The primary and secondary synchronization signals (PSS and SSS) are defined to facilitate the cell search. Since there are not any prior knowledge before the UE performs the cell search, the synchronization signals are always transmitted in the center 6 resource blocks of the system bandwidth, so that the UE always knows where it is. The PSS is transmitted in the last OFDM symbol of slots 0 and 10 for frame structure type 1 and the third OFDM symbol in slots 2 and 12 for frame structure type 2. The SSS is transmitted in the second last OFDM symbol of slots 0 and 10 for frame structure type 1 and the last OFDM symbol in slots 1 and 11 for frame structure type 2.

The PSS transmitted in the two slots of one radio frame is same. This transmission is used to acquire the symbol synchronization and part of the cell ID information. There are three different ZC root sequences with length 63 defined for PSS to represent the three unique IDs within one physical layer cell ID group. The middle element of the sequence is punctured when mapped onto the center 6 resource blocks due to the existence of direct current (DC) subcarrier. The mapping results in the symmetric property of PSS in time domain which can reduce the computation complexity at the UE [31]. Once the PSS is acquired, UE can detect SSS based on the time relationship between PSS and SSS. As the time relationship between PSS and SSS is different for frame structure type 1 and 2, the detection of SSS can also obtain the information of the frame structure and cyclic prefix. Due to the close distance between PSS and SSS symbol, the PSS may be used for coherent detection of SSS, which is up to UE implementation. The SSS is represented by two short sequences with length 31 and the 168 unique combinations of two short sequences represent the physical layer cell ID groups. For all the combinations of the two short sequences, the index of the first short sequence is always less than that of the second short sequence, which can reduce the ambiguity of cell group ID detection [32]. The positions of two short sequences of SSS are switched across the two slots of one radio frame, which can be used to acquire the subframe synchronization.

After the cell search is done, UE needs to acquire the system broadcast information conveyed on PBCH and SIBs. The PBCH is detected to obtain the system bandwidth, the number of CRS antenna ports, SFN and PHICH duration described in Sect. 2.1.4.1. The SIBs are mapped to the PDSCH in physical layer, and the corresponding PDCCH with CRC scrambled by the system information radio network temporary identifier (SI-RNTI) is transmitted in the common search space for all the UE's detection. The PDCCH indicates the scheduled resources, MCS, etc. of SIBs. There is a periodicity to the transmission of SIBs with different periodicity. The SIB1 uses a fixed periodicity of 80 ms and there are four repetitions within 80 ms. The first transmission of SIB1 is scheduled in subframe 5 of radio frames for which the SFN mod 8 = 0, and the repetitions are scheduled in subframe 5 of all other radio frames for which SFN mod 2 = 0 [20]. The SIB1 contains the information to determine whether a UE is allowed to access this cell and defines the scheduling of other system information (SI). The SI message is used to convey one or more SIBs other than SIB1. All the SIBs included in one SI message are transmitted with the same periodicity, and one SIB can only be mapped to one SI message. The SIB2 is always mapped to the first SI configured in SIB1. The SIB2 contains radio resource configuration information that is common for all UEs, e.g., MBSFN sub-frame configuration and PRACH configuration information. The detail information of each SIBs can be found in [20].

The UE needs to perform random access to achieve uplink synchronization for the PUSCH transmission and establish RRC connection after obtaining the system information related to random access including PRACH configuration, frequency position, the root sequences, cyclic shift, set type (restricted or unrestricted), and so on. The transmission of PRACH occupies 6 resource blocks in a subframe or more consecutive subframes, which is scalable to different system bandwidth. Contention base random access procedure is used during initial access phase because the UE is in idle state, which have four steps as shown in Fig. 2.38 [33].

Step 1: Random Access Preamble Transmission

The UE first selects one preamble sequence from the configured preamble sequence set(s), and transmits it in the configured frequency resource with the desired power.

Step 2: Random Access Response (RAR)

Upon the successfully detected preamble sequences, the eNB will transmit the RAR in downlink in response to these transmitted preamble sequences. The RAR is transmitted in the scheduled PDSCH indicated by a PDCCH marked with random access RNTI (RA-RNTI). The RAR conveys at least the identity of the detected random access preamble, timing alignment information, Random Access Response Grant (RARG) [12], and the assignment of Temporary cell RNTI (Temporary C-RNTI). If there are multiple random access preambles detected simultaneously, the intended information in RAR can be contained in one PDSCH.

Fig. 2.38 Contention based random access procedure

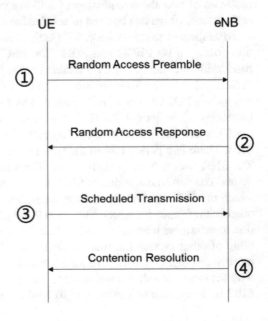

Step 3: PARG-based UL Transmission

After the reception of RAR, the UE would be synchronized in uplink using the signaled timing alignment information. However, the UE does not establish the connection with the cell yet. The PUSCH scheduled by RARG conveys some request information from higher layer, e.g., RRC Connection Request and NAS UE identifier [33]. The PUSCH is scrambled by the Temporary C-RNTI.

Step 4: Contention Resolution on Downlink

It may happen that multiple UEs transmit the same random access preambles simultaneously and then receive the same RAR, which results in the collision of different UEs using the same Temporary C-RNTI. To solve the collision, the eNB transmits a downlink message in PDSCH scheduled by a PDCCH marked with the Temporary C-RNTI in Step 2. The UEs will check whether the UE identity signaled in the downlink message is the same as the identity conveyed in Step 3. If it is matched, it implies that the random access is successful and the Temporary C-RNTI will be promoted to C-RNTI for the UE; otherwise, the UE needs to restart the random access procedure from Step 1.

In addition to the contention-based random access procedure, there is also non-contention-based random access procedure applicable to only handover and DL data arrival during RRC_CONNECTED requiring random access procedure [33].

There are two states defined for the UE which are RRC_IDLE and RRC_CONNECTED (Fig. 2.39) [33]. In RRC_IDLE state, there is no RRC context and specific cell attachment. The UE under this state does not have data to receive/transmit and can be in sleep mode for power saving. Once it is triggered to establish RRC connection, the random access procedure as described above is performed and establish the RRC context to move to RRC_CONNECTED state. In case of RRC_CONNECTED state, the normal communication between the UE and eNB can be performed under the RRC context known by both UE and eNB. If there is no data to transfer, the RRC connection can be released to go to RRC_IDLE state.

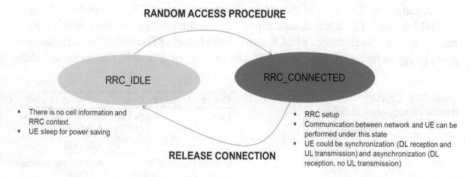

Fig. 2.39 State machine of RRC_IDLE and RRC_CONNECTED

2.2   5G-NR Design of Carrier and Channels

2.2.1   Numerology for the Carrier

NR spans much wider frequency range compared to LTE, which currently defines two frequency ranges (FR) as FR1 and FR2. The corresponding frequency range for FR1 and FR2 are 450–6,000 MHz and 24,250–52,600 MHz respectively, and a set of operating bands are defined for FR1 and FR2 [34]. For the spectrum beyond 3GHz, there is larger spectrum bandwidth available (see Sect. 2.3.1), which can be utilized to fulfill the high data rate requirement of IMT-2020 [35]. For each operating band, a UE or base station can support a number of carriers which is dependent on the bandwidth of a carrier and the UE capability. The carrier bandwidth is related to base station and UE processing capability. The supported carrier bandwidths for FR1 and FR2 are summarized in Table 2.18. However, from UE perspective, the supported transmission bandwidth (i.e., UE channel bandwidth) may be smaller than the carrier bandwidth due to the restriction of UE capability. In this case, the network can configure a part of contiguous spectrum being equal to or smaller than UE channel bandwidth from the carrier for the UE, which is also called as bandwidth part (BWP). A UE can be configured with up to four BWPs in the downlink and uplink, but there is only one BWP being active at a given time. The UE is not expected to receive or transmit outside an active BWP, which is beneficial for UE power saving because it does not have to transmit or receive on the entire system bandwidth.

Since CP-OFDM-based waveform is applied for both downlink and uplink transmission, see Sect. 2.5.1, the design of numerology for a carrier is similar to LTE includes subcarrier spacing (SCS) and CP. The key factor to determine the SCS is the impact of Doppler shift which is related to the carrier frequency and mobility. The frequency range of LTE operating bands is within that of NR. The 15 kHz SCS design has been well verified in the practical networks. Hence 15 kHz SCS is supported in NR as well. Considering the mobility requirement supporting up to 500 km/h and the wider frequency range, having only 15 kHz SCS is not enough and multiple larger SCS values with $2^\mu \times 15$ kHz, where $\mu = 0, 1, 2, 3, 4$ were introduced. Larger SCS results in the shorter time duration of OFDM symbol which is helpful for short TTI transmission especially for the latency sensitive service, e.g., remote control. Such design of SCS also has the benefit of reducing implementation complexity in terms of the number of subcarriers for a certain carrier bandwidth.

Table 2.18 Carrier bandwidth supported in NR

Frequency range	Carrier bandwidth (MHz)[a]
FR1	5, 10, 15, 20, 25, 30, 40, 50, 60, 70, 80, 90, 100
FR2	50, 100, 200, 400

[a]Carrier bandwidth is dependent on the subcarrier spacing and operating band [34]

The utilization of SCS is dependent on the frequency range. Of course, it is not necessary to support all the SCS values for each and every frequency range.

The CP length was determined to well mitigate the impact of the delay spread and have a reasonable overhead. In the case 15 kHz SCS of LTE, the ratio of CP length over the duration of one OFDM symbol is $144/2048 = 7.03\%$ for the OFDM symbols other than the first one in each slot. The CP length for the first OFDM symbol in each slot is slightly longer as $160/2048 = 7.81\%$ due to the restriction of 0.5 ms slot and to help with the settling of the automatic gain adjustment. This CP length has been shown to be a good trade-off between the mitigation of delay spread and the CP overhead, and is, therefore, reused for NR in the case of 15 kHz SCS and normal CP. As the duration of one OFDM symbol is equal to the reciprocal of SCS, the CP length for other SCS will be proportionally reduced by 2^μ, $\mu = 0, 1, 2, 3, 4$ which keeps the same ratio of CP overhead as 15 kHz. This can also enable the OFDM symbols of different SCS within one slot to align, which is beneficial for the coexistence of carriers with different SCS especially for the TDD networks because of the need of synchronization. It is noted that the CP length of the first and $7 * 2^\mu$th OFDM symbols within one subframe is slightly longer, and the detail of frame structure can be found in Sect. 2.2.2. Extended CP is only supported for 60 kHz SCS, and the length of extended CP is scaled from that of LTE. The 60 kHz SCS is envisioned to be applicable for URLLC service because of the resulting shorter TTI. When the URLLC service using 60 kHz SCS is deployed in sub 3 GHz, the normal CP may not be able to mitigate the inter-symbol interference in case of high delay spread scenario, and the extended CP is useful in this scenario.

Due to the proportional reduction of CP for larger SCS, it is clear that the CP will not be proper for scenarios with large delay spread. The larger SCS is, thus, typically utilized for high frequency which has smaller delay spread [36]. There is the frequency range restriction for different SCS as shown in Table 2.19. The SCS and CP for one BWP can be obtained from the higher layer parameter *subcarrierSpacing* and *cyclicPrefix*, respectively. In one carrier, multiple numerologies can be configured.

Table 2.19 Supported transmission numerologies

μ	SCS $\Delta f = 2^\mu \cdot$ 15 [kHz]	Cyclic prefix Type	Length of CP within one subframe $l = 0$ or $7 \cdot 2^\mu$ (normal CP)	others	Applicable frequency range FR1 Sub 1 GHz	1–3 GHz	3–6 GHz	FR2
0	15	Normal	$144 * \Delta + 16 * \Delta$	$144 * \Delta$	√	√	√	–
1	30	Normal	$144 * \Delta/2 + 16 * \Delta$	$144 * \Delta/2$	√	√	√	–
2	60	Normal	$144 * \Delta/4 + 16 * \Delta$	$144 * \Delta/4$	–	√	√	√
2	60	Extended	$512 * \Delta$		–	√	√	√
3	120	Normal	$144 * \Delta/8 + 16 * \Delta$	$144 * \Delta/8$	–	–	–	√
4[a]	240	Normal	$144 * \Delta/16 + 16 * \Delta$	$144 * \Delta/16$	–	–	–	–

Note: Δ is the oversampling factor over 2048
[a]240 KHz is defined in the numerology, but not used for FR1 and FR2 currently in Rel-15

Given a certain SCS, a resource block is defined as 12 consecutive subcarriers in the frequency domain. It is different from LTE which always has 7 or 6 OFDM symbols within one resource block. Here there is no restriction on the number of symbols within one resource block to facilitate the multiple short TTIs transmission per SCS. The smallest resource unit within the resource block is called resource element (RE) indicated by (k, l), where k is the subcarrier index in the frequency domain and l represents the OFDM symbol index relative to a starting position in the time domain.

Two kinds of resource blocks are defined: common resource block (CRB) and physical resource block (PRB). The CRB is defined from the system perspective, where the CRBs are numbered from 0 and upwards in the frequency domain for subcarrier spacing configuration μ, i.e., $n_{CRB}^{\mu} = k / 12$, where k is the subcarrier index relative to a frequency starting position (i.e., reference point) of a carrier common for all SCS. The PRB is defined within one BWP which is a set of contiguous common resource blocks. The relation between the CRB and RRB is $n_{CRB} = n_{PRB} + N_{BWP,i}^{start}$, where $N_{BWP,i}^{start}$ is the starting position of BWP i, $i = 0, 1, 2, 3$ expressed by CRB.

The actual number of resource blocks per carrier depends on the carrier bandwidth, SCS, and the frequency band [34].

2.2.2 Frame Structure

One radio frame with 10 ms duration consists of ten subframes and each subframe is of 1 ms duration. Each radio frame is divided into two equally sized half-frames of five subframes, which are half-frame 0 consisting of subframe 0–4 and half-frame 1 consisting of subframes 5–9. The number of slots within one subframe is 2^{μ}, $\mu = 0, 1, 2, 3, 4$ which is dependent on SCS. In one slot, there are always 14 OFDM symbols regardless of the SCS configuration in the case with normal CP. The relationship among the number of OFDM symbol, slot, subframe, and radio-frame is demonstrated in Fig. 2.40.

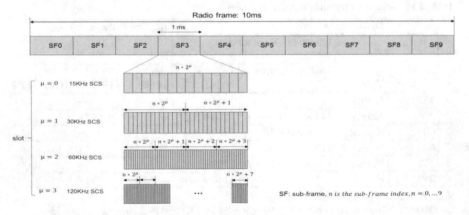

Fig. 2.40 Frame, subframe and slot

NR has very flexible frame structure compared to LTE, which has only two frame structures which are for FDD and TDD, respectively. The slots in one radio frame can be flexibly configured for downlink or uplink transmission. The 14 OFDM symbols in one slot can be classified as "downlink," "flexible," or "uplink." A slot format includes downlink symbols, uplink symbols, and flexible symbols. The downlink and uplink symbols can only be used for downlink and uplink transmission, respectively, however the flexible symbols can be used for downlink transmission, uplink transmission, GP, or reserved resources. The slots can be classified into five different types: as downlink only, uplink only, downlink dominant, uplink dominant, and full flexible (Fig. 2.41). For Type 3 or Type 4, the downlink PDSCH scheduling and the corresponding ACK/NACK or the UL grant and PUSCH transmission is contained in one slot, which can be considered as a type of self-contained transmission to reduce latency. All the OFDM symbols in Type 5 slot are flexible and can be reserved for future to achieve the forward compatibility.

The flexible frame structure is beneficial to adapt the service traffic in downlink and uplink, which can be found in Sect. 2.3.1. The signaling of slot formats needed for UE to obtain the frame structure includes cell-specific higher layer configuration, UE-specific higher layer configuration, UE-group DCI, and UE-specific DCI as below.

2.2.2.1   Cell-Specific Higher Layer Configuration

The cell-specific higher layer parameter *TDD-UL-DL-ConfigurationCommon* is used by the UE to set the slot format per slot over a number of slots indicated by the parameter. The higher layer parameter includes [37]:

{

- reference SCS
- pattern 1

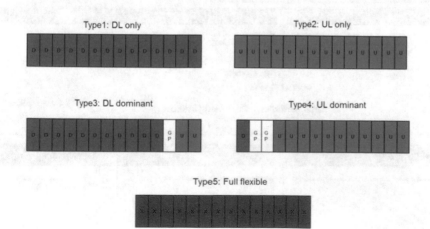

Fig. 2.41 Slot types

- DL/UL transmission periodicity (P_1)
- Number of downlink slots (X_1): Number of consecutive full DL slots at the beginning of each DL-UL pattern
- Number of downlink OFDM symbols(x_1): Number of consecutive DL symbols in the beginning of the slot following the last full DL slot
- Number of uplink slots (Y_1) : Number of consecutive full UL slots at the end of each DL-UL pattern,
- Number of uplink OFDM symbols(y_1): Number of consecutive UL symbols in the end of the slot preceding the first full UL slot

}

Additionally, a second pattern, i.e., pattern 2, can be configured in the *TDD-UL-DL-ConfigurationCommon*, which has a different configuration, e.g., the second pattern has more uplink slots configuration for uplink capacity and coverage. In this case, it is known as dual periodicity slot configuration (Fig. 2.42). The configuration of the second pattern is optional. It is noted that the remaining slots and symbols not addressed by *TDD-UL-DL-ConfigurationCommon* during the DL/UL transmission period are default to be "flexible" and can be further configured by other signaling.

2.2.2.2   UE-Specific Higher Layer Configuration

The UE-specific higher layer parameter *TDD-UL-DL-ConfigDedicated* can be additionally configured to override only the flexible symbols provided by *TDD-UL-DL-ConfigurationCommon*. The *TDD-UL-DL-ConfigDedicated* provides a set of slot configurations, where each slot configuration includes the slot index and the symbol

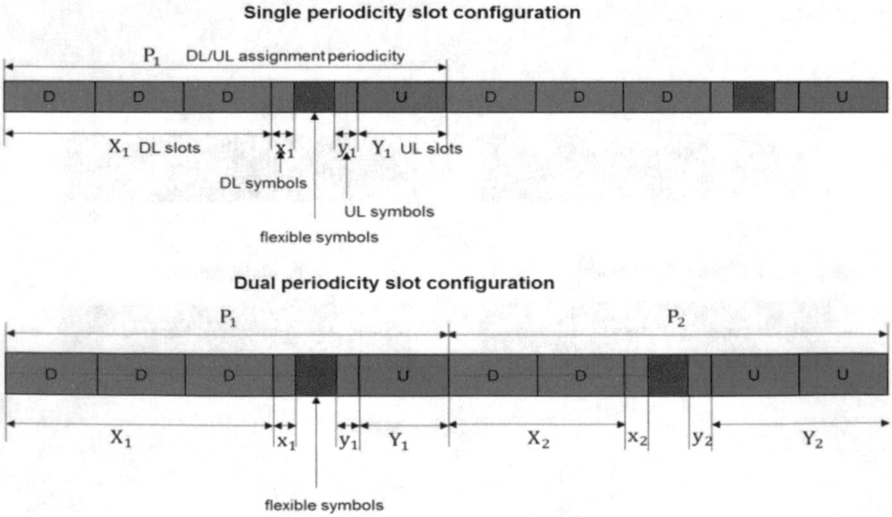

Fig. 2.42   Illustration of single and dual periodicity slot format configuration

configuration for the indicated slot. The symbols in the indicated slot can be configured for all downlink, all uplink or a number of downlink first symbols in the slot and a number of uplink last symbols in the slot [38].

2.2.2.3 Group Common PDCCH

In case the symbols in a set of consecutive slots in time domain indicated to a UE as flexible by higher layer parameters *TDD-UL-DL-ConfigurationCommon* and *TDD-UL-DL-ConfigDedicated*, the layer 1 signaling DCI format 2_0 can be used to provide the slot format(s) for the set of consecutive slots. There are total number of 56 slot formats defined in the specification [38].

DCI format 2_0 is a group common PDCCH which can be detected by a group of UEs. Whether a UE needs to detectDCI format 2_0 is configured by the network. When a UE is configured by higher layers signaling with parameter *SlotFormatIndicator*, the UE is provided with a slot format indicator (SFI-RNTI) and the payload size of DCI format 2_0. In addition, the location of SFI-index field for a cell in DCI format 2_0 is provided. The SFI-index represents a slot format combination. The mapping between a list of slot format combinations and the corresponding SFI-index is configured by the parameter *SlotFormatIndicator*. The slot format combination refers to the one or more slot formats that occur in the set of consecutive slots in time domain.

2.2.2.4 DL/UL Dynamic Scheduling

If an SFI-index field value in DCI format 2_0 indicates that the set of symbols of the slot as flexible, the set of symbols can be dynamically scheduled for downlink or uplink transmission, which depends on the received DCI format. When the UE receives a DCI format (e.g., DCI format 1_0, 1_1, 0_1 [39]) for scheduling downlink transmission PDSCH or CSI-RS in the set of symbols of the slot, the set of symbols will be used for downlink transmission. In case the UE receives a DCI format (e.g., DCI format 0_0, 0_1) [39] for scheduling PUSCH, PUCCH, or SRS transmission in the set of symbols, it will be used for uplink transmission. If there is not any scheduling information for the set of symbols, the UE does not transmit or receive in these resources.

In summary, there are four levels signaling to configure the frame structure which provides much flexibility as shown in Fig. 2.43. However, considering the potential complexity and the cost of cross-link interference mitigation in the practical system, it may be enough to use only a few frame structure. Regarding the signaling, it is not mandatory for the UE to receive all the four levels of signaling. The UE can receive one or more of these signaling to obtain the frame structure.

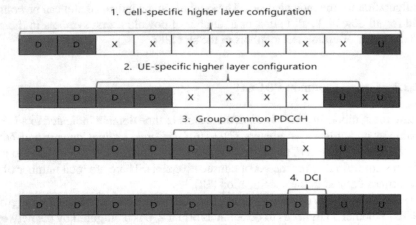

Fig. 2.43 Signaling of frame structure configuration

2.2.3   *Physical Layer Channels*

After the initial access procedure (see Sect. 3.1.1), gNB starts to perform normal communication with the UE. The information to be transmitted in both the downlink and uplink is conveyed on different physical channels. The physical channels correspond to a set of resource elements carrying the information originating from higher layers. According to the transmission direction and the characteristics of the carried information of physical channel, a set of downlink and uplink physical channels are defined [40]. In this section, the functions of different physical channels in the DL and UL transmission are introduced. The physical random access channel does not convey the information from higher layer, and transmits the preamble to build RRC connection, uplink synchronization, etc., which is discussed in Sect. 3.1.1.

2.2.3.1   Physical Broadcast Channel (PBCH)

The PBCH is detected by UE and is used to acquire the essential system information during the initial access described in section of IAM. The system information from higher layer is divided into MIB (Master Information Block) and a number of SIBs (System Information Block). The MIB is always carried on the PBCH with 80 ms periodicity and made repetitions within 80 ms. The MIB includes the following information [37]:

- System Frame Number (SFN): the six most significant bit (MSB) of the 10-bit SFN
- Subcarrier spacing for SIB1, Msg.2/4 for initial access and broadcast SI-messages
- SSB Subcarrier Offset: the frequency domain offset between SSB and the overall resource block grid in number of subcarriers

2.2   5G-NR Design of Carrier and Channels   95

- The time position of the first DM-RS for the downlink or uplink
- The resource configuration for SIB1 detection, e.g., a common control resource set, a common search space and necessary PDCCH parameters for SIB1
- The cell is barred or not
- Intra frequency reselection is allowed or not

In addition to the information from the MIB transmitted in the PBCH, there are additionally 8 bits information from physical layer as four least significant bit (LSB) of 10-bit SFN, 1 bit half-frame indication, 3 MSB of SS/PBCH block index in case with 64 candidate SS/PBCH blocks (otherwise there is one bit as MSB of PRB offset and two reserved bits). The total payload size of PBCH is 56 bits including 24 bits CRC, which is encoded to 864 bits with very low coding rate to guarantee the reliability of PBCH detection.

The PBCH is transmitted with PSS and SSS together as the SS/PBCH block (SSB). The same subcarrier spacing and cyclic prefix is applied for PSS, SSS, and PBCH. An SSB consists of four consecutive OFDM symbols in the time domain and 240 consecutive subcarriers forming 20 SSB resource blocks in the frequency domain (Fig. 2.44). For each SSB, the PSS and SSS are located at the first and the third OFDM symbol, respectively, and the length of PSS and SSS sequence is 127 symbols mapped onto the center resource elements of each located symbol; PBCH and the associated DM-RS is mapped onto the second, the fourth OFDM symbol and partial of the third OFDM symbol. There is one DM-RS resource element every four resource elements giving a PBCH DM-RS density of 1/4. For the third OFDM symbol within one SSB, PBCH and the associated DM-RS is mapped onto the 48 resource elements of both sides.

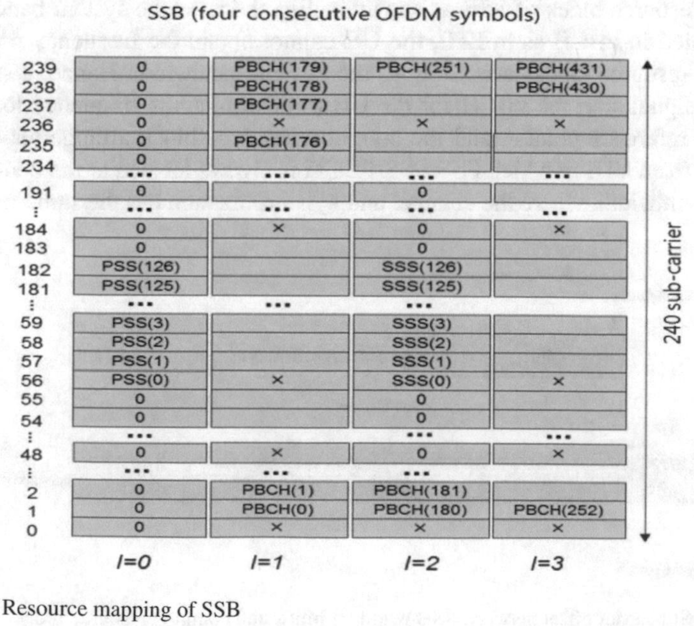

SSB (four consecutive OFDM symbols)

sub-carrier	l=0	l=1	l=2	l=3
239	0	PBCH(179)	PBCH(251)	PBCH(431)
238	0	PBCH(178)		PBCH(430)
237	0	PBCH(177)		
236	0	×	×	×
235	0	PBCH(176)		
234	0			
...	...	...	...	...
191	0		0	
⋮	...	...	...	...
184	0	×	0	×
183	0		0	
182	PSS(126)		SSS(126)	
181	PSS(125)		SSS(125)	
⋮	...	...	...	
59	PSS(3)		SSS(3)	
58	PSS(2)		SSS(2)	
57	PSS(1)		SSS(1)	
56	PSS(0)	×	SSS(0)	×
55	0		0	
54	0		0	
⋮	...	...	...	...
48	0	×	0	×
⋮	...	...	...	...
2	0	PBCH(1)	PBCH(181)	
1	0	PBCH(0)	PBCH(180)	PBCH(252)
0	0	×	×	×

240 sub-carrier

Fig. 2.44 Resource mapping of SSB

In NR, beam-based access mechanism was adopted for cell coverage. Multiple SSBs covering different beam directions are used for performing beam sweeping. Since more SSBs are needed in the high-frequency scenario. The maximum number of candidate SSBs within a half frame is 4/8/64 corresponding to the sub 3 GHz, 3–6 GHz and above 6 GHz scenario, respectively. The transmission time instances of candidate SSBs, i.e., SSB pattern, within one-half frame are predefined which is dependent on the subcarrier spacing and the applicable frequency range [38]. As SSB is used to determine the subframe timing of the associated PRACH resources, etc., the candidate SSBs need to be identified. The candidate SSBs in a half frame are indexed from 0 to L-1, where $L = 4, 8, 64$. The index of candidate SSB is implicitly associated with its corresponding PBCH DM-RS sequence when $L = 4$ and 8. In case $L = 64$, if there are 64 DM-RS sequences to associate with the SSB index, it will significantly increase the SSB blind detection complexity. To help with the blind detection, the index of candidate SSB is determined by obtaining the three MSB from the PBCH and three LSB from the PBCH DM-RS sequences. Once the UE successfully detects one SSB, the UE will obtain the essential system information from PBCH, frame timing according to the SFN and the index of detected SSB. For the initial access UE, the UE can assume that the periodicity of SSBs in a half frame is 20 ms.

The channel raster and synchronization raster in NR is decoupled in order to have a sparse synchronization raster for reducing search complexity. In this case, the SSB resource block may not be aligned with the common resource block, and there is a subcarrier offset between them. The subcarrier offset is from subcarrier 0 in the common resource block with the subcarrier spacing provided by MIB to subcarrier 0 of the SSB, where this common resource block overlaps with the subcarrier 0 of the first resource block of SSB as shown in Fig. 2.45. As the system bandwidth is not signaled in PBCH as in LTE, the UE cannot obtain the frequency position of common resource blocks according to the system bandwidth. Hence, a subcarrier offset is signaled in the PBCH for the UE to determine the frequency location of common reference point A and the common resource block arrangement. This is different from LTE in which PSS/SSS/PBCH is always located at the center of the system bandwidth, where the channel and synchronization has the same raster.

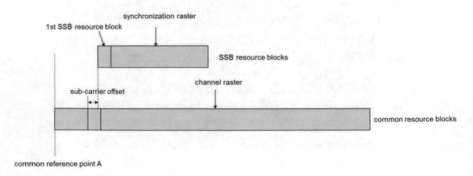

Fig. 2.45 Subcarrier offset between SSB resource block and common resource block

2.2.3.2   Physical Shared Data Channel (PDSCH)

There is only one transmission scheme defined for the PDSCH transmission, which is based on DM-RS similar to TM9 in LTE. From the UE perspective, up to eight layers transmission is supported. There is no open loop transmit diversity scheme as SFBC used in LTE defined for PDSCH transmission, which means NR could be less robust in certain environments (such as high speeds). Hence, link adaptation in NR plays a crucial role in the network's choice of transmission parameters such as spatial precoder, frequency domain granularity of spatial precoder, and time-domain granularity of spatial precoder.

The maximum number of codeword per UE for the PDSCH transmission is 2 and there is a one to one mapping between transport block and the codeword. Similar to LTE, there are the corresponding MCS, HARQ-ACK and CQI for each codeword. However, the codeword to layer mapping is slightly different from LTE discussed in Sect. 2.5.3.1.1. There is one codeword if the number of transmission layer is less than or equal to 4; otherwise the number of codeword is 2. Such codeword to layer mapping has the benefit of reducing the signaling overhead of the MCS, HARQ, and CQI when the rank is no more than 4, but it cannot enable to use the successive interference cancellation at the UE side in case with single codeword transmission. The relationship between the number of codeword and layer is a trade-off of performance and signaling overhead. The general transmission structure of NR PDSCH is shown in Fig. 2.46. As NR only supports DM-RS-based transmission scheme, the precoding operation is transparent and the data of each layer is directly mapped onto the DM-RS ports.

The channel coding scheme used for PDSCH transmission is LDPC (see Sect. 2.5.2.1). The scrambling and modulation operation is similar to LTE, and the modulation schemes for PDSCH are QPSK, 16QAM, 64QAM, and 256QAM.

In NR, since low latency is a very important requirement [35, 41], shorter TTI is a key feature supported to fulfill the requirement. To enable the shorter TTI, the resource allocation for the PDSCH is more flexible especially in the time domain. The number of OFDM symbols allocated for PDSCH can be {3, 4,…14} or {2,4,7} which is dependent on PDSCH mapping type. PDSCH mapping type A is used for starting the PDSCH in the first three symbols of a slot with duration of three symbols or more until the end of a slot. PDSCH mapping type B is used for starting the PDSCH anywhere in the slot, with duration of 2, 4 or 7 OFDM symbols. In addition,

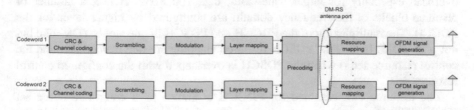

Fig. 2.46 Transmission structure of NR PDSCH

the slot aggregation in time domain for PDSCH transmission is supported to improve the coverage. In this case, the same symbol allocation is used across a number of consecutive slots. The number of aggregated slots can be 2, 4, or 8, which is configured by higher signaling, but there is only one downlink control channel for the slot aggregation to reduce the signaling overhead. The frequency domain resource allocation supports two types, and it is relative to a bandwidth part. Type 0 allocates resources with a bitmap of resource blocks groups (RBG), where each RBG is a set of consecutive virtual resource blocks. Type 1 allocates resources as a set of contiguous non-interleaved or interleaved virtual resource blocks. The network can also indicate to the UE how many physical resource blocks (PRB) are bundled with the same precoder, which constitutes precoding resource block groups (PRG) (see Sect. 2.5.3.1.2).

On the downlink, the UE receives a DL DCI over PDCCH (see Sect. 2.2.3.3). DCI format 1_1 offers the most scheduling flexibility, while DCI format 1_0 is more robust and can be used for fallback. A DCI scheduling PDSCH can also indicate a rate-matching pattern that allows reserving resource elements where PDSCH and DM-RS cannot be mapped. Such resource reservation signaling allows, for example, rate matching around entire OFDM symbols or PRBs, or rate matching around LTE common reference signals (CRS), or rate matching around ZP CSI-RS resources. This signaling also allows dynamically mapping of the PDSCH in REs that are semi-statically allocated for PDCCH, when the network decided not to use those resources for PDCCH in a certain slot.

2.2.3.3   Physical Downlink Control Channel (PDCCH)

The PDCCH carriers downlink control information (DCI) for PDSCH scheduling, PUSCH scheduling, or some group control information, e.g., power control information for PUSCH/PUCCH/SRS and slot format configuration. The defined set of DCI formats are shown in Table 2.20 [39].

Similar to LTE, to minimize the PDSCH decoding latency, the PDCCH is normally located at the beginning 1/2/3 OFDM symbols of a slot in time domain. However PDCCH does not span the entire carrier bandwidth in the frequency domain as in LTE. The rational is that the UE channel bandwidth may be smaller than carrier bandwidth, as well as the resource granularity of the PDCCH spanning the entire carrier bandwidth is rough which could result in increasing the resource overhead especially for larger bandwidth, e.g., 100 MHz. Hence, a number of resource blocks in the frequency domain are configured by higher layer for the PDCCH. The multiplexing of the PDCCH and PDSCH in one slot is TDM-like but not pure TDM. In NR, the PDSCH resource mapping are rate matched around the control resource set(s) when the PDSCH is overlapped with the configured control resource sets.

The resource unit assigned for PDCCH is known as a control resource set (CORESET). A control resource set consists of $N_{\mathrm{RB}}^{\mathrm{CORESET}}$ resource blocks in the frequency domain and $N_{\mathrm{symb}}^{\mathrm{CORESET}}$ symbols in the time domain, where the resource

Table 2.20 DCI formats

DCI format	Usage
0_0	Scheduling of PUSCH in one cell
0_1	Scheduling of PUSCH in one cell
1_0	Scheduling of PDSCH in one cell
1_1	Scheduling of PDSCH in one cell
2_0	Notifying a group of UEs of the slot format
2_1	Notifying a group of UEs of the PRB(s) and OFDM symbol(s) where UE may assume no transmission is intended for the UE
2_2	Transmission of TPC commands for PUCCH and PUSCH
2_3	Transmission of a group of TPC commands for SRS transmissions by one or more UEs

blocks are configured by a bitmap. These two parameters are configured by the higher layer parameter *ControlResourceSet* IE [37]. The assigned resource blocks are in the form of a number of resource block group (RBG) consisting of six consecutive resource blocks each. Up to three control resource sets can be configured for one UE to reduce the PDCCH blocking probability.

Given the configured PDCCH resources, the PDCCH is mapped onto these resources and transmitted. A PDCCH is formed by aggregating a number of control channel elements (CCEs), which depends on the aggregation level of that particular PDCCH. The aggregation level may be 1,2,4,8, or 16. One CCE consists of six resource-element groups (REGs) where a REG equals one resource block during one OFDM symbol. There is one resource element for PDCCH DM-RS every four resource elements in one REG, and therefore the number of available resource elements of one CCE is 48. The REGs within a control resource set are numbered in increasing order in a time-first manner, starting with 0 for the first OFDM symbol and the lowest-numbered resource block in the control resource set as demonstrated in Fig. 2.47.

The PDCCH transmission is also based on DM-RS as PDSCH, and the transparent precoding cycling can be applied to achieve transmit diversity gain. More precoders for one PDCCH is beneficial for performance from diversity perspective, but it impacts on channel estimation performance because channel interpolation cannot be performed across the resources using different precoders. To balance the precoding gain and channel estimation performance, the concept of REG bundle was introduced. An REG bundle consists of L consecutive REGs, and the same precoding is used within a REG bundle. One CCE consists of *6/L* REG bundles. For the two modes of CCE to REG mapping in one control resource set: non-interleaving and interleaving. The non-interleaving mapping is such that the REG bundles of one CCE are consecutive and located together. For the interleaving mode, the REG bundles of one CCE are interleaved and distributed to achieve frequency diversity gain and randomize the inter-cell interference [40]. The REG bundle size depends on the CCE to REG mapping mode, which is summarized in Table 2.21. The bundle size enables the REGs with the same resource block index spanning over different OFDM symbols to belong to one REG bundle as illustrated in Fig. 2.48. The bundle size in the case of interleaving mode is configurable.

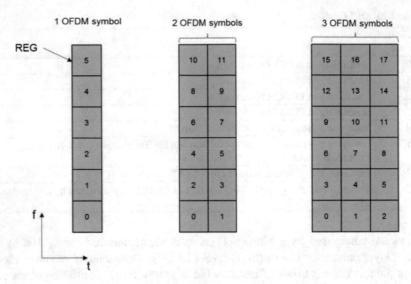

Fig. 2.47 Example of REG numbering in the control resource set

Table 2.21 REG bundle size

Number of OFDM symbols in a CORESET	REG bundle size	
	Non-interleaving	Interleaving
1	6	2, 6
2	6	2, 6
3	6	3, 6

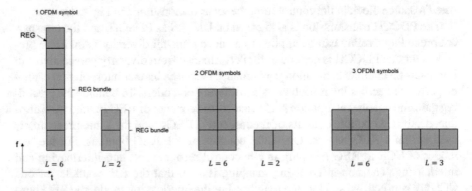

Fig. 2.48 Illustration of REG bundle for different number of OFDM symbols

At the receiver side, PDCCH is blindly detected by UE. To reduce the blind detection complexity, the number of candidate PDCCHs for blind detection is restricted, while reducing the PDCCH blocking probability. A set of PDCCH candidates for a UE to monitor is defined as the PDCCH search space set. A search space set can be a common search space set or UE-specific search space set. The common search space normally conveys some common control information for many UEs, e.g., SIB, paging, RACH response, slot format configuration, and TPC command. Hence, the supported aggregation levels in the common search space are high as 4/8/16 to guarantee the performance. DCI format 0_0, 1_0, 2_0, 2_1, 2_2 and 2_3 can be signaled in the common search space. DCI format 0_1 and 1_1 for UE-specific scheduling information are signaled in the UE-specific search space. The aggregation levels of UE-specific search space is 1/2/4/8/16.

The PDCCH blind detection procedure is similar to LTE, but the difference is some related blind detection parameters are configured by the network for flexibility. The number of search space sets for UE to monitor is configured by higher layer signaling, and each search space set is associated with one control resource set. In addition, the information related to each configured search space set is also signaled, e.g., it is common or UE-specific search space set, the number of PDCCH candidates for each aggregation level, etc. [12].

2.2.3.4   Physical Uplink Shared Data Channel (PUSCH)

For PUSCH, both DFT-S-OFDM and CP-OFDM-based transmission are supported (see Sect. 2.5.1). Two transmission schemes are supported for the PUSCH: codebook-based transmission and non-codebook-based transmission. The difference of these two schemes is whether the precoding operation is transparent or not. For codebook-based transmission scheme, the UE is configured to use one or more SRS resources for SRS transmission. More than one SRS resource is better for higher frequency scenario because beam management is easier. Based on the transmitted SRS, the network selects the preferred SRS resource, the transmission rank and precoder corresponding to the preferred SRS resource. Then the network signals them in the uplink grant DCI format 0_1. The signaled information is used by the UE for PUSCH precoding. The codebook used for DFT-S-OFDM and CP-OFDM are defined in [40]. In the case of DFT-S-OFDM, only rank 1 codebook was defined for both two and four antenna ports. The reason is that DFT-S-OFDM is normally used for the coverage limited scenario due to low cubic metric. In this scenario, rank 1 transmission is the typical case. For non-codebook-based transmission scheme, the UE first determines a number of precoders based on the measurement from CSI-RS in downlink. Then a number of SRS is precoded by these precoders and transmitted by the UE. The network selects the preferred precoded SRS. After the selection, the network signals the index of the preferred precoded SRS to UE. The UE uses the precoder corresponding to the signaled precoded SRS for the PUSCH transmission, but the precoding is transparent. The non-codebook-based transmission scheme is more applicable for TDD operation with channel reciprocity.

PUSCH transmission is based on DM-RS and up to four layers transmission for one UE is supported. The number of transmission layer is determined by the network regardless of codebook- or non-codebook-based transmission scheme. The same codeword to layer mapping as downlink is used, and, therefore, there is only one codeword for PUSCH. However, from network perspective, it can support up to 12 layers transmission in the form of MU-MIMO because of DM-RS capacity. In NR, downlink and uplink use the same DM-RS structure in order to keep the orthogonality of downlink and uplink reference signal, which can ease the cross-link interference mitigation in the case of flexible duplex.

2.2.3.5 Physical Uplink Control Channel (PUCCH)

The PUCCH conveys uplink control information (UCI) including scheduling request (SR), HARQ-ACK, and periodic CSI. Based on the control content and payload size of PUCCH, multiple PUCCH formats are supported shown in Table 2.22.

The PUCCH has a short duration transmission with 1 or 2 OFDM symbols and a long duration transmission with 4–14 OFDM symbols within one slot. Within one slot, intra-slot frequency hopping can be configured to take advantage of the frequency diversity gain in case the number of OFDM symbols is more than 1. If it is configured, the number of symbols in the first hop is the floor of half of the length of OFDM symbols. The support of duration less than four symbols PUCCH is for low latency transmission. The number of symbols for PUCCH is configured by higher layer signaling based on the requirement of both latency and coverage. The PUCCH formats with 1 or 2 symbols are normally applied in the DL dominant slot, which is transmitted in the last 1 or 2 symbol of one slot. Other PUCCH formats can be used in the UL dominant slot.

2.2.3.5.1 PUCCH Format 0

The transmission of PUCCH format 0 is constrained to one resource block in the frequency domain. The 1 or 2 bits information on PUCCH format 0 is transmitted by performing sequence selection. The used sequences are the cyclic shifts of one

Table 2.22 PUCCH formats

PUCCH format	Length of OFDM symbols L	Number of bits	UCI type
0	1–2	1 or 2	HARQ-ACK, SR, HARQ-ACK/SR
1	4–14	1 or 2	HARQ-ACK, SR, HARQ-ACK/SR
2	1–2	>2	HARQ-ACK, HARQ-ACK/SR, CSI, HARQ-ACK/SR/CSI
3	4–14	>2	HARQ-ACK, HARQ-ACK/SR, CSI, HARQ-ACK/SR/CSI
4	4–14	>2	CSI, HARQ-ACK/SR, HARQ-ACK/SR/CSI

low PAPR base sequence [40]. One cyclic shift is selected from a set of 2 or 4 cyclic shifts according to the 1 or 2 bits information to be transmitted. The selected cyclic shift is used to generate PUCCH sequence and then transmitted. For SR, a positive SR uses a predefined cyclic shift and the sequence corresponding to the predefined cyclic shift is transmitted, and nothing is transmitted for negative SR. In case there is simultaneous transmission of HARQ-ACK and positive SR, a different set of 2 or 4 cyclic shift values are used for selection. It should be noted that for PUCCH format 0 there is no DM-RS sequence for coherent detection. The network can detect the received sequence to obtain the information.

2.2.3.5.2   PUCCH Format 1

The one or two information bits are modulated as one BPSK or QPSK symbol, respectively. The modulation symbol is spread by a sequence with length 12, and then is block-wise spread with an orthogonal code covering (OCC). Two-dimensional spreading is used, which is similar to the PUCCH format 1a/1b in LTE. The multiplexing of PUCCH data and DM-RS are interlaced in the time domain. The length of the PUCCH data depends on the number of OFDM symbols for PUCCH format 1, which is equal to $\lfloor L/2 \rfloor$. The remaining $L - \lfloor L/2 \rfloor$ symbols are used for DM-RS. The multiplexing capacity is determined by the number of available cyclic shifts and OCC. The transmission structure of PUCCH format 1 with 14 OFDM symbols is illustrated in Fig. 2.49.

2.2.3.5.3   PUCCH Format 2

The PUCCH format 2 is for moderate payload size UCI transmission due to the restriction of the number of OFDM symbols. As the typical scenario of PUCCH format 2 with 1 or 2 OFDM symbols is not the coverage limited case, PUCCH format 2 transmission is similar to OFDM-based PUSCH. The coded information bits after channel coding are scrambled and QPSK modulated. The modulation symbols are then mapped onto the resource elements of the physical resource blocks

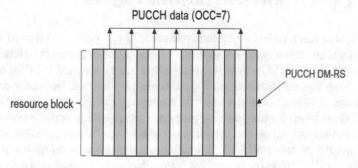

Fig. 2.49 PUCCH format 1 with 14 OFDM symbols

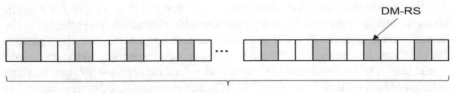

Fig. 2.50 Multiplexing of data and DM-RS for PUCCH format 2

configured for PUCCH format 2. In each of the symbols configured for PUCCH format 2, DM-RS and PUCCH UCI information are FDM-based multiplexed, and there is one resource element for DM-RS every three resource elements as shown in Fig. 2.50. The number of resource block and symbol for PUCCH format 2 is configured by higher layer signaling.

2.2.3.5.4   PUCCH Format 3 and 4

PUCCH format 3 is for large payload size of UCI. Both PUCCH format 3 and 4 transmission are based on DFT-S-OFDM which has low PAPR to enhance coverage, which is similar to LTE PUCCH format 4 and 5, respectively. The difference is that the number of symbol, ranging from 4 to 14, is configured. The number of subcarrier assigned for PUCCH format 3 shall fulfill the requirement of $2^{\alpha_2}.3^{\alpha_3}5^{\alpha_5}$ as PUSCH to reduce the implementation complexity. PUCCH format 4 is, however, constrained in one resource block, and block-wise spreading across the frequency domain within one resource blocks is used. The length of block-wise spreading is 2 or 4. The PUCCH format 4 of different UEs can be multiplexed by using different orthogonal sequence. The orthogonal sequence is configured by the higher layer signaling. To preserve the single carrier property, the multiplexing of PUCCH data and DM-RS are TDM as in Table 2.23 [40].

2.2.4   Physical Layer (PHY) Reference Signals

For a wireless communication system, reference signal (aka pilot signal) is one of the key elements of the system design. Reference signal in general carriers multiple fundamental functionalities to ensure proper and highly efficient PHY layer performances. Such functionalities include synchronization in time, frequency and phase between the transmitters and receivers, conveying channel characteristics (long term and short term) for transmission property determination at the transmitter side and the channel estimation and feedback at the receiver side, access link identification and quality measurement, etc. Though designing a single reference (for downlink) to fulfill all of these functionalities is possible, as in the case of the first release

Table 2.23 DM-RS positions for PUCCH format 3 and 4

| PUCCH length | DM-RS position *l* within PUCCH span | | | |
| | No additional DM-RS | | Additional DM-RS | |
	No hopping	Hopping	No hopping	Hopping
4	1	0, 2	1	0, 2
5	0, 3		0, 3	
6	1, 4		1, 4	
7	1, 4		1, 4	
8	1, 5		1, 5	
9	1, 6		1, 6	
10	2, 7		1, 3, 6, 8	
11	2, 7		1, 3, 6, 9	
12	2, 8		1, 4, 7, 10	
13	2, 9		1, 4, 7, 11	
14	3, 10		1, 5, 8, 12	

of LTE, it will largely limit the performance and flexibility of the system. Therefore, fabricating a set of small number of reference signals to jointly fulfill these functionalities is a better choice. While on the surface NR may appear to have quite some similarity in terms of reference signal design to that of LTE, learning from the advantages and shortcomings of LTE, NR reference signal adopted a new design framework from the beginning and took into account additional design requirements that are unique for 5G networks. In the following, NR reference signal design framework and considerations is first elaborated followed by design details for each type of reference signals including demodulation reference signal (DM-RS), channel state information reference signal (CSI-RS), sounding reference signal (SRS), and phase-tracking reference signal (PT-RS). Quasi co-location (QCL) and transmission configuration indicator (TCI) are then described as the linkage between different reference signals.

2.2.4.1 Reference Signal Design Framework and Considerations [42]

As of the design framework, one of the fundamental change of NR is to remove cell common reference signal (CRS) as in LTE system. In LTE, CRS carries several important functionalities including cell identification, time and frequency synchronization, RRM measurement, data and control channel demodulation, and CSI measurement. It is (almost) the single reference signal for downlink in LTE release 8 and is to transmit in every downlink subframe with certain time and frequency density to meet the minimum requirements of the most stringent functionality. CRS transmission is always-on regardless of the presence and absence of data transmission except for the case with cell on/off and license assisted access (LAA) introduced in its later releases. An always-on signal imposes persistent interference and overhead even when there is no data traffic and degrades the system performance significantly especially when the network deployment is dense [43]. In addition to interference and overhead issue, always-on CRS also largely limits the flexibility of the system design and forward compatibility which is a problem that became obvious for LTE during the discussions of small cell enhancements and new carrier type in 3GPP Release-12. Therefore, eliminating always-on CRS was adopted as a basic design assumption for NR.

The functionalities carried by CRS thus should be distributed among other reference signals or newly designed NR reference signals. Figure 2.51 shows the overview of LTE functionalities and the corresponding reference signals.

As shown in Fig. 2.51, in LTE, UE acquires coarse time/frequency synchronization and cell identification through the detection of PSS/SSS (SS). Cell identification is also partially carried out through CRS as PBCH demodulation reference signal. CRS also carries the functionalities of digital AGC, fine time/frequency synchronization, and RRM measurement. CRS provides data demodulation reference signal for transmission modes 1–6 and control demodulation reference signal for all of the transmission modes (except for EPDCCH demodulation). In CSI measurement, CRS is used for deriving the signal part for transmission mode 1–8 and the interference part for transmission mode 1–9. CSI-RS is utilized in transmission modes 9 and 10 for measuring signal quality. DM-RS is the data demodulation reference signal for transmission mode 7–10 and EPDCCH control channel

Fig. 2.51 LTE reference signal set and functionalities

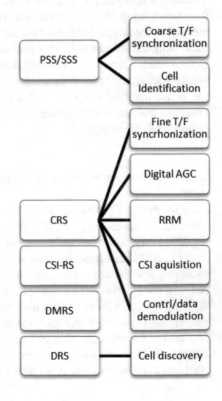

demodulation reference signal. DRS consists of SS, the windowed CRS, and optionally CSI-RS and carries the functionalities of cell discovery and RRM measurement.

In LTE, QCL is defined with respect to delay spread, Doppler spread, Doppler shift, average gain, and average delay. LTE defines several QCL assumptions among reference signals:

- CRS ports of serving cell are assumed to be quasi co-located with respect to delay spread, Doppler spread, Doppler shift, average gain, and average delay
- CSI-RS, DM-RS ports, and their tied CRS ports are assumed to be quasi co-located with respect to Doppler spread, Doppler shift, delay spread, and average delay
- SS and CRS ports are assumed to be quasi co-located with respect to Doppler shift and average gain

The long-term channel characteristics derived from CRS can be utilized to facilitate the reception of the target reference signals when they can be assumed to be quasi co-located.

With the removal of CRS, all of the functionalities (including cell identification (partial), fine time/frequency tracking, RRM measurements, Digital AGC, CSI acquisition (partial), control channel demodulation) and long-term channel property estimation for QCL assumptions carried by it should be distributed to other reference signals or newly design NR reference signals.

In addition, NR design should support frequency up to 100 GHz. To combat the significantly higher path loss at these high-frequency bands, hybrid (analog/digital) beamforming as a good trade-off between complexity and performance needs to be supported. In addition to the RS functionalities for the low-frequency band, NR design should consider the additional RS design to facilitate the analog beam acquisition, tracking, and feedback. Another issue for communication system in the high carrier frequency band is the phase noise. Compared to low-frequency band, the phase noise is considerably larger in the high-frequency band. Performance degradation caused by phase noise could be mitigated by increasing the subcarrier spacing to some extent. However, larger subcarrier spacing means shorter symbol length and larger CP length overhead. Thus, with the possible maximum subcarrier spacing limitation, a specific RS to fulfill this purpose may be introduced.

The overall reference signal design framework is summarized as follows (Fig. 2.52):

- SSB enhanced for T/F synchronization
- CSI-RS configurations to multiple functionalities such as fine T/F synchronization and main QCL assumption source (as TRS), RRM (jointly with SSB), CSI acquisition, discovery, and beam management
- DM-RS for demodulation of PDSCH/PUSCH and PDCCH
- PT-RS is introduced for phase noise compensation
- Uplink SRS

One key aspect that worth mention is that all of the above reference signals are UE-specifically configurable and hence effectively no cell common reference signal exists in NR.

Fig. 2.52 NR downlink reference signal framework

2.2.4.2 Demodulation Reference Signal

As only one transmission scheme based on DM-RS is defined in NR, its design needs to consider different scenarios and various and sometimes conflicting requirements including:

- Good channel estimation performance: as the reference signal for demodulation, good channel estimation performance is a most important criteria for DM-RS design. This demands sufficient signal density in time and frequency matching the underline radio channel characteristics and the configured numerology.
- DM-RS time and frequency pattern design needs to support a large range of carrier frequency from below 1 GHz to up to 100 GHz and various velocity of up to 500 km/h while also considering that numerology may generally scale with carrier frequency.
- Total number of orthogonal DM-RS ports and its multiplexing scheme(s) with relatively small DM-RS overhead while ensuring good demodulation performance to support the large number of data layers for massive SU/MU-MIMO transmissions.
- In addition to MBB traffic, NR also needs to support URLLC type of services. DM-RS design should then also enable very small channel estimation and demodulation processing time at the receiver.
- Flexible and configurable numerology and frame structure is a key property of the NR design. DM-RS design needs to fit into these vast number of possible configurations. One implementation of the flexible frame structure is to enable flexible duplexing of downlink and uplink transmission which may introduce severe cross-link interference. To mitigate the impact of such interference to channel estimation performance, common downlink and uplink DM-RS design that allow configuration of orthogonal downlink and uplink DM-RS transmissions is desirable.

In the following, overall design of NR DM-RS is described followed by details of Type-1 and Type-2 DM-RS configurations.

2.2.4.2.1 Overall Design of NR DM-RS

For DM-RS time and frequency pattern, two types (Type-1 and Type-2) of configurations are introduced in NR. Type-1 DM-RS supports up to four orthogonal DM-RS ports when 1 symbol is configured for DM-RS transmission and up to eight orthogonal DM-RS ports when 2 symbols are configured. Type-2 DM-RS supports up to six orthogonal DM-RS ports when 1 symbol is configured for DM-RS transmission and up to 12 orthogonal DM-RS ports when 2 symbols are configured. These orthogonal DM-RS ports are multiplexed in time domain and frequency domain by OCC. Both types of DM-RS configurations are configurable for downlink and for uplink and it can be configured such that the DM-RS for downlink and uplink are orthogonal to each other.

Two 16 bits configurable DM-RS scrambling IDs are supported. The configuration is by RRC and in addition the scrambling ID is dynamically selected and indicated by DCI. Before RRC configuring the 16 bits DM-RS scrambling IDs, cell ID is used for DM-RS scrambling.

When mapping to symbol locations of a PDSCH/PUSCH transmission within a slot, front-loaded DM-RS symbol(s) only or front-loaded DM-RS plus additional DM-RS symbol(s) can be configured. The additional DM-RS when presents should be the exact copy of the front-loaded DM-RS for the PDSCH/PUSCH transmission, i.e., the same number of symbols, antenna ports, and sequence.

With front-loaded only DM-RS, channel estimation can only rely on these 1 or 2 symbols in the early part of data transmission duration in order to speed up demodulation and reduce overall latency. However, without additional DM-RS symbol to enable time domain interpretation/filtering, the channel estimation and hence overall performance will degrade for scenarios with just moderate mobility.

For PDSCH/PUSCH mapping Type-A, the front-loaded DM-RS starts from the third or fourth symbols of each slot (or each hop if frequency hopping is supported). For PDSCH/PUSCH mapping Type-B, the front-loaded DM-RS starts from the first symbol of the transmission duration. Number of additional DM-RS can be 1, 2, or 3 per network configurations. The location of the each additional DM-RS depends on the transmission duration (i.e., number of OFDM symbols) of the PDSCH/ PUSCH transmission and follows a set of general rules for better channel estimation performance. These rules include no more than 2 OFDM symbols for PDSCH/ PUSCH after the last DM-RS, 2–4 symbols between neighboring DM-RSs, and DM-RSs almost evenly distributed in time. Figures 2.53 and 2.54 show some examples.

2.2.4.2.2 DM-RS Type-1 Configuration

For DM-RS type-1 configuration, as shown in Fig. 2.55, comb of every other tone in frequency and a cyclic shift in frequency is allocated to a DM-RS antenna port. With two combs and two cyclic shifts, four orthogonal DM-RS ports are supported when one OFDM symbol is configured. When two OFDM symbols are configured for DM-RS type-1 configuration, time domain OCC of size 2 is further used to generate orthogonal DM-RS ports which gives total 8 orthogonal ports.

DM-RS type-1 can be configured for CP-OFDM PDSCH and PUSCH. For DFT-s-OFDM PUSCH, only type-1 DM-RS is used. Type-1 DM-RS with a specific configuration is also used before RRC as default DM-RS pattern.

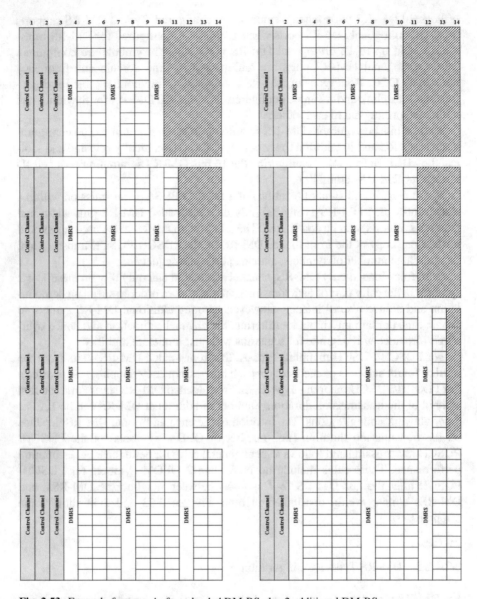

Fig. 2.53 Example for type-A: front-loaded DM-RS plus 2 additional DM-RSs

2.2.4.2.3   DM-RS Type-2 Configuration

For DM-RS type-2 configuration, as shown in Fig. 2.56, frequency domain OCC of size 2 over adjacent 2 REs and FDM are used to support 6 orthogonal DM-RS ports when 1 OFDM symbol is configured. When 2 OFDM symbols are configured for DM-RS type-2 configuration, time domain OCC of size 2 is further used to generate orthogonal DM-RS ports which gives total 12 orthogonal ports. With more orthogonal ports, type-2 configuration can potentially provide high system throughput for massive MIMO where a large number data streams of MU-MIMO is desired. DM-RS type-2 can only be configured for CP-OFDM PDSCH and PUSCH by RRC configuration.

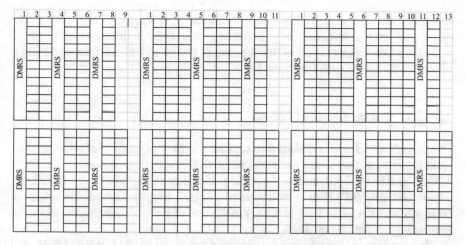

Fig. 2.54 Example for type-B: front-loaded DM-RS plus 2 additional DM-RSs for PUSCH

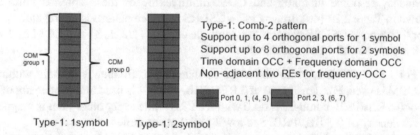

Fig. 2.55 DM-RS type 1

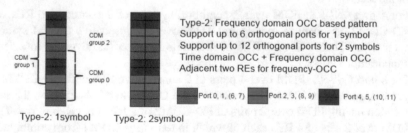

Fig. 2.56 DM-RS type-2

2.2.4.3   CSI-RS

CSI reference signal (CSI-RS) for NR generally takes the LTE CSI-RS as starting point, but expended further to support, in addition to CSI acquisition, beam management, time and frequency tracking, and RRM measurement. In order to support these variety of functionalities which naturally come with different performance target and design requirements, NR CSI-RS overall design needs to have a lot more flexibility and configurability in terms of number of antennas ports, time and frequency resource pattern, density in frequency, periodicity in time, etc.

In the following, we first describe the overall design of CSI-RS followed by specifics of CSI-RS for each functionality.

2.2.4.3.1   General Design of CSI-RS

Overall, NR CSI-RS was designed to support a variety of number of antenna ports from single port to up to 32 orthogonal ports, different frequency domain density of $\{1/2, 1, 3\}$ REs per PRB, and configurable periodicity of $\{4, 5, 8, 10, 16, 20, 32, 40, 64, 80, 160, 320, 640\}$ slots as well as aperiodic (i.e., triggered based) transmission. Note that here the periodicity in time is in the unit of slot and hence depends on the numerology of the carrier/BWP.

The numbers of CSI-RS antenna ports supported are $\{1, 2, 4, 8, 12, 16, 24,$ and $32\}$. Extensive discussions and considerations had been taken to design the time and frequency resource mapping and CDM multiplexing of these antenna ports as shown in Table 2.24 [40]. In general, a CSI-RS resource pattern is aggregated with 2 or 4 or 8 ports CSI-RS patterns and multiple density $\{1/2, 1, 3\}$ and CDM of 2, 4, or 8 are supported. More specifically:

– For 1 port CSI-RS, comb-like resource mapping in frequency domain without CDM is used. For density (ρ) of 3 REs/PRB, comb-4 is used and for density of 1 or 0.5 REs/PRB, comb-12 is used within a PRB while every other PRB is mapped for density of 0.5 REs/PRB. See row 1 and 2 of the table.
– For 2 port CSI-RS, frequency domain CDM of 2 (FD-CDM2) over 2 consecutive REs is used. For density of 0.5 REs/PRB, every other PRB is mapped. See row 3 of the table.
– For 4 port CSI-RS, FDM over 2 neighboring pairs of 2 consecutive REs with FD-CDM2 (as given in row 4) and TDM over 2 consecutive OFDM symbols each with FD-CDM2 of 2 REs of the same location (as given in row 5) are supported.
– For 8 port CSI-RS, FDM over 4 pairs of 2 consecutive REs with FD-CDM2 (as given in row 6), or TDM over 2 consecutive OFDM symbols each on the same location having FDM over 2 pairs of FD-CDM2 of 2 REs (as given in row 7), or FDM over 2 sets of 4 REs each of which in term uses CDM4 from combination of FD2 (over 2 consecutive REs) and TD2 (over 2 consecutive OFDM symbols) (as shown in row 8) are supported.
– For 12 port CSI-RS, FDM over 6 pairs of 2 consecutive REs with FD-CDM2 (as given in row 9), or FDM over 3 sets of 4 REs each of which in term uses CDM4 from combination of FD2 (over 2 consecutive REs) and TD2 (over 2 consecutive OFDM symbols) (as shown in row 10) are supported.
– For 16 port CSI-RS, TDM over 2 consecutive OFDM symbols each on the same location having FDM over 4 pairs of FD-CDM2 of 2 REs (as given in row 11), or FDM over 4 sets of 4 REs each of which in term uses CDM4 from combination

Table 2.24 CSI-RS locations (time and frequency patterns) within a slot

Row	Number of ports	Density	cdm-type	Time and frequency locations $(\bar{k}, \bar{l})$	CDM group index	k'	l'
1	1	3	No CDM	$(k_0, l_0), (k_0 + 4, l_0), (k_0 + 8, l_0)$	0,0,0	0	0
2	1	1, 0.5	No CDM	(k_0, l_0)	0	0	0
3	2	1, 0.5	FD-CDM2	(k_0, l_0),	0	0, 1	0
4	4	1	FD-CDM2	$(k_0, l_0), (k_0 + 2, l_0)$	0,1	0, 1	0
5	4	1	FD-CDM2	$(k_0, l_0), (k_0, l_0 + 1)$	0,1	0, 1	0
6	8	1	FD-CDM2	$(k_0, l_0), (k_1, l_0), (k_2, l_0), (k_3, l_0)$	0,1,2,3	0, 1	0
7	8	1	FD-CDM2	$(k_0, l_0), (k_1, l_0), (k_0, l_0 + 1),$ $(k_1, l_0 + 1)$	0,1,2,3	0, 1	0
8	8	1	CDM4 (FD2,TD2)	$(k_0, l_0), (k_1, l_0)$	0,1	0, 1	0, 1
9	12	1	FD-CDM2	$(k_0, l_0), (k_1, l_0), (k_2, l_0),$ $(k_3, l_0), (k_4, l_0), (k_5, l_0)$	0,1,2,3,4,5	0, 1	0
10	12	1	CDM4 (FD2,TD2)	$(k_0, l_0), (k_1, l_0), (k_2, l_0)$	0,1,2	0, 1	0, 1
11	16	1, 0.5	FD-CDM2	$(k_0, l_0), (k_1, l_0), (k_2, l_0),$ $(k_3, l_0), (k_0, l_0 + 1), (k_1, l_0 + 1),$ $(k_2, l_0 + 1), (k_3, l_0 + 1)$	0,1,2,3, 4,5,6,7	0, 1	0
12	16	1, 0.5	CDM4 (FD2,TD2)	$(k_0, l_0), (k_1, l_0), (k_2, l_0), (k_3, l_0)$	0,1,2,3	0, 1	0, 1
13	24	1, 0.5	FD-CDM2	$(k_0, l_0), (k_1, l_0), (k_2, l_0),$ $(k_0, l_0 + 1), (k_1, l_0 + 1),$ $(k_2, l_0 + 1), (k_0, l_1), (k_1, l_1),$ $(k_2, l_1), (k_0, l_1 + 1), (k_1, l_1 + 1),$ $(k_2, l_1 + 1)$	0,1,2,3,4,5, 6,7,8,9,10,11	0, 1	0
14	24	1, 0.5	CDM4 (FD2,TD2)	$(k_0, l_0), (k_1, l_0), (k_2, l_0), (k_0, l_1),$ $(k_1, l_1), (k_2, l_1)$	0,1,2,3,4,5	0, 1	0, 1
15	24	1, 0.5	CDM8 (FD2,TD4)	$(k_0, l_0), (k_1, l_0), (k_2, l_0)$	0,1,2	0, 1	0, 1, 2, 3
16	32	1, 0.5	FD-CDM2	$(k_0, l_0), (k_1, l_0), (k_2, l_0),$ $(k_3, l_0), (k_0, l_0 + 1), (k_1, l_0 + 1),$ $(k_2, l_0 + 1), (k_3, l_0 + 1), (k_0, l_1),$ $(k_1, l_1), (k_2, l_1), (k_3, l_1),$ $(k_0, l_1 + 1), (k_1, l_1 + 1),$ $(k_2, l_1 + 1), (k_3, l_1 + 1)$	0,1,2,3, 4,5,6,7, 8,9,10,11, 12,13,14,15	0, 1	0
17	32	1, 0.5	CDM4 (FD2,TD2)	$(k_0, l_0), (k_1, l_0), (k_2, l_0),$ $(k_3, l_0), (k_0, l_1), (k_1, l_1), (k_2, l_1),$ (k_3, l_1)	0,1,2,3,4,5,6,7	0, 1	0, 1
18	32	1, 0.5	CDM8 (FD2,TD4)	$(k_0, l_0), (k_1, l_0), (k_2, l_0), (k_3, l_0)$	0,1,2,3	0,1	0,1, 2, 3

of FD2 (over 2 consecutive REs) and TD2 (over 2 consecutive OFDM symbols) (as shown in row 12) are supported.
- For 24 port CSI-RS, TDM over 2 set of 2 consecutive OFDM symbols (i.e., total 4 OFDM symbols) each on the same location having FDM over 3 pairs of FD-CDM2 of 2 REs (as given in row 13), or TDM of 2 sets of 12 ports where each set uses FDM over 3 sets of 4 REs each of which in term uses CDM4 from combination of FD2 (over 2 consecutive REs) and TD2 (over 2 consecutive OFDM symbols) and these 2 sets of 12 ports are aligned in frequency location (as given in row 14), or FDM over 3 sets of 8 REs each of which in term uses CDM8 from combination of FD2 (over 2 consecutive REs) and TD4 (over 4 consecutive OFDM symbols) (as shown in row 15) are supported.
- For 32 port CSI-RS, TDM over 2 set of 2 consecutive OFDM symbols (i.e., total 4 OFDM symbols) each on the same location having FDM over 4 pairs of FD-CDM2 of 2 REs (as given in row 16), or TDM of 2 sets of 16 ports where each set uses FDM over 4 sets of 4 REs each of which in term uses CDM4 from combination of FD2 (over 2 consecutive REs) and TD2 (over 2 consecutive OFDM symbols) and these 2 sets of 16 ports are aligned in frequency location (as given in row 17), or FDM over 4 sets of 8 REs each of which in term uses CDM8 from combination of FD2 (over 2 consecutive REs) and TD4 (over 4 consecutive OFDM symbols) (as shown in row 18) are supported.

Among the above time and frequency pattern and multiplexing configurations, a subset of them may be used for different functionalities.

2.2.4.3.2 CSI-RS for CSI Acquisition

The prominent usage of CSI-RS, as indicated by its name, is for UE to perform measurements for CSI report. CSI-RS related resource configurations for CSI reporting are described below. A very high level description is given here. In general, UE needs to perform channel measurement(s) for reporting CSI over some NZP (none zero power) CSI-RS resource(s). For certain cases, UE also needs to perform interference measurement(s) over additional resource(s) in order to derive proper CSI report(s) and these additional resource(s) can be CSI-IM (channel state information—interference measurement) resources or they can also be NZP CSI-RS resources or combination of both.

The resources for CSI reporting, named CSI resource settings, are configured via higher layer signaling. Each CSI Resource Setting *CSI-ResourceConfig* contains a configuration of a list of CSI Resource Sets (given by higher layer parameter *csi-RS-ResourceSetList*), where the list comprises references to either or both of NZP CSI-RS resource set(s) and SS/PBCH block set(s) or the list comprises references to CSI-IM resource set(s). Each NZP CSI-RS resource set consists of a number of NZP CSI-RS resource(s). The UE can be configured with one or more CSI-IM resource set configuration(s) as indicated by the higher layer parameter *CSI-IM-ResourceSet*. Each CSI-IM resource set consists of a number of CSI-IM resource(s).

A CSI-IM resource configuration is basically a set of RE resources with time and frequency pattern of (2, 2) or (4, 1). The UE is to measure interference level on CSI-IM resource but not any specific signal. On the other hand, when NZP CSI-RS resource is configured for interference measurement, the UE is to measure interference level associated with specific signals transmitted there.

It is worth to clarify here what is a ZP (zero power) CSI-RS resource in NR. UE can be configured with one or more ZP CSI-RS resource set configurations, with each ZP CSI-RS resource set consisting of a number of ZP CSI-RS resources. A ZP CSI-RS resource configuration consists of a set of RE resources with time and frequency pattern as of a CSI-RS resource and, with corresponding configuration and indication from the network, the UE will then assume the RE resources associated with the ZP CSI-RS resource(s) are not mapped for PDSCH transmission.

2.2.4.3.3   CSI-RS for Beam management

For downlink beam management as well as uplink beam management when beam correspondence is supported by the UE, CSI-RS resource set is configured for the UE to select preferred downlink transmit beam(s) (i.e., spatial domain transmission filter) via L1 RSRP reporting.

Downlink receiving beam sweeping can be implemented by gNB configures and transmits a CSI-RS resource set with the higher layer parameter *repetition* set to "on," and the UE may assume that the CSI-RS resources within the set are transmitted with the same downlink transmission beam and hence try different receiving beam. Downlink transmission beam sweeping can be implemented by gNB configures and transmits a CSI-RS resource set with the higher layer parameter *repetition* set to "off," and the UE shall not assume that the CSI-RS resources within the set are transmitted with the same downlink transmission beam. The UE may use the same receiving beam to receive and measure these CSI-RS resources and then report the best CSI-RS resources which corresponds to the best transmission beams at the gNB side. With the combination of multiple CSI-RS resource sets with higher layer parameter *repetition* set to "on" or "off" and proper configuration of UE reporting, different beam management procedure can be implemented. Note that for CSI-RS resource set used for beam management purpose, it should have only the same number (1 or 2) of ports for all the CSI-RS resources within the set.

2.2.4.3.4   CSI-RS for Time and Frequency Tracking

Tracking functionalities in NR system (as in LTE) includes coarse time/frequency tracking, fine time/frequency tracking, and delay/Doppler spread tracking. In addition to the absence of CRS, NR supports large variety of system bandwidth, subcarrier spacing and carrier frequency. In addition to meeting the tracking performance requirements, signaling overhead, configuration flexibility, and performance robustness should also be considered. In order to satisfy tracking performance, based on evaluation for various applicable scenarios, the reference signal should meet certain minimum design criterion, for example, time and frequency density and spacing, periodicity, and bandwidth.

Based on the above considerations, instead of introducing a separate type of reference signal for tracking purposes, NR defines a set of CSI-RS resources for tracking which is labeled with higher layer parameter trs-Info as tracking reference signal. To guarantee UE baseband performance, though CSI-RS for tracking is UE-specifically configurable, a UE in RRC connected mode is expected to receive higher layer configuration of at least one such type of CSI-RS resource set. The CSI-RS for tracking when configured should satisfy the following restriction:

- 2 or 4 CSI-RS resources of one antenna port in the CSI-RS resource set. For FR1 and FR2, 4 CSI-RS resources can be configured within 2 consecutive slots with 2 CSI-RS resources per slot. In addition, for FR2, 2 CSI-RS resources can be configured within 1 slot.
- The time-domain locations of the two CSI-RS resources in a slot is spaced by 4 OFDM symbols.
- Each CSI-RS resource is of a single port with frequency density of 3, i.e., mapping to every fourth tones.
- The bandwidth of the CSI-RS for tracking is configurable but should be at least of 52 PRB or equal to the number of resource blocks of the associated BWP.
- The periodicity of the periodic CSI-RS for tracking is configured as {10, 20, 40, or 80} ms. Periodic CSI-RS for tracking should be always available for a UE in connected mode. In addition, aperiodic CSI-RS for tracking is also supported in association with a periodic CSI-RS for tracking when configured.

Due to its design purpose and property, periodic CSI-RS for tracking is used by the UE to estimate long-term properties of its observed channels including timing, carrier frequency, and Doppler. It is then natural for the UE to use these long-term properties to derive channel estimator for receiving subsequent control and data channels. Therefore, periodic CSI-RS for tracking serves the main source reference signal for QCL assumptions at the UE.

2.2.4.3.5   CSI-RS for Mobility Measurement

CSI-RS is also used for mobility measurement when it is configured with the higher layer parameter *CSI-RS-Resource-Mobility* either independently or jointly with an associated SSB indicated by the higher layer parameter *associatedSSB* where the UE obtains the timing of the CSI-RS resource from the timing of the corresponding serving cell. And only CSI-RS resource with one antenna port can be configured for mobility measurement.

2.2.4.4   Sounding Reference Signal

Sounding reference signal (SRS) is supported in NR transmitting from the UE at the uplink. Comparing to that of the LTE, several new aspects are considered in NR to augment the SRS design.

The most prominent use case for uplink SRS is to obtain downlink CSI information in TDD system and enable outstanding downlink MIMO performance by utilizing channel reciprocity. Since uplink SRS is an UE-specific signal, with potentially large number of active UEs in the cell as well as mobility of these UEs, SRS capacity may become a bottleneck. In addition, UE transmission power is generally much lower than that of a gNB, SRS can be power limited and received at a relatively low SINR level resulting in poor estimation quality of CSI. Therefore, increasing SRS capacity as well as coverage performance is a critical issue to solve.

Another issue is the unbalanced UE capability for downlink reception vs. uplink transmission. In general, most UEs can receive with more antennas at a number of downlink aggregated carriers but may only be able to transmit with a subset of these antennas at a smaller number of uplink carrier(s) at the same time which then hinders the gNB's capability to obtain full CSI knowledge of the downlink channels. In order to address these issues without dramatically increasing the complexity and cost of the UEs, switching SRS transmission from antenna to antenna and from carrier to carrier is specified in NR.

SRS can of course also be used to obtain uplink CSI and enable uplink MIMO schemes. In addition to codebook-based uplink MIMO as in LTE, NR also introduced non-codebook-based MIMO scheme for uplink where potential uplink precoding choice(s) (determined by the UE) is conveyed through precoded SRS(s). In addition, to support FR2 with beamforming, uplink beam management using SRS is needed at least for UE without the capability of beam correspondence.

In the following, general SRS design is described following by SRS specifics for different use cases.

2.2.4.4.1   General Design of SRS

In NR, an SRS resource with 1, 2, or 4 antenna ports are supported which can be mapped to 1, 2, or 4 consecutive OFDM symbols. Comb transmission on every 2 or 4 REs in frequency domain is supported. In addition, cyclic shift is supported with the maximum number of cyclic shifts equal to 12 (or 8) when Comb of size 4 (or 2) is used. SRS sequence ID is configured by higher layer parameter. Up to 6 OFDM symbols from the end of a slot may be used for SRS transmission which is a significant increase from that of LTE. A SRS resource may be configured for periodic, semi-persistent, aperiodic SRS transmission. In frequency domain, SRS allocation is aligned with 4 PRB grid. Frequency hopping is supported as in the case of LTE. With same design approach, NR SRS bandwidth and hopping configuration is designed to cover a larger span of values comparing to that of LTE.

For a given SRS resource, the UE is configured with repetition factor 1, 2, or 4 by higher layer parameter to improve coverage. When frequency hopping within an SRS resource in each slot is not configured, i.e., the number of repetition is equal to the number of SRS symbols, each of the antenna ports of the SRS resource in each slot is mapped in all the SRS symbols to the same set of subcarriers in the same set of PRBs. When frequency hopping within an SRS resource in each slot is configured without repetition, each of the antenna ports of the SRS resource in each slot is mapped to different sets of subcarriers in each OFDM symbol, where the same transmission comb value is assumed for different sets of subcarriers. When both frequency hopping and repetition within an SRS resource in each slot are configured, i.e., with 4 SRS symbols and repetition of 2, each of the antenna ports of the SRS resource in each slot is mapped to the same set of subcarriers within each pair of 2 adjacent OFDM symbols, and frequency hopping is across the two pairs.

A UE may be configured single symbol periodic or semi-persistent SRS resource with inter-slot hopping, where the SRS resource occupies the same symbol location in each slot. A UE may be configured 2 or 4 symbols periodic or semi-persistent SRS resource with intra-slot and inter-slot hopping, where the N-symbol SRS resource occupies the same symbol location(s) in each slot. For case of 4 symbols, when frequency hopping is configured with repetition 2, intra-slot and inter-slot hopping is supported with each of the antenna ports of the SRS resource mapped to different sets of subcarriers across two pairs of 2 adjacent OFDM symbol(s) of the resource in each slot. Each of the antenna ports of the SRS resource is mapped to the same set of subcarriers within each pair of 2 adjacent OFDM symbols of the resource in each slot. For the case where number of SRS symbols equal to repetition factor, when frequency hopping is configured, inter-slot frequency hopping is supported with each of the antenna ports of the SRS resource mapped to the same set of subcarriers in adjacent OFDM symbol(s) of the resource in each slot.

The UE can be configured with one or more SRS resource sets. For each SRS resource set, a UE may be configured with a number of SRS resources. The use case (such as beam management, codebook-based uplink MIMO, and non-codebook-based uplink MIMO, and antenna switching which actually is for general downlink CSI acquisition) for a SRS resource set is configured by the higher layer parameter.

2.2.4.4.2   SRS for DL CSI Acquisition

For downlink CSI acquisition, the usage of the SRS resource set is configured as "antenna switching" to cover not only the cases where SRS transmission antenna switching happens but also the case where the number of transmit and receive antennas are the same and no antenna switching happens. As stated above, in the case that UE supports less number of uplink carrier than downlink carrier, SRS switching between component carriers is also supported in NR.

SRS Antenna Switching

For SRS antenna switching, the following cases are supported: 1 (or 2) TX (transmission) to 2 (or 4) RX (receiving) antenna switching denoted as "1T2R," "2T4R," "1T4R," and "1T4R/2T4R" where a UE supports both "1T4R" and "2T4R" switching. In addition, "T=R" where number of transmission and receiving are equal is also supported.

To support antenna switching, an SRS resource set is configered with two (for "1T2R" or "2T4R") or four (for "1T4R") SRS resources transmitted in different symbols. Each SRS resource consists of one (for "1T2R" or "1T4R") or two (for "2T4R") antenna port(s) and the SRS port(s) of each SRS resource is associated with different UE antenna port(s). In addition, for the case of "1T4R," two aperiodic SRS resource sets with a total of four SRS resources transmitted in different symbols of two different slots can be configured. The SRS port of each SRS resource in given two sets is associated with a different UE antenna port. The two sets are each configured with two SRS resources, or one set is configured with one SRS resource and the other set is configured with three SRS resources.

Due to hardware limitation, the UE is configured with a guard period of a number of symbols in-between the SRS resources of a set for antenna switching. During these symbols, the UE does not transmit any other signal, in the case the SRS resources of a set are transmitted in the same slot. The number of symbols depends on the subcarrier spacing.

SRS Carrier Switching

For a UE with capability of supporting more downlink carriers than uplink carrier(s), in TDD system, a carrier with slot formats comprised of DL and UL symbols may not be configured for PUSCH/PUCCH transmission. In such a case, normally no uplink transmission including that of SRS is expected. In order to obtain downlink CSI and take advantages of TDD reciprocity via SRS, the UE can be configured with SRS resource(s) in this carrier. Since the UE may not have RF capability to transmit SRS on carrier(s) without PUSCH/PUCCH and at the same time to transmit on carrier(s) configured with PUSCH/PUCCH, the UE needs to borrow some RF capability from a carrier configured with PUSCH/PUCCH (denoted as switch-from carrier) and to switch the associated RF chain for transmission of SRS on this carrier (denoted as switch-to carrier). The switch-from carrier is configured by the network according to the UE's reported capability. During SRS transmission on switching-to carrier including any interruption due to uplink and downlink UE RF retuning time as reported by the UE, the UE temporarily suspends the uplink transmission on the switching-from carrier.

In the case of collision between SRS transmission (including any interruption due to uplink and downlink RF retuning) on the switch-to carrier and uplink or downlink transmission(s) on the switch-from carrier, priority rules are defined for collision handling.

Both periodic/semi-persistent and aperiodic SRS are supported on the switch-to carrier. For triggering aperiodic SRS transmission, in addition to DL and UL DCI, group-based DCI format is also introduced.

2.2.4.4.3   SRS for Codebook and Non-codebook-Based Uplink MIMO

In NR uplink, UE can be configured by the network for codebook-based transmission or non-codebook-based transmission.

For codebook-based uplink PUSCH transmission, one or two SRS resources can be configured for the UE. When multiple SRS resources are configured, an SRS resource indicator field in uplink DCI is used to indicate the selected SRS resource over whose antenna ports are used by the network to derive the precoding for uplink transmission.

For non-codebook-based uplink PUSCH transmission, multiple (up to 4) SRS resources can be configured for the UE. The UE can calculate the precoder used for the transmission of precoded SRS based on measurement of an associated NZP CSI-RS resource. The network determines the preferred precoder for PUSCH transmissions based on the reception of these precoded SRS transmissions from the UE and indicates to the UE via the wideband SRS resource indicator either in DCI for dynamic scheduling or in RRC for configured grant.

2.2.4.4.4   SRS for UL Beam Management

In the case that beam correspondence is not supported by the UE, the UE will not be able to determine its transmission beam for uplink based on downlink beam management outcomes. SRS with higher layer parameter usage set to "BeamManagement" will then be used for uplink beam management. To enable beam sweeping, when the higher layer parameter *usage* for SRS is set to "BeamManagement," only one SRS resource in each of multiple SRS sets can be transmitted at a given time instant. The SRS resources in different SRS resource sets can be transmitted simultaneously. SRS resource indicator is used to indicate the selected beam from the network.

2.2.4.5   Phase Tracking Reference Signal

Phase tracking reference signal (PT-RS) is introduced in NR for high-frequency band (FR2) to compensate phase noise for both downlink and uplink data transmissions (PDSCH/PUSCH). When PT-RS is configured and dynamically indicated (implicitly via DCI) by the network, the UE shall assume PT-RS being present only in the resource blocks used for the PDSCH/PUSCH with corresponding time/frequency density and location determined by several factors of the associated data transmission format. Specifics of PT-RS design are different for PDSCH, PUSCH with CP-OFDM, and PUSCH with DFT-s-OFDM.

2.2.4.5.1 PT-RS for PDSCH

The presence of PT-RS is configurable by the network. In addition, the time and frequency density of the PT-RS is dependent on the scheduled MCS and bandwidth of its associated PDSCH transmission for good trade off of PT-RS overhead and PDSCH demodulation performance as shown in Table 2.25.

The thresholds in the tables below are configured by higher layer signaling. For the network to derive the values of these threshold to determine the density and location of the PT-RS inside PDSCH transmission resource blocks, a UE reports its preferred MCS and bandwidth thresholds based on the UE capability at a given carrier frequency, for each subcarrier spacing applicable to data channel at this carrier frequency, assuming the MCS table with the maximum modulation order as it reported to support.

The DL DM-RS port(s) associated with PT-RS port are assumed to be quasi co-located. For PDSCH transmission of a single codeword, the PT-RS antenna port is associated with the lowest indexed DM-RS antenna port for the PDSCH.

For PDSCH transmission of two codewords, the PT-RS antenna port is associated with the lowest indexed DM-RS antenna port assigned for the codeword with the higher MCS. If the MCS indices of the two codewords are the same, the PT-RS antenna port is associated with the lowest indexed DM-RS antenna port assigned for codeword 0.

2.2.4.5.2 PT-RS for PUSCH of CP-OFDM

PT-RS for PUSCH of CP-OFDM is generally similar to that of PDSCH which is of course uses the same CP-OFDM waveform with the difference described in the following.

For uplink MIMO transmission, depending on implementation architecture, antennas and RF chains at the UE may differ in terms of supporting of coherent transmission from these antennas and RF chains. If a UE has reported the capability of supporting full-coherent UL transmission, the UE shall expect the number of UL PT-RS ports to be configured as one if UL-PTRS is configured. For partial-coherent

Table 2.25 Configuration of PT-RS

Scheduled MCS	Time density (L_PTRS)
$I_{MCS} <$ ptrsthMCS$_1$	PT-RS is not present
ptrsthMCS1 $\leq I_{MCS} <$ ptrsthMCS2	present on every fourth symbol
ptrsthMCS2 $\leq I_{MCS} <$ ptrsthMCS3	present on every second symbol
ptrsthMCS3 $\leq I_{MCS}$	present on every symbol

Scheduled bandwidth	Frequency density (K_PTRS)
$N_{RB} <$ ptrsthRB0	PT-RS is not present
ptrsthRB0 $\leq N_{RB} <$ ptrsthRB1	present on every second RB
ptrsthRB1 $\leq N_{RB}$	present on every fourth RB

and non-coherent codebook-based UL transmission, the actual number of UL PT-RS port(s) is determined based on TPMI and/or TRI in DCI. The maximum number of configured PT-RS ports is given by the higher layer parameter and the UE is not expected to be configured with a larger number of UL PT-RS ports than it has reported need for.

For codebook or non-codebook-based UL transmission, the association between UL PT-RS port(s) and DM-RS port(s) is signaled by DCI when needed. For non-codebook-based UL transmission, the actual number of UL PT-RS port(s) to transmit is determined based on SRI(s). A UE may be configured with the PT-RS port index for each configured SRS resource by the higher layer parameter.

2.2.4.5.3   PT-RS for PUSCH of DFT-s-OFDM

For PUSCH of DFT-s-OFDM waveform, DFT transform precoding is enabled. When PT-RS is configured for PUSCH transmission of DFT-s-OFDM, PT-RS samples are inserted into data transmission before DFT transform as shown in Fig. 2.57. PT-RS before DFT transform is divided into a number of (2, 4, or 8) PT-RS groups and each PT-RS group consists of a number of (2 or 4) samples. An orthogonal sequence of the same size is applied within a PT-RS group.

The group pattern of PT-RS depends on the scheduled bandwidth according to Table 2.26 [44] and the thresholds are configured by the network.

A scaling factor is applied according to the scheduled modulation order per the following table [44].

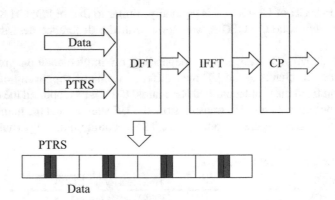

Fig. 2.57 Pre-DFT PT-RS for DFT-s-OFDM PUSCH

Table 2.26 PT-RS group pattern as a function of scheduled bandwidth

Scheduled bandwidth N_{RB}	Number of PT-RS groups	Number of samples per PT-RS group
$N_0 \leq N_{RB} < N_1$	2	2
$N_1 \leq N_{RB} < N_2$	2	4
$N_2 \leq N_{RB} < N_3$	4	2
$N_3 \leq N_{RB} < N_4$	4	4
$N_4 \leq N_{RB}$	8	4

Table 2.27 PT-RS scaling factor

Scheduled modulation	PT-RS scaling factor
$\pi/2$-BPSK	1
QPSK	1
16QAM	$\dfrac{3}{\sqrt{5}}$
64QAM	$\dfrac{7}{\sqrt{21}}$
256QAM	$\dfrac{15}{\sqrt{85}}$

Note that since PUSCH of DFT-s-OFDM only support 1 layer transmission, its associated PT-RS has also only 1 port and the mapping is straightforward.

2.2.4.6   Quasi-co-Location and Transmission Configuration Indicator

As discussed in section for reference signal design framework, there is no common and always-on CRS type reference signal in NR and the necessary functionalities are distributed into a set of downlink reference signals. Therefore, these distributed functionalities need to be integrated properly at the receiver to facilitate good performance of data and control channel demodulation, measurements, etc. Quasi co-location between antenna ports is used as the venue for such integration.

The definitions of antenna port and quasi co-location are given in [40] as following:

> "An antenna port is defined such that the channel over which a symbol on the antenna port is conveyed can be inferred from the channel over which another symbol on the same antenna port is conveyed.
>
> Two antenna ports are said to be quasi co-located if the large-scale properties of the channel over which a symbol on one antenna port is conveyed can be inferred from the channel over which a symbol on the other antenna port is conveyed. The large-scale properties include one or more of delay spread, Doppler spread, Doppler shift, average gain, average delay, and spatial Rx parameters."

Via quasi co-location relationship between antenna ports of a source reference signals and a target reference signal with respect to a set of large-scale channel properties, the UE receiver can utilize channel properties estimated/derived from the source reference signal for channel estimation and measurement performed on the target reference signal.

Quasi co-location relationship between reference signals are configured and signaled through TCI (transmission configuration indicator) state(s). The UE shall not assume that two antenna ports are quasi co-located with respect to any QCL type unless specified otherwise. A TCI state contains information for configuring quasi co-location relationship including one or two downlink reference signals as the source reference signal and the QCL types associated with the one or two source reference signals. QCL types is defined in the following to categorize a subset of QCL parameters:

– 'QCL-TypeA': {Doppler shift, Doppler spread, average delay, delay spread}
– 'QCL-TypeB': {Doppler shift, Doppler spread}
– 'QCL-TypeC': {Doppler shift, average delay}
– 'QCL-TypeD': {Spatial Rx parameter}

As different reference signals are designed and optimized for different purpose(s), a single type of reference signal is not feasible to derive all QCL parameters of a channel between the TRP and the UE with sufficient accuracy. For example, CSI-RS for tracking is specifically designed to obtain and maintain fine time and frequency synchronization and Doppler estimation for a channel on certain spatial (or beam-forming) direction and should be used to derive parameters in QCL-TypeA in most of the situations. However, CSI-RS for tracking may not be a good option to obtain spatial Rx parameter via beam sweeping which can result in very large overhead and long delay. On the other hand, CSI-RS for beam management (i.e., CSI-RS resource configured with higher layer parameter *repetition*) is generally used to derive QCL-TypeD for FR2. Therefore, it needs both CSI-RS for tracking and CSI-RS for beam management for the UE to obtain a full set of quasi co-location parameters. For the case of two source DL RSs, the QCL types should not be the same, regardless of whether the references are to the same DL RS or different DL RSs.

The possible configuration of quasi co-location relationship through TCI states are given in Table 2.28 below. Note that QCL-TypeD may not always be applicable.

In the absence of CSI-RS configuration, for example, before RRC configuration, the UE may assume PDSCH/PDCCH DM-RS and SS/PBCH block to be quasi co-located with respect to Doppler shift, Doppler spread, average delay, delay spread, and, when applicable, spatial Rx parameters.

2.3   5G-NR Spectrum and Band Definition

2.3.1   5G Spectrum and Duplexing

2.3.1.1   IMT-2020 Candidate Spectrum

The IMT spectrum identified in the ITU's World Radiocommunication Conferences (WRC) 2015 and 2019 (which are below 6 GHz and above 24 GHz, respectively) are applicable for 5G deployments. 3GPP defines frequency bands for the 5G-NR interface according to the guidance both from ITU and from the regional regulators, with prioritization according to the operators' commercial 5G plan. According to [45], three frequency ranges are identified for 5G deployments for both eMBB and IoT applications, including the new frequency ranges of 3–5 and 24–40 GHz, as well as the existing LTE bands below 3 GHz.

As illustrated in Fig. 2.58, generally, a triple-layer concept can be applied to the spectral resources based on different service requirements. An "over-sailing layer" below 2 GHz is expected to remain the essential layer for extending the 5G mobile broadband coverage both to wide areas and to deep indoor environments. This is espe-

Table 2.28 Configurations of quasi co-location relationship via TCI states

Target reference signal	Source reference signal(s) and QCL type(s)	Notes
Periodic CSI-RS for tracking	SS/PBCH block I QCL-TypeC and QCL-TypeD	
	SS/PBCH block I QCL-TypeC CSI-RS for BM I QCL-TypeD	
Aperiodic CSI-RS for tracking	Periodic CSI-RS for tracking I QCL-TypeA and QCL-TypeD	
CSI-RS for CSI acquisition	CSI-RS for tracking I QCL-TypeA and QCL-TypeD	
	CSI-RS for tracking I QCL-TypeA SS/PBCH block I QCL-TypeD	
	CSI-RS for tracking I QCL-TypeA CSI-RS for BM I QCL-TypeD	
	CSI-RS for tracking I QCL-TypeB	When QCL-TypeD is not applicable
CSI-RS for BM	CSI-RS for tracking I QCL-TypeA and QCL-TypeD	
	CSI-RS for tracking I QCL-TypeA CSI-RS for BM I QCL-TypeD	
	SS/PBCH block I QCL-TypeC and QCL-TypeD	
DM-RS for PDCCH	CSI-RS for tracking I QCL-TypeA and QCL-TypeD	
	CSI-RS for tracking I QCL-TypeA CSI-RS for BM I QCL-TypeD	
	CSI-RS for CSI acquisition I QCL-TypeA	When QCL-TypeD is not applicable
DM-RS for PDSCH	CSI-RS for tracking I QCL-TypeA and QCL-TypeD	
	CSI-RS for tracking I QCL-TypeA CSI-RS for BM I QCL-TypeD	
	CSI-RS for CSI acquisition I QCL-TypeA and QCL-TypeD	

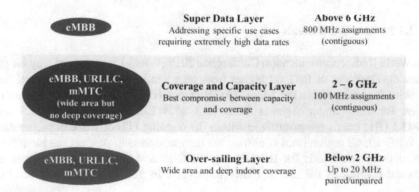

Fig. 2.58 Multi-layer approach for 5G scenarios

cially important for mMTC and URLLC applications. On the other hand, the "coverage and capacity layer" spanning from 2 to 6 GHz can be used for striking a compromise between capacity and coverage. However, compared to the range below 2 GHz, these bands suffer from a higher penetration loss and propagation attenuation. The "super data layer" above 6 GHz can be invoked for use-cases requiring extremely high data rates but relaxed coverage. Given this triple-layer concept, the eMBB, mMTC and URLLC services that require different coverage and rate-capability can be accommodated in the appropriate layer. However, a service-based single-layer operation would complicate the 5G deployments and it is inefficient in delivering services that simultaneously require both good coverage and high data rate as well as low latency, etc. To accommodate these diverse services, the employment of joint multiple spectral layers becomes a "must" for a meritorious 5G network.

2.3.1.1.1   C-Band (3,300–4,200 and 4,400–5,000 MHz)

Spectrum availability for IMT in the 3,300–4,200 and 4,400–5,000 MHz ranges is increasing globally. The 3400–3600 MHz frequency band is allocated to Mobile Service on a co-primary basis in almost all countries throughout the world. Administrations will make available different portions of the 3,300–4,200 and 4,400–5,000 MHz ranges at different times, incrementally building large contiguous blocks.

The 3GPP 5G-NR specification support 3,300–3,800 MHz from the start, using a TDD access scheme. In line with the release plans from many countries, the 3,300–3,800 MHz band will be the primary 5G band with greatest potential for global harmonization over time: it is recommended that at least 100 MHz of contiguous bandwidth from this band be allocated to each 5G network. In order to take advantage of the harmonized technical specification—the 3GPP 5G NR specification for 3,300–3,800 MHz band, regulators are recommended to adopt a frequency arrangement with an aligned lower block edge of usable spectrum and harmonized technical regulatory conditions, at least in countries of the same Region. The 5G-NR ecosystem of 3,300–3,800 MHz is expected to be commercially ready in 2018 in all the three regions [46] (Fig. 2.59).

2.3.1.1.2   Mm-Wave Bands

The World Radiocommunication Conference 2015 (WRC-15) paved the way for the future development of IMT on higher frequency bands by identifying several frequencies for study within the 24.25–86 GHz range (Fig. 2.60) for possible identification for IMT under Agenda Item 1.13 of WRC-19. The 24.25–27.5 and 37–43.5 GHz bands are prioritized within the ongoing ITU-R work in preparation for WRC-19; all regions and countries are recommended to support the identification of these two bands for IMT during WRC-19 and should aim to harmonize technical conditions for use of these frequencies in 5G. The frequency band of

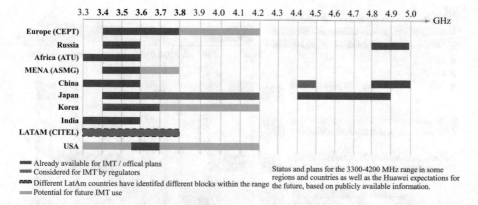

Fig. 2.59 Global availability and planning of the 3,300–4,200 MHz and 4,400–5,000 MHz frequency ranges

Group 30 (GHz)	Group 40 (GHz)	Group 50 (GHz)	Group 70/80 (GHz)
24.25-27.5 31.8-33.4	37-40.5 40.5-42.5 42.5-43.5	45.5-47 47-47.2 47.2-50.2 50.4-52.6	66-71 71-76 81-86

Fig. 2.60 Candidate frequency bands of WRC-19 Agenda Item 1.13

27.5–29.5 GHz, though not included in the WRC-19 Agenda Item 1.13, is considered for 5G in the USA, South Korea and Japan.

The 24.25–29.5 and 37–43.5 GHz ranges are the most promising frequencies for the early deployment of 5G millimeter wave systems, and several leading markets are considering portions of these two ranges for early deployments (Fig. 2.61), and the two ranges are also being specified in 3GPP Release 15 based on a TDD access scheme. It is recommended that at least 400 MHz of contiguous spectrum per network from these ranges be assigned for the early deployment of 5G.

2.3.1.1.3   Sub-GHz Frequency Bands for 5G

Sub-3 GHz candidate frequency bands are given in Fig. 2.62.

The S-band (2,496–2,690 MHz) is another 5G candidate band that may have early commercial deployment. Currently, there are LTE TDD networks deployed in this band partially in both China and US, while EU deploys LTE FDD in the two edge of this band instead. It is high likely that those regions may deploy 5G with TDD duplex mode on the left spectrum of that band, with maximizing eco-system sharing.

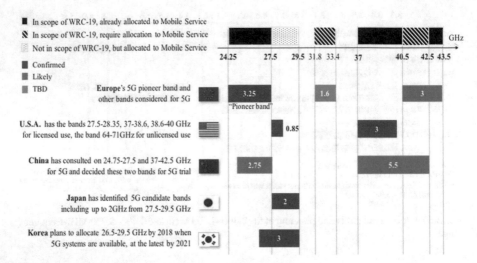

Fig. 2.61 Mm-wave frequency bands for early deployment of 5G

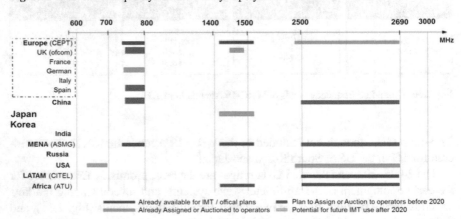

Fig. 2.62 Sub-3GHz frequency bands for early deployment of 5G

The L-band (1,427–1,518 MHz) is one 5G candidate band that has the potential to be allocated to mobile in most countries in the world. CEPT and CITEL regions have adopted the SDL (Supplemental Down Link) scheme for this band. The requirement for standalone operation in the band (both UL and DL transmissions) has emerged in some other regions. In the case of standalone 5G systems, a TDD access scheme is a potentially appropriate option, which can accommodate traffic asymmetry in the UL/DL directions with good potential for economies of scale. The same 5G-NR equipment can serve both the TDD and SDL markets. In addition, the SDL band can be paired with an individual SUL band (as described in Sect. 2.3.2) as well, following the traditional way of FDD operation.

The 700 MHz band has already been harmonized for mobile in most countries. Europe plans to use this band for 5G. Over the long term, the other frequencies of UHF band (470–694/698 MHz) could also be used for mobile, while the USA has already started the process of transferring the band from broadcasting to mobile service.

2.3.1.2 5G Duplexing Mechanisms

2.3.1.2.1 5G Candidate Band Types and Duplex Modes

Duplexing is another key factor affecting the network operation. There are two typical spectrum types available for IMT systems, i.e., paired spectrum and unpaired spectrum. At present, FDD and TDD are two dominant duplex modes, which are adopted to paired spectrum and unpaired spectrum, respectively, as introduced in following:

- **FDD over paired spectrum:** FDD is more mature in 2G/3G/4G telecommunication systems, whose spectrum are mainly located in low-frequency range below 3 GHz. Due to the limit frequency resource in low-frequency range, generally the bandwidth of a FDD band is quite limited. In addition, common RF filter design for FDD band is needed for the convenience of eco-system sharing, thus a fixed duplex distance between DL and UL spectrum are defined for each FDD band, which is also the obstacle of expanding the bandwidth of a FDD band.
- **TDD over unpaired spectrum:** As the increasing of telecommunication traffic load, TDD attracts more attention due to more available wideband unpaired spectrum in middle and high-frequency range. LTE TDD successfully operates the wide-coverage public networks for eMBB service in 2.6 GHz (Band 41) and 3.5 GHz (Band 42). Moreover, the DL and UL channel reciprocity in TDD system can well accommodate the optimized multi-antenna operation with efficient sounding design instead of redundant feedback of channel state information, which brings a significant throughput gain especially for multiple user MIMO mechanisms. The channel statement information feedback overhead required for multi-antenna mechanism is un-neglectable for a spectrum with very large DL transmission bandwidth, thus TDD operation with UL sounding is the must for the multi-antenna solutions in C-band and mm-Wave bands. This is also one of the main reasons that the current 5G new spectrum in C-band and mm-wave choose TDD mode.
- **SDL over unpaired spectrum:** LTE also defines a supplementary DL (SDL) band type, which also fits to unpaired spectrum. However, SDL band type can only operate jointly with a normal FDD or TDD band which has UL transmission resources for feedback. A SDL carrier is aggregated to a normal FDD or TDD primary carrier, following the normal access, scheduling and HARQ processes, which is quite in line with a normal FDD DL carrier.

In addition to the tradition band types, 5G introduces a new supplementary UL (SUL) type.

- **SUL over unpaired spectrum:** SUL band is combined with a normal 5G TDD (or SDL and FDD) band, to provide the UL control signal and feedback transmission as well as UL data transmission. The original introduction of SUL band type is to compensate the UL coverage shortage of the 5G TDD spectrum in C-band and mm-wave bands, where SUL bands are generally overlapped with the UL spectrum region of the existing LTE FDD band, which provide the possibility of an operator to reuse their spared UL frequency resources in LTE network for

5G-NR UL transmission for cell-edge users. The detailed concept, specifications and benefits of SUL band combinations are introduced in Sect. 2.4.

2.3.1.2.2   Flexible Duplex: Convergence of FDD and TDD

In the future, more and more operators will own multiple spectrum bands, and most likely both FDD and TDD bands. Consequently, there are, currently, many LTE networks operating in both FDD and TDD bands jointly to accommodate the eMBB and IoT services to end users. It is foreseen that convergence of TDD and FDD will be the evolution trend of mobile communication systems.

Joint Operation of FDD and TDD

There are several FDD/TDD joint operation mechanisms specified in 3GPP, as below:

- **Common network with multi-RATs of FDD and TDD:** One public cellular network contains two radio access layers in a FDD band and a TDD band individually, with the authentication, access, and mobility control by a common core network. Which radio access layer is selected by a user is dependent on the coverage and the available radio resource units of the two layers. Switching between FDD and TDD layers is generally triggered by the inter-RAT handover based on the UE reference signal receiving power (RSRP) measurement of the FDD cell and TDD cell, respectively. In such a joint FDD/TDD multi-RAT operation, the FDD or TDD radio access layer provide independent scheduling and transmission procedure.
- **FDD/TDD carrier aggregation:** LTE-Advanced system introduced FDD/TDD carrier aggregation mechanism since Release-12, which enables the physical-layer scheduling of parallel data transmission in both FDD and TDD carriers for a single UE, which requires either co-site FDD/TDD eNodeB (eNB) or ideal backhaul between FDD and TDD eNBs. Either a FDD carrier or TDD carrier can be the anchor carrier to provide the basic mobility functions such as handover and cell-reselection, while it can activate and deactivate one or more (up to four in DL and two in UL) secondary component carriers for parallel data transmission. In LTE-Advanced system, each component carrier has its own scheduling control signaling and HARQ process, while the HARQ feedback of the SCCs can be transmitted only on the UL carrier of PCC. Accordingly 3GPP defines detailed specifications on how to deal with HARQ feedback timing and potential confliction cases for different TDD DL/UL configurations. 5G-NR goes further by enabling the SCC scheduling via PCC DL control signaling. However, 3GPP Release-15 only specifies the cross-carrier scheduling for the component carriers with the same numerology, while 5G-NR TDD carriers in mm-wave band and C-band generally have different numerologies from that of low-frequency FDD band.

- **FDD/TDD dual connectivity (DC)**: FDD/TDD DC provides the capability of parallel transmit multiple streams over FDD and TDD to a single UE semi-statically configured at high layer. The higher layer (RLC/PDCH or upward) parallel transmission makes it applicable to those deployment scenarios without ideal almost zero latency backhaul, such as inter-site FDD/TDD stream aggregation for a single UE.

Synchronization for TDD Band

Network synchronization is a basic requirement for the TDD cellular systems. It requires multiple cells synchronize the TDD frame structure, including DL and UL switching point, as well as the DL/UL configurations.

There are multiple types of synchronization requirements as the following:

- **Intra-frequency inter-cell synchronization**

  - Figure 2.63 describes the inter-cell interference, while (a) is the given network topology with (b) for the scenario with two neighboring cells are configured with two different TDD DL/UL configurations, and (c) is for the scenario with two neighboring cells are asynchronous in DL/UL switching points. It can be seen that for the period with cell 1 schedules DL transmission while cell 2 schedules the UL transmission at the same time, the high transmit power from cell 1 base station (BS) ends to a strong interference at the receiver of the cell 1 BS, thus may block the cell 1 BS reception during that time period.
  - To avoid such inter-cell BS-to-BS interference, it is recommended for operator to deploy a TDD network with frame synchronization. Guard period between DL and UL transmission is usually chosen corresponding to the interference-protection distance from an aggregator-BS to the victim-BS.

- **Intra-band Inter-frequency inter-cell synchronization**

  - When one operator deploys TDD system on two or more carriers in the same band, network synchronization is also necessary. Figure 2.64 describes the inter-frequency BS-to-BS interference, for both neighboring carrier and two carriers with a certain frequency distance. Since the BS-to-BS interference comes from the same network and also the same BS, the intra-band transmission signals will fall into the RF filter of the same receiver filter range thus block the BS receiver totally.

- **Intra-band inter-operator synchronization with regional regulation coordination**

  - For different operators who deploy TDD systems on their own spectrum located in the same band, the similar inter-frequency BS-to-BS interference is suffered from the victim TDD network that with the UL time widow overlap with the aggressor TDD network with the DL transmission. There was a proposal to introduce a guard band between the two neighboring TDD networks in the same frequency band, however for cellular network with macro BSs, a

quite big guard band is required in addition to the inter-operator special isola-
tion to avoid the BS receiver blocking and basic receiving performance of the
victim TDD network. According to the regulation evaluation in both China
MIIT [47] and EU ECC [48], a guard band of 5~10 MHz is needed for two
2.6 GHz LTE TDD systems with operating bandwidth of 20 MHz individually,
and more than 25 MHz guard band is required for two 3.5 GHz NR TDD systems with operating bandwidth of 100 MHz individually as shown in Fig. 2.65.
Such big guard band is a big waste of the cherished spectrum resources, and
thus is un-acceptable for any regional regulator as well as operators.

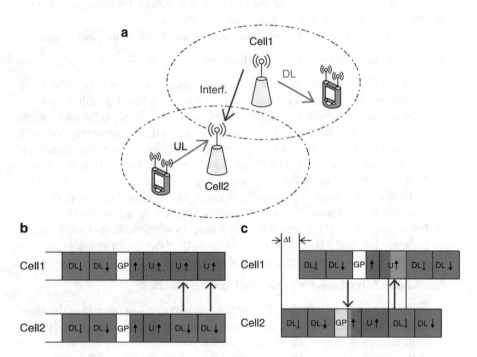

Fig. 2.63 TDD network asynchronous inter-cell interference. (**a**) Network topology. (**b**)
Asynchronous in TDD configuration. (**c**) Asynchronous in DL/UL switching points

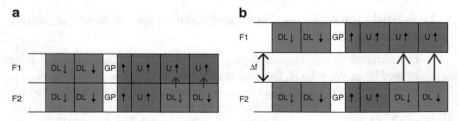

Fig. 2.64 TDD network asynchronous inter-frequency interference. (**a**) Neighboring carrier. (**b**)
Two carriers with guard band in between

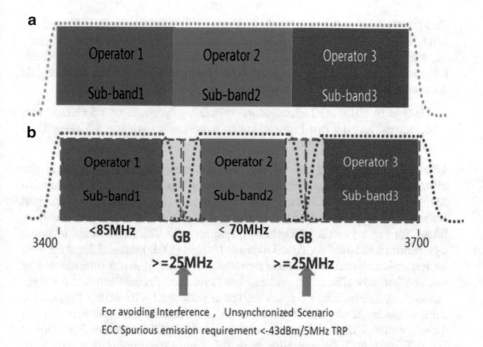

Fig. 2.65 Intra-band Inter-operator TDD operation: synchronous vs. asynchronous. (**a**) Inter-operator synchronous TDD operation. (**b**) Inter-operator asynchronous TDD operation

To reach the intra-band inter-operator TDD synchronization, the regional regulator generally specify a common TDD DL/UL frame structure according to the DL/UL traffic load ratio statistics from multiple operators. Few country also ask operators to do coordination if they have to deploy TDD network in the same band.

- **China:** For the first 2.6 GHz TDD network in the world, China operators contributed many efforts to coordinate the network. Finally, under the guideline of MIIT (Ministry of Industry and Information Technology) of China, the synchronization among operators' networks at the band 2.6 GHz was implemented based on the same frame structure TDD configuration 2 and the same DL/UL traffic ratio 4:1. China MIIT is actively organizing MNOs and relevant stake holders to negotiate a single frame structure for synchronization of 5G networks in 3.5 GHz band.

- **Japan:** Public open hearing of potential operators for Japan 3.4–3.6 GHz band was held by MIC (Ministry of Internal Affairs and Communications) on 23 January 2014. All operators presented a clear position of TDD, and advocated the necessity of operator's consensus for collaboration for realizing TDD synchronization including the DL/UL configuration ideally, in order to achieve No Guard Band for efficient usage of spectrum resources. All the operators have the same opinion of DL heavy frame configuration, by referring to the heavy data traffic in downlink side. Japan Ministry of Internal Affairs and Communications

(MIC) issued the guideline for the introduction of 4G for comments in September 2014, which includes 3,480–3,600 MHz should be assigned for three operators (40 MHz per operator) for TDD use, and licensees are obliged to agree with each other in advance about the matters for TDD synchronized operation, where frame structure configuration 2 is also chosen.

- On 6 April 2018, MIC allocated the remaining spectrum of 3.5 GHz to two operators, and the two licenses are to synchronize existing 3.5 GHz TDD network.

- **UK:** On 11 July 2017, UK Ofcom issued auction regulations [49] for the award of the 2.3 and 3.4 GHz spectrum bands and also updated Information Memorandum [50] alongside the regulations (Fig. 2.66). The updated Information Memorandum sets out the conditions of the licenses that will be issued for the 2.3 and 3.4 GHz bands, respectively. The licenses are technology neutral and based on Time Division Duplex (TDD) mode. Licensees will be required to synchronize their networks in order to avoid interference to one another, so traffic alignment and the "Preferred Frame Structure configuration2" as in Fig. 2.67a for transmission with the limits of the Permissive Transmission Mask are mandated to implement the synchronization. Timeslots must have the duration of 1 ms. TD-LTE frame configuration 2 (DL/UL ratio 4:1) is compatible with this frame structure. Other details of the preferred frame structure can be found in paragraph 12 in the Information Memorandum.

  - Early this year, Ofcom conducted the auction, and the 3.4 GHz band plan based on final auction results[5] is shown in Fig. 2.66.

Vodafone	H3G	UK Broadband	Telefónica	EE	UK Broadband
3.410 – 3.460 GHz	3.460-3.480GHz	3.480-3.500 Ghz	3.500-3.540 GHz	3.540-3.580 GHz	3.580-3.600 Ghz

Fig. 2.66 UK 3.4 GHz band plan based on final auction results

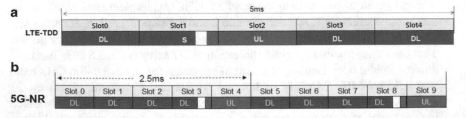

Fig. 2.67 TDD frame structure with DL/UL ratio of 4:1. (**a**) LTE-TDD Frame structure of configuration 2, 5 ms-single DL/UL switching period. (**b**) 5G-NR Frame structure of 2.5 ms-single DL/UL switching period

---

[5] https://www.ofcom.org.uk/spectrum/spectrum-management/spectrum-awards/awards-in-progress/2-3-and-3-4-ghz-auction

– As described above, almost all the commercial LTE TDD systems globally adopt TDD configuration 2, i.e., 5 ms period with "DSUDD" pattern and the DL/UL ratio around 4:1 as in Fig. 2.67a. For the TDD band with only 5G-NR deployment, the same DL/UL ratio of 4:1 can be chosen as well, but with 2.5 ms period with "DDDSU" pattern due to wider subcarrier spacing and shorter slot length, as shown in Fig. 2.67b. As described in Sect. 2.2.2, 3GPP specifies a very flexible frame structure for 5G-NR, therefore other TDD DL/UL configurations can also be adopted according to regional regulation coordination among operators, while both the traffic statistics and the implementation complexity should be taken into account.

- **Intra-band LTE/NR synchronization**

  For both 2.6 GHz Band 41 and 3.5 GHz Band 42, there are countries, e.g., China and Japan, that have plan to deploy both LTE TDD and 5G-NR system within the same band. Same as the intra-band inter-operator TDD synchronization, the regulators mandates the same DL/UL switching period and switching points. LTE and 5G-NR usually have different numerologies, i.e., 15 kHz subcarrier spacing for LTE and 30 kHz subcarrier spacing for NR. For LTE TDD, the frame configuration 2 is the most widely used frame structure, i.e., 3:1 DL/UL ratio with 5 ms DL/UL switching period. Then, 5G-NR is recommended to operate with a TDD DL/UL switching period of 5 ms with the pattern of "DDDDDDDSUU" and the DL/UL ratio of 8:2 to attain synchronization with LTE. In addition, the slot format configuration in NR is very flexible which can match all the configurations of special subframe in LTE. The only modification is to adjust the starting point of the frame as shown in Fig. 2.68.

It is worth noting that network synchronization is not an exclusive feature of communication systems that work in Time Division Duplex (TDD) mode, while it could also be applied to the communication systems working in Frequency Division Duplex (FDD) mode if performance gain is expected by utilizing techniques such as

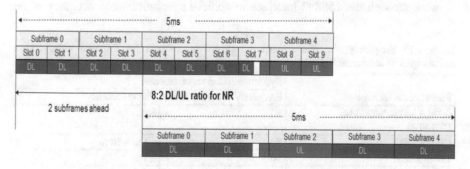

Fig. 2.68 Intra-band TDD synchronization between LTE and 5G-NR: frame structure configuration

interference cancellation (IC). The difference lies in the requirement of synchronization accuracy precision. GSM, UMTS, WCDMA, and LTE-FDD mobile technologies require only frequency synchronization with accuracy within 50 parts per billion (ppb) at the radio interface. CDMA2000, TD-SCDMA, and LTE-TDD services have the same frequency requirements as the other earlier 2G/3G network, but also specify requirements for phase and time. Different with FDD systems where frequency synchronization is generally sufficient, it is essential that the time and phase reference in TDD system is traceable to Coordinated Universal Time (UTC). Without the common UTC time reference cell sites cannot operate as intended. In the LTE-TDD system, two phase accuracy granularities are specified, 1.5 and 5 μs, corresponding to two cell radius sizes differentiated by 3 km. Table 2.29 presents a summary of the synchronization requirements of different mobile network modes.

The mechanisms to guarantee the TDD network synchronization are mature since 3G TD-SCDMA system, and already widely used in LTE TDD systems globally. Currently, the main solutions include:

- **Type1: distributed synchronization scheme based on GNSS systems.** GNSS signal receivers are directly deployed on the terminal and BSs, each BS acquire satellite time signals (GPS, Beidou, Glaness, etc.) directly to achieve the time synchronization between different BSs and to ensure the maximum deviation of any two of the base station no more than 3 μs. Generally macro BSs locate in open areas, which is easy to install GPS antenna with good satellite signal receiving performance. However, for those BSs deployed indoor or outdoor but surrounded by tall objects that easily block GPS signals, it'll be difficult to receive GPS signal correctly.
- **Type2: centralized synchronization scheme based on IEEE1588v2** system maintained by the cellular network backhaul. IEEE 1588v2 specifies accurate time transfer protocol, which can achieve sub-microsecond precision time synchronization like the current GPS. The clock synchronization information of the main time source is transmitted through the 1588v2 protocol packet on the transmission network. The BSs can obtain time information from transmission network through the 1588v2 interface to achieve synchronization accuracy of ns

Table 2.29 Frequency and phase synchronization requirements for different radio access technologies as delineated in [48]

Radio access technology	Synchronization requirements	
	Frequency accuracy (ppb)	Phase accuracy
GSM, UMTS, WCDMA, LTE-FDD	50	NA
CDMA2000	50	±3 to ±10 μs
TD-SCDMA	50	±3 μs
LTE-TDD	50	±1.5 μs (for cell radius ≤ 3 km)
	50	±5 μs (for cell radius > 3 km)
5G-NR	50	±1.5 μs

level. It requires all nodes of the bearer network to support PTP, and additionally the clock synchronization quality is affected by network QoS.

The above two types of synchronization solutions are quite complemented to each other. Both of them are widely used in the commercial LTE networks already. It's up to each operator to choose their own network synchronization solutions.

Dynamic TDD and Flexible Duplex

Traditional TDD network has a static TDD DL/UL configuration, which is usually decided by the statistical UL/DL traffic load ratio among multiple operators in a specific country or region. However, in the practical telecommunication networks, the DL traffic constitutes a large portion of the entire tele-traffic, as shown in the practical network traffic statistics in Fig. 2.69a. With the popularity of video stream- ing increasing, it is forecast that the proportion of DL content will grow even further in the future as in Fig. 2.69b [4], hence it is natural that more resources should be allocated to the DL. Therefore, a smaller proportion of the resources is left for the UL, which will further affect the UL coverage performance.

Telecommunication industry never gives up on improving the total spectrum uti- lization efficiency [51]. If the network can configure the DL/UL radio resource ratio according to the actual DL/UL traffic ratio geographically and also timely, assum- ing can resolve the serious DL to UL interference in the wide-area TDD network, the spectrum utilization can be improved a lot. There are two steps in the duplex evolving roadmap as below, additionally there were academia papers propose full duplex [52, 53] as the final duplex evolution step, which is however far immature due to many implementation challenges.

Dynamic TDD:

LTE-Advanced system introduced the dynamic configuration of TDD DL/UL ratio from 3GPP Release-12 and onwards, called as enhanced interference man- agement traffic adaption (eIMTA) (Fig. 2.70) [117]. In the theoretical analyses and simulation [54], eIMTA can be applied to an isolated area such as indoor hotspot scenario like football game with a very special DL/UL traffic statistics. One drawback of eIMTA is that the newly introduced interference between DL and UL signals will reduce Signal to Interference plus Noise Ratio (SINR) when the inter-cell interference is present, thus cause the communication quality deg- radation over that of the conventional TDD system. As interpreted in the follow- ing figure, the interference suffered by Cell0 base station receiver increases since the DL signals from neighbor base stations with higher transmit power than UL signals cause strong interference.

The commercial LTE TDD networks are mainly wide-coverage cellular network; thus, eIMTA has not yet been deployed in practical systems due to its severe inter- carrier and intra-carrier interference of the normal transceiver implementation in Macro eNB.

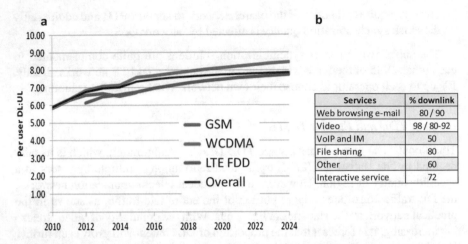

Fig. 2.69 DL/UL traffic load ratio statistics of IMT systems and eMBB services. (**a**) IMT-system DL/UL traffic statistics and prediction. (**b**) DL traffic proportion of different eMBB services

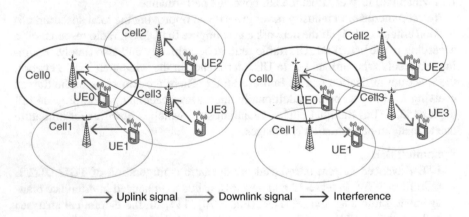

Fig. 2.70 Dynamic TDD

Flexible Duplex:

- **Flexible configuration on DL/UL ratio**: 5G standardization takes into account the future trend of FDD/TDD convergence from day one, which *defines* a very flexible physical-layer design as well as the symmetric DL/UL air-interface. 5G-NR inherits the dynamic TDD DL/UL ratio from LTE eIMTA, while goes further with both static cell-specific DL/UL configurations and the DL/UL switching period, as well as the additional potential semi-static or dynamic UE-specific DL/UL configurations. As shown in Fig. 2.71, 3GPP Release-15 specifies up to 62 slot configurations with each OFDM-symbol marked as DL, UL or unknown direction, while the unknown-direction symbols can be configured specifically for each UE receivers. Among those 62 slot configurations, there are four typical configurations as DL only slot, DL dominant slot, UL only slot and UL dominant slot [38]. Within one 5G-NR radio frame, each slot can have its own DL//UL configurations out of the total 62 candidates. Theoretically, 5G-NR can support a large number of candidate DL/UL configurations for the frame structure seen from UE side, while keeping the basic cell-level TDD synchronization on those basic DL or UL symbols which does not necessary occupy the whole time of a frame.

- **Other application scenarios of flexible duplex:** Theoretically, such flexible slot direction can be configured for paired spectrum as well. Due to the more and more serious imbalance between DL and UL traffic load, the symmetric DL/UL bandwidths in FDD band end to more and more spared UL radio resources when FDD DL resources are fully occupied by the eMBB traffic. Accordingly, there is a potential solution to make use of the spared UL for the low power DL transmission of pico cells, as shown in Fig. 2.72a. The flexible duplex mode can also be applied to some new FDD DL spectrum and SDL band as in Fig. 2.72b, where there is possibility to introduce SRS in the DL carriers to enable efficient multi-antenna mechanism based on the DL/UL channel reciprocity. For flexible duplex application in FDD DL or UL spectrum or SDL bands, the main limit would be regional spectrum regulation restrictions.

  Flexible Duplex is also applicable to access-integrated wireless backhaul and D2D etc., as in Fig. 2.73. In reality, it is very costly for operators to widely deploy high-speed and low-latency backhaul (e.g., fiber), considering the cost and difficulty to find suitable sites; while the tradition wireless backhauling or relay requires additional spectrum to avoid the interference to-and-from the access link [55, 56], which is very low efficient. With flexible duplex, the same resource may be allocated to the backhaul link and access link, with MU-MIMO type of advanced receiver to mitigate the inter-link interference. The similar mechanism is for in-band D2D system as well.

- **Flexible Duplex interference mitigation mechanisms:** For cellular network with flexible duplex configuration, the main issue is still the DL to UL interference among cells. To reach a good inter-cell inter-direction interference mitigation performance with an acceptable advanced receiver implementation complexity, 5G-NR air-interface is designed to have a symmetrical DL/UL trans-

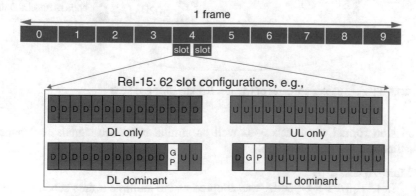

Fig. 2.71 5G-NR slot DL/UL configurations in 3GPP release-15

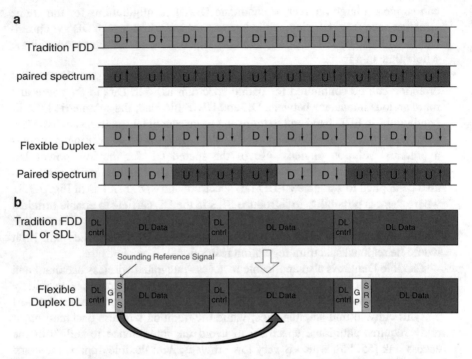

Fig. 2.72 Flexible duplex use cases in FDD or SDL band. (**a**) Flexible duplex in FDD UL carrier. (**b**) Flexible duplex in FDD DL carrier or SDL band

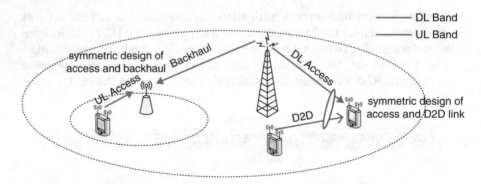

Fig. 2.73 Flexible duplex use cases of wireless Backhauling and D2D

mission format, multi-access as well as similar reference signals for channel estimation and demodulation.

A. Multi-Access aspect

In LTE, DFT-S-OFDMA was adopted as the multiple access scheme for UL due to low Peak-to-Average Power Ratio (PAPR) for better coverage, while OFDMA is adopted for DL for more efficient wideband frequency-domain selective scheduling. Accordingly, the subcarrier mapping schemes are different for

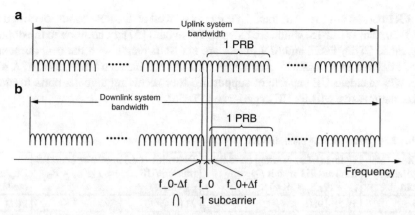

Fig. 2.74 LTE DL and UL subcarrier mapping with 7.5 kHz shift

DL and UL signals, leading half-subcarrier offset between the two kinds of signals, as interpreted in Fig. 2.74. Such design works well in conventional systems, but cannot support new applications in 5G. In order to support flexible duplex, symmetrical design is proposed for multiple access schemes with both DL and UL signals adopting OFDMA scheme. Subcarrier mapping scheme should also align with each other, so as to avoid inter-carrier interference between DL and UL signals. Based on such symmetry, current processes to receive two DL (UL) signals can be reused to receive DL and UL signals simultaneously.

B. Reference signal aspect

In LTE, the reference signal (RS) patterns were designed quite differently for DL and UL. In order to well support simultaneous reception of DL and UL signals, it is preferred to avoid interference between data and DM RSs and ensure DL and UL DM RSs to be orthogonal with each other, so that the receiver can estimate DL and UL fading channels for both signals accurately to ensure decoding afterward. In 5G, the DL/UL symmetric air-interface enables the orthogonal resources among reference signals or between reference signal and data transmissions in different transmission direction. By such design, the inter-direction interference will be seen from the advanced receiver as the orthogonal streams in multi-user MIMO of the actual scheduled signal. Virtual MIMO advanced receiver is well implemented in LTE network already, which can be the basis of the 5G flexible duplex advanced receiver.

2.3.2   3GPP 5G-NR Band Definition

2.3.2.1   3GPP Rel.15 5G-NR Band Definition

The 5G-NR candidate spectrums specified in 3GPP Release-15 are classified into two frequency ranges. FR1 represents for the sub-6 GHz spectrum, ranges from 450 MHz to 6 GHz; while FR2 is for millimeter wave, from 24.25 to 52.6 GHz.

FR1 consists of the traditional LTE bands as well as the new bands identified for IMT-2020 in WRC'15, which are listed in Table 2.30 [57]. In addition to the duplexing mode FDD, TDD, and SDL, there are six SUL bands with the new duplexing mode defined in Release-15. Among FR1 bands, NR bands n7, n38, n41, n77, n78, and n79 mandate UE capable of supporting four receiving antenna ports to maximize the massive MIMO UE experience [57, 58].

Table 2.30 NR operating bands in FR1

NR operating band	Uplink (UL) *operating band* BS receive/UE transmit $F_{UL\_low} - F_{UL\_high}$ (MHz)	Downlink (DL) *operating band* BS transmit/UE receive $F_{DL\_low} - F_{DL\_high}$ (MHz)	Duplex mode
n1	1920–1980	2110–2170	FDD
n2	1850–1910	1930–1990	FDD
n3	1710–1785	1805–1880	FDD
n5	824–849	869–894	FDD
n7	2500–2570	2620–2690	FDD
n8	880–915	925–960	FDD
n12	699–716	729–746	FDD
n20	832–862	791–821	FDD
n25	1850–1915	1930–1995	FDD
n28	703–748	758–803	FDD
n34	2010–2025	2010–2025	TDD
n38	2570–2620	2570–2620	TDD
n39	1880–1920	1880–1920	TDD
n40	2300–2400	2300–2400	TDD
n41	2496–2690	2496–2690	TDD
n50	1432–1517	1432–1517	TDD[a]
n51	1427–1432	1427–1432	TDD
n66	1710–1780	2110–2200	FDD
n70	1695–1710	1995–2020	FDD
n71	663–698	617–652	FDD
n74	1427–1470	1475–1518	FDD
n75	N/A	1432–1517	SDL
n76	N/A	1427–1432	SDL
n77	3300–4200	3300–4200	TDD
n78	3300–3800	3300–3800	TDD
n79	4400–5000	4400–5000	TDD
n80	1710–1785	N/A	SUL
n81	880–915	N/A	SUL
n82	832–862	N/A	SUL
n83	703–748	N/A	SUL
n84	1920–1980	N/A	SUL
n86	1710–1780	N/A	SUL

[a]UE that complies with the NR Band n50 minimum requirements in this specification shall also comply with the NR Band n51 minimum requirements

3GPP Release-15 only specifies four millimeter wave bands within FR2 as listed in Table 2.31 [59]. All of the Release-15 FR2 bands adopt TDD mode for the convenience of efficiently utilizing the DL/UL channel reciprocity for multi-antenna transmission.

2.3.2.2   3GPP 5G-NR Band Combination

Due to the large path-loss and penetration-loss of middle and high-frequency 5G new bands, C-band (n77, n78, n79) and millimeter wave bands (n257, n258, n260, n261) have very limited coverage and thus limit the cell-edge user experience throughput of the UEs that access the operated NR base station. Therefore typical network operation will combine the new wideband middle or high-frequency 5G band with the low-frequency band, while the latter band will serve as an anchor carrier to provide a continuous coverage and seamless mobility. Different operator may choose different band combination(s) for their 5G network deployment according their owned spectrum.

2.3.2.2.1   5G-NR Band Combination Mechanisms

3GPP specified two band combination mechanisms for LTE-Advanced: carrier aggregation (CA) since Release-10, and dual-connectivity (DC) since Release-12. For 5G-NR, 3GPP defines two additional band combination mechanisms, which are LTE/NR DC with typically NR non-standalone operation, and NR band combination of the normal TDD (or FDD/SDL) band with a SUL band. In total, there are three potential band combination mechanisms for 5G-NR standalone deployment and two potential band combination mechanisms for 5G-NR non-standalone deployment, described as below.

For 5G-NR standalone deployment, the following three band combination mechanisms can be adopted. However, only NR-NR DL CA and SUL specifications are completed within Release-15, while NR-NR DC only specify few mm-wave and C-band band combinations partially.

– NR-NR CA: An operated network aggregates multiple NR component carriers (CC) to serve a single UE in order to increase the potential scheduled bandwidth and therefore increase the experience throughput (Fig. 2.75). The data transmission in the secondary CC (SCC) can be scheduled by the control signaling of the primary CC (PCC), and the HARQ feedback of the SCC can be transmitted over the PCC as well. There are three types of NR-NR CA, i.e., intra-band CA with continuous CC, intra-band CA with non-continuous CA, and inter-band CA, as shown in Fig. 2.75a, b, and c, respectively.

– NR-NR DC: Similar to NR-NR CA, the data streams of single UE can be transmitted over multiple CCs simultaneously to reach a higher bit rate. However, NR-NR DC forces a self-contained data scheduling, transmission and HARQ

Table 2.31 NR operating bands in FR2 [57]

Operating band	Uplink (UL) operating band (MHz) $f_{\text{UL\_low}} - f_{\text{UL\_high}}$	Downlink (DL) operating band (MHz) $f_{\text{DL\_low}} - f_{\text{DL\_high}}$	Duplex mode
n257	26,500–29,500	26,500–29,500	TDD
n258	24,250–27,500	24,250–27,500	TDD
n260	37,000–40,000	37,000–40,000	TDD
n261	27,500–28,350	27,500–28,350	TDD

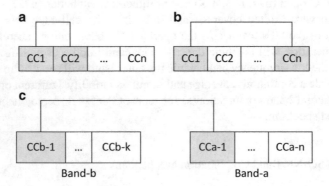

Fig. 2.75 NR-NR CA. (**a**) Intra-band contiguous-CC CA. (**b**) Intra-band non-contiguous-CC CA. (**c**)Inter-band CA

procedure within a CC, and requires longer activating and de-activating period for a CC.

– NR Band combination for SUL [16]: This band combination combines the NR SUL CC with the normal NR CC(s) that contains DL, which can be one or multiple TDD CC(s), FDD CC(s), or SDL CC(s). The typical operation combinations for SUL is to combine with a TDD CC. Different from NR-NR CA, the combined normal NR CC and SUL CC forms a single cell, and all the UL time-frequency resources in the normal CC and SUL CC form a single pool for a single-cell scheduling, as well as share the same HARQ process. Such a mechanism has the benefit of supporting both simultaneous UL carrier switching and good balance between DL capacity and UL coverage, as described in Sect. 2.4.

For 5G-NR non-standalone deployment, the following two band combination mechanisms were specified within Release-15.

– LTE-NR DC: One network can allow single-UE data transmission of multiple streams over LTE and NR in parallel. The typical operation mode is called EN-DC, i.e., the non-standalone NR deployment with LTE as the anchor network to provide the basic coverage and mobility layer, accessing to the EPC.
– LTE-FDD and NR SUL band combination: One special case of the LTE-NR DC operation mode is to configure the NR cell as NR band combination with SUL,

while the SUL carrier can be either orthogonal from LTE UL carrier or shared
with LTE UL carrier. More details will be introduced in Sect. 2.4.

2.3.2.2.2   5G Band Combination Definition

3GPP Release-15 only completed the specifications of some prioritized 5G band
combinations for LTE-NR DC, NR band combination for SUL, NR-NR CA, and
NR-NR DC, as shown in Tables 2.32, 2.33, and 2.34, respectively.

2.3.2.2.3   Multi-Band Coexistence Requirements and Solutions:
Intermodulation and Harmonics

For the aforementioned band combination mechanisms (Table 2.35), the coexistence of multiple bands may suffer from severe receiver desensitization due to spurious emissions, and thus the strict RF requirements as well as the corresponding
solutions for each band combinations individually has to be defined. Spurious emissions are emissions which are caused by unwanted transmitter effects such as harmonics emission and intermodulation products. Receiver desensitization is specified
in terms of Maximum Sensitivity Deduction (MSD) in UE specifications [60].

The intermodulation is the unwanted emission within the downlink transmission
bandwidth of the FDD band, generated by the dual uplink transmissions at different
frequency via the nonlinear transceivers at UE side. Low-order intermediation
causes severe interference for some band combinations of LTE/NR DC, CA and
LTE/NR SUL band combinations. For those band combinations, single UL transmission at UE side is recommended to avoid the FDD DL performance degradation
due to intermodulation.

Harmonic emission is generated by the active uplink in a lower frequency band
within a specified frequency range, whose transmitter harmonics from the UE side
fall within the downlink transmission bandwidth assigned in a higher band at the

Table 2.32   LTE/NR EN-DC band combinations [60]

LTE bands	NR bands
Band 25, 26, 41	2.496–2.69 GHz
Band 1, 3, 8, 11, 18, 19, 20, 21, 26, 28, 40, 41, 42	3.3–4.2 GHz
Band 1, 2, 3, 5, 7, 8, 11, 18, 19, 20, 21, 26, 28, 38, 39, 41, 42, 66	3.3–3.8 GHz
Band 1, 3, 8, 11, 18, 19, 21, 26, 28, 39, 41, 42	4.4–5.0 GHz
Band 1, 3, 5, 7, 8, 11, 18, 19, 21, 26, 28, 41, 42, 66	26.5–29.5 GHz
Band 3, 7, 8, 20, 28, 39, 41	24.25–27.5 GHz
Band 2, 5, 12, 30, 66	37–40 GHz
Band 5, 66	27.5–28.35 GHz

Table 2.33 NR band combinations for SUL (data from [57, 60, 61])

NR Standalone				NR Non-standalone
SUL combinations		SUL band (UL) (MHz)	NR band (UL/DL)	LTE/NR SUL combinations
Release-15	SUL_n78-n80	1710–1785	3.3–3.8 GHz	DC_1-SUL_n78-n80
				DC_3_SUL_n78-n80
				DC_7_SUL_n78-n80
				DC_8_SUL_n78-n80
				DC_20_SUL_n78-n80
				DC_1-3-SUL_n78-n80
				DC_3-7-SUL_n78-n80
				DC_3-8-SUL_n78-n80
				DC_3-20-SUL_n78-n80
	SUL_n78-n81	880–915		DC_3-SUL_n78-n81
				DC_8_SUL_n78-n81
	SUL_n78-n82	832–862		DC_3-SUL_n78-n82
				DC_20_SUL_n78-n82
	SUL_n78-n83	703–748		DC_8-SUL_n78-n83
				DC_20_SUL_n78-n83
				DC_28_SUL_n78-n83
	SUL_n78-n84	1920–1980		DC_20-SUL_n78-84
				DC_1_SUL_n78-n84
				DC_3_SUL_n78-n84
	SUL_n78-n86	1710–1780		DC_8-SUL_n78-n86
				DC_66_SUL_n78-n86
	SUL_n79-n80	1710–1785	4.4–5.0 GHz	DC_8-SUL_n79-n80
				DC_3_SUL_n79-n80
	SUL_n79-n81	1710–1780		DC_8-SUL_n79-n81
	SUL_n79-n84	1920–1980		DC_1-SUL_n79-n84
	SUL_n75-n81	880–915	1432–1517 MHz (SDL)	DC_8-SUL_n75-n81
	SUL_n75-n82	832–862		DC_20-SUL_n75-n82
	SUL_n76-n81	880–915	1427–1432 MHz (SDL)	DC_8-SUL_n76-n81
	SUL_n76-n82	832–862		DC_20-SUL_n76-n82
	SUL_n41-n80	1710–1785	2496–2690 MHz	DC_3-SUL_n41-n80
	SUL_n41-n81	880–915		DC_8-SUL_n41-n81
	SUL_n77_n80	1710–1785	3.3–4.2 GHz	DC_1-SUL_n77-n80
				DC_3_SUL_n77-n80
	SUL_n77_n84	1920–1980		DC_1_SUL_n77-n84
				DC_3_SUL_n77-n84
Release-16	DL_n78(2A)_UL_n78-n86	1710–1780	3.3–3.8 GHz	DC_66_SUL_n78(2A)-n86

Table 2.34 NR-NR CA {55}

Intra-band contiguous CA in FR1		Intra-band contiguous CA in FR1		FR1/FR2 inter-band CA	
NR CA band	NR band (MHz)	NR CA Band	NR band	NR CA band	NR band
CA_n77	3300–4200	CA_n3A-n77A	FDD: 1805–1880 MHz & 1710–1785 MHz; TDD: 3300–4200 MHz	CA_n8-n258	FDD: 880–915 MHz & 925–960 MHz; TDD: 24.25–27.5 GHz
CA_n78	3300–3800	CA_n3A-n78A	FDD: 1805–1880 MHz & 1710–1785 MHz; TDD: 3300–3800 MHz	CA_n71-n257	FDD: 663–698 MHz & 617–652 MHz; TDD: 26.5–29.5 GHz
CA_n79	4400–5000	CA_n3A-n79A	FDD: 1805–1880 MHz & 1710–1785 MHz; TDD: 4400–5000 MHz	CA_n77-n257	TDD: 3.3–4.2 GHz; TDD: 26.5–29.5 GHz
		CA_n8A-n75A	FDD: 880–915 MHz & 925–960 MHz; SDL: 1432–1517 MHz	CA_n78-n257	TDD: 3.3–3.8 GHz; TDD: 26.5–29.5 GHz
		CA_n8-n78A	FDD: 880–915 MHz & 925–960 MHz; TDD: 3300–3800 MHz	CA_n79-n257	TDD: 4.4–5.0 GHz; TDD: 26.5–29.5 GHz
		CA_n8A-n79A	FDD: 880–915 MHz & 925–960 MHz; TDD: 4400–5000 MHz		
		CA_n28A_n78A	FDD: 703–748 MHz & 758–803 MHz; TDD: 3300–3800 MHz		
		CA_n41A-n78A	TDD: 2496–2690 MHz; TDD: 3300–3800 MHz		
		CA_n75A-n78A[1]	SDL: 1432–1517 MHz; TDD: 3300–3800 MHz		
		CA_n77A-n79A	TDD: 3300–4200 MHz; TDD: 4400–5000 MHz		
		CA_n78A-n79A	TDD: 3300–3800 MHz; TDD: 4400–5000 MHz		

[1] Applicable for UE supporting inter-band carrier aggregation with mandatory simultaneous Rx/Tx capability.

Table 2.35 Band combinations NR-DC (two bands)

NR-DC band	NR band (GHz)
DC_n77-n257	TDD: 3.3–4.2
	TDD: 26.5–29.5
DC_n78-n257	TDD: 3.3–3.8
	TDD: 26.5–29.5
DC_n79-n257	TDD: 4.4–5.0
	TDD: 26.5–29.5

UE receiver. Harmonic can be mitigated by harmonic rejection filter at lower band or avoided by limiting the spectrum resources of the UL transmission in the low-frequency band, and coordination with the DL scheduled resource allocation in the high-frequency band.

Unlike the usual harmonic issue which concerns more on UE self-desensitization problem when low-frequency band UL harmonics falls onto high-frequency band DL carriers, some band combinations may be subjected to low-band desensitization when high-band UL carrier is located at third or fifth harmonic of low-band DL carrier, due to the known harmonic mixing problem.

Intermodulation Avoidance with Single UL Transmission

The intermodulation is the generation of signals in its nonlinear elements caused by the presence of two or more signals at different frequency reaching the transmitter via the antenna, as shown in Fig. 2.76. For a UE configured with some band combinations, intermodulation emission may occur, which is in spite of what band

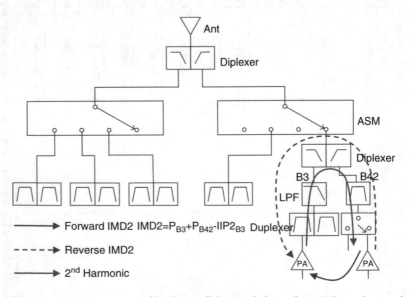

Fig. 2.76 Intermodulation generated by the parallel transmissions of two different frequencies

Table 2.36 IMD for LTE NR band combinations for NR Band n78

LTE band f_x	NR band f_y	Source of IMD	Note for victim band(s)
B1	n78 (3.3–3.8 GHz)	IMD2, IMD4, IMD5	IMD2: Band1, $f_y - f_x$ IMD4: Band1, $3*f_x - f_y$ IMD5: Band1, $2*f_y - 3*f_x$
B3		IMD2, IMD4, IMD5	IMD2: Band3, $f_y - f_x$ IMD4: Band3, $3*f_x - f_y$ IMD5: Band3, $2*f_y - 3*f_x$
B5		IMD4	Band 5, $f_y - 3*f_x$
B7		IMD4	Band Z, $3*f_x - f_y$
B8		IMD4	Band 8, $f_y - 3*f_x$
B20		IMD4	Band 20, $f_y - 3*f_x$
B28		IMD5	Band 28, $f_y - 4*f_x$
B39		N/A	
B41		N/A	
B42		N/A	

combination mechanism including LTE/NR DC, LTE/NR UL sharing, and the inter-band CA or DC either in LTE or in NR system. Low-order Intermodulation can be so severe that it can cause serious damage for the low-frequency band DL reception. For example, for Band 3 (FDD: DL/1,805–1,880 MHz, UL/1,710–1,785 MHz) and Band 42 (TDD: DL&UL/3,400–3,600 MHz) LTE-CA [62], the LTE DL sensitivity would decrease by 29.8 dB due to the second-order intermodulation interference. Such serious intermodulation issue and the challenge to UE implementation was acknowledged by 3GPP industries for a long time, and was hotly discussed during Release-15 5G-NR standardization period.

Taking the example of the NR core band n78 (3.3–3.8 GHz) related band combi-nations, Table 2.36 gives the intermodulation analysis, where IMDn stands for the nth-order intermodulation. The analysis can also be extended to 3.3–4.2 GHz. From the table, it can be seen that among all the band combinations involving C-band, combination with B1 and B3 will suffer from very strong IMD interference at sec-ond order, while B5, B7, B8, and B20 will suffer from the fourth order of IMD, and B28 suffers from the fifth order of IMD. Intermodulation at second order and third order will cause unacceptable performance degradation at tens of dB level in the corresponding emission frequency range, while even the fourth-order and fifth-order IMD impact are also quite obvious.

Due to the serious degradation to the low-frequency FDD DL performance, 3GPP specified a single-UL solution for those band combinations with intermodulation issue: limiting the UL transmission is a single band. That is, the UE operates on only one of the carriers at a given time among a pair of the low-fre-quency and high-frequency carriers.

For all the band combination mechanisms, which band to choose for the UL transmission at a time is dependent on the RSRP measurements. Usually the high-frequency UL carrier is selected for the UEs at cell center, while low-frequency UL carrier is selected for the UEs at cell edge (Fig. 2.77).

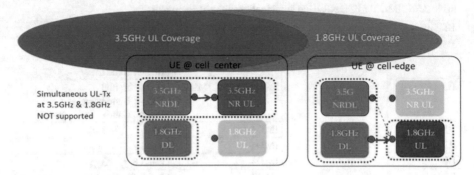

Fig. 2.77 Single UL transmission for B3 + n78 band combination

Table 2.37 Harmonic mixing for LTE NR band combinations with 3.5 GHz band

LTE band f_x	NR band f_y	Order of harmonic mixing
B2	3.3–3.8 GHz (band n78)	Second
B3		Second
B8		Fourth
B20		Fourth
B26		Fourth
B28		Fifth
B66		Second
B1	3.3–4.2 GHz (band n77)	Second
B3		Second
B8		Fourth
B18		Fourth, fifth
B19		Fifth
B20		Fourth
B26		Fourth
B28		Fifth

Harmonics Distortion Avoidance by Cross-Band Scheduling Coordination

Different from intermodulation generated by the transmissions from different frequency jointly, harmonic emission is generated by a single active uplink in a lower frequency band, whose transmitter harmonics from the UE side fall within the downlink transmission bandwidth assigned in a higher band at the UE receiver.

Taking Band n78 again as the example, Table 2.37 provides the harmonic analyses for these band combinations. It shows that for C-band n78 involved band combinations, the combinations with B2, B3, and B66 suffer a second-order harmonics issue, while the combinations with B8, B20, and B26 suffer a fourth-order harmonics issue, and the combination with B28 suffer a fifth-order harmonics issue.

Besides utilizing harmonic rejection filter at lower band to mitigate the MSD issue, Harmonic can largely be avoided by limit the spectrum resources of the UL transmission in the low-frequency band, and coordination with the DL scheduled resource allocation in the high-frequency band. Since the high-frequency band has a very wide bandwidth, e.g., 100 MHz carrier at C-band, while the harmonic issue only falls at a much narrower frequency range than the whole carrier bandwidth in the high-frequency band, it is, thus, straight forward to schedule the DL transmission for the UE to avoid the frequency range which may suffer the harmonics issue from its simultaneous UL transmission in the low-frequency FDD band.

2.4   4G/5G Spectrum Sharing (aka LTE/NR Coexistence)

The 3GPP standardization of standalone and non-standalone 5G-NR was frozen on June 2018 and December 2017, respectively. The C-band (3,000–4,200 and 4,400–5,000 MHz) is emerging as the primary frequency band for the introduction of 5G by 2020, with the availability of at least 100 MHz channel bandwidth per 5G network. It is crucial for 5G-NR commercial deployment to provide an optimal balance between coverage and capacity, and preferable to deploy 5G-NR BSs co-site with the exiting LTE BSs for cost-efficient implementation. 5G-NR with the adoption of massive MIMO will boost DL throughput at both cell center and cell edge with affordable complexity. Beamforming can be used to reach similar coverage as the LTE network deployed at 1.8 GHz band. However, due to limited transmit power at terminal and limited UL transmission slots, it's difficult for 5G-NR to improve its UL coverage only based on C-Band TDD carrier. Accordingly, 4G/5G spectrum sharing, also called as LTE/NR coexistence, was proposed to reuse the UL spare resources in the lower frequencies already licensed for mobile use (e.g., 700, 800, 900, 1,800, and 2,100 MHz) as the coverage compensation to 5G-NR network, scheduled as a supplementary UL in combination with 3,300–3,800 MHz. 4G/5G sharing allows operators to benefit from faster and cost-efficient deployment of C-band (can expand to SDL and mm-wave in the future), and thus delivers enhanced capacity without incurring network densification costs.

2.4.1   Motivation and Benefit

5G-NR were developed for supporting diverse services, such as enhanced mobile broadband (eMBB), massive machine type communication (mMTC) and ultra-reliable low latency communication (URLLC). The new spectrum released for 5G deployments, primarily above 3 GHz, unfortunately has a relatively high path-loss, which limits the coverage, especially for the uplink (UL). The high propagation loss, the limited number of UL slots in a TDD frame and the limited user-power gravely limit the UL coverage, but the available spectrum bandwidth is very rich. Moreover, the stringent requirements of the 5G diverse applications lead to a number of 5G challenges, such as ensuring seamless coverage, high spectrum efficiency, and low latency. This section addresses some of those challenges with the aid of a unified spectrum sharing mechanism, and by means of an UL/DL decoupling solution based on 4G/5G frequency sharing, which has already been specified in 3GPP Release-15. The key concept is to accommodate the UL resources in an LTE FDD frequency band as a supplemental UL carrier in addition to the NR operation in the TDD band above 3 GHz. The performance of the spectrum sharing has been confirmed with field trials (see Sect. 5.4.5).

2.4.1.1 NR Coverage on New Spectrum

Coverage is a very essential performance criterion for wireless communication system and is affected by numerous factors, including the transmission power, propagation loss, and receiver sensitivity. Since the propagation loss varies with the frequency, the coverage substantially differs in different frequency bands. Therefore, the provision of a good performance in all frequency bands remains a key challenge for 5G deployments. Furthermore, due to the limited UL transmission power and higher path-loss in NR than in LTE, the UL coverage is usually the bottleneck in 5G deployments.

2.4.1.1.1 Link Budget

In Fig. 2.78, the coverage performance of the 3.5 GHz TDD band and compare it to that of the 1.8 GHz FDD band is portrayed. Partial list of the parameters assumed for this comparison are shown in the Figure, while the rest are given in Table 2.38.

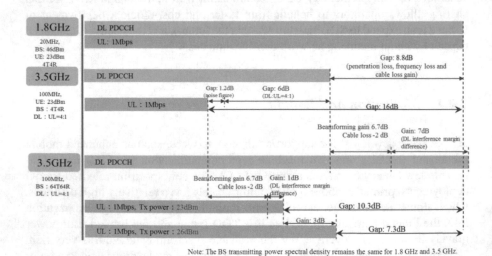

Fig. 2.78 Link budgets of 1 Mbps throughput for different frequency bands

Table 2.38 Parameters assumed in the link budgets

Parameters	1.8 GHz with 4T4R		3.5 GHz with 4T4R		3.5 GHz with 64T64R	
	PDCCH	PUSCH	PDCCH	PUSCH	PDCCH	PUSCH
Tx antenna gain G_{Ant}^{TX} (dBi)	17	0	17	0	10	0
Tx cable loss L_{CL}^{TX} (dB)	2	0	0	0	2	0
Rx antenna gain G_{Ant}^{RX} (dBi)	0	18	0	18	0	10
Rx cable loss L_{CL}^{RX} (dB)	0	2	0	0	0	2
Penetration loss L_{pe} (dB)	21	21	26	26	26	26
Receiver sensitivity γ (dBm)	−129.44	−134.3	−129.44	−134.3	−141.02	−141.23
Shadowing loss L_{SF} (dB)	9	9	9	9	9	9
Propagation loss due to frequency L_f (dB)	0	0	5.78	5.78	5.78	5.78
Interference margin I_m (dB)	14	3	14	3	7	2
Thermal noise per subcarrier N_{RE} (dBm)	−132.24	−132.24	−129.23	−129.23	−129.23	−129.23
Noise figure N_F (dB)	7	2.3	7	3.5	7	3.5

In the link budget, the UL coverage is calculated when the UL data rate is set to 1 Mbps for supporting typical uplink video traffic. By contrast, the DL coverage is usually limited by the physical downlink control channel (PDCCH) quantified in terms of the block error rate of the primary PDCCH. It can be observed that the UL coverage and DL coverage are balanced over the 1.8 GHz FDD band with the aid of four transmit and four receive antennas. For the 3.5 GHz TDD band using the same transmit and receive antennas as that of 1.8 GHz scenario, in excess of 10 dB coverage gap is observed. This is mainly due to the large propagation loss, the penetration loss and the limited number of UL transmission slots in a frame of the 3.5 GHz TDD band. By comparison, for the 3.5 GHz TDD band using 64 transmit and 64 receive antennas, a similar DL coverage performance can be achieved to that of 1.8 GHz, due to the beamforming gain provided by massive MIMOs and by the DL interference margin difference. Explicitly, since massive MIMOs also reduce the inter-cell interference, they reduce the DL interference margin. However, the UL coverage is poorer compared to the DL of 3.5 GHz, even when massive MIMOs are employed, because the UL power spectral density of the 3.5 GHz TDD band is lower than that of the 1.8 GHz FDD band at the same maximum device transmission power. This is partly due to having less UL slots in a TDD-frame than in an FDD-frame, which means that more frequency resources per slot should be allocated for a given UL throughput of say 1 Mbps. Therefore, how to improve the UL coverage is indeed an important issue for 5G deployments.

2.4.1.1.2   UL/DL Assignment Impact on NR Coverage

As described in Sect. 2.3.1.2, the UL/DL traffic ratio in a TDD system is usually DL dominant adaptive to the eMBB DL/UL traffic load statistics, which is 4:1 for almost all the commercial LTE TDD systems, and is most likely the chosen operating point for 5G-NR TDD configuration as well. Therefore, a smaller proportion of the resources is left for the UL, which will further affect the UL coverage performance.

   In contrast, the UL coverage is pretty good for the low-frequency FDD band, due to both the low path-loss and penetration loss as well as the continuous available UL slots all the time for the potential transmission of terminals at the cell edge.

   Moreover, for LTE FDD band, the same bandwidth is allocated to both the UL and DL, which means that the UL spectrum resources are under-utilized. While in practical networks, the corresponding UL and DL resource utilization rates in LTE FDD systems are provided in Fig. 2.79 by several operators. Based on the statistics of four operators in 2016 Q4, the average downlink resource utilization rate is from 40 to 60%, while the average uplink resource utilization rate is about 10%. Such situation gives the possibility to utilize the FDD UL spared spectrum resources for 5G-NR UL transmission for the cell-edge users, to compensate for its transmission in the high-frequency TDD band.

2.4.1.1.3   5G-NR Deployment Challenges Due to the Coverage

In the following, we will discuss a few challenging issues that have to be considered in 5G deployments, particularly for the TDD mode and in higher frequency bands.

- **5G Band selection: Wideband spectrum availability vs. coverage.**
  The availability of the bands below 3 GHz remains limited for 5G-NR in the near-future and the lower bands fail to support high data-rates due to their limited bandwidth. On the other hand, the wider NR bands above 3 GHz experience increased propagation losses, leading to a limited coverage. Therefore, independent usage of the spectrum below and above 3 GHz fails to strike a compelling trade-off between a high data rate and large coverage.

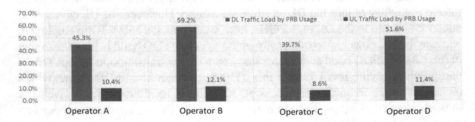

Fig. 2.79 DL/UL frequency resource usage of four operated LTE FDD Network in 2016Q4

- **TDD DL/UL ratio: Spectrum utilization efficiency vs. DL/UL coverage balance vs. Multi-service operation**

  As discussed, the NR TDD operation is usually configured for a limited number of UL transmission slots (e.g., DL:UL = 4:1) in a frame due to the heavy DL traffic load, even though more slots should be allocated to the UL for improving the UL coverage. This can increase the UL data-rates, when the bandwidth cannot be further increased due to the maximum transmission power constraint. Since the DL spectral efficiency is usually higher than that of the UL, having more UL slots would, then, further reduce the spectral utilization efficiency. Therefore, there is a clear trade-off between the UL coverage and spectral utilization efficiency.

  In addition, 5G system is required to provide diverse services including eMBB, mMTC and URLLC, etc. As analyses in Sect. 2.3, it is hard for a single network TDD configuration fits to both eMBB and IoT type of services. Efficient eMBB transmission requires high DL resources proportion, while IoT services for low power wide area scenario (LPWA) are highly dependent on the UL coverage and thus needs continuous UL transmission and preferably as low path-loss as possible. Therefore, efficient 5G spectrum exploitation mechanism is desired to support all kinds of services using a unified TDD configuration.

- **TDD DL/UL switching period: Transmission efficiency vs. latency.**

  For the TDD operation, frequent DL/UL switching is required for URLLC services to provide very quick DL/UL scheduling and almost immediate ACK/NACK feedback. However, a certain guard period is also needed at each DL/UL switching point (e.g., 130 μs is used in TD-LTE networks) for avoiding serious blocking of the UL receiver due to the strong DL interference emanating from other cells. Frequent DL/UL switching would lead to a high idle-time (14.3% vs. 2.8% for 1 and 5 ms switch period). Such a short DL/UL switching period in a single TDD band network operation will lead to an unacceptable degradation of spectrum efficiency for eMBB transmission, which is generally in favor of a larger DL/UL switching period, e.g., 2.5 or 5 ms.

- **Site planning: seamless coverage vs. deployment investment and mobility.**

  For early 5G-NR deployment, co-site installation with the existing LTE networks would be cost-effective and convenient. However, due to the higher propagation loss above 3 GHz, one has to introduce denser cells and new sites. Otherwise, 5G-NR cannot attain the same seamless UL coverage as that of LTE. In the following, we will discuss a few challenging issues that have to be considered in 5G deployments, particularly for the TDD mode and in higher frequency bands.

In summary, for a fast and cost-effective 5G network deployment, to accommodate multi-services of both eMBB and IoT applications efficiently, and to balance the spectrum efficiency, coverage as well as low latency, 5G network has to operate on both high-frequency wideband TDD band for high capacity and low-frequency band with good coverage

2.4.1.2   UL/DL Decoupling Enabled by 4G/5G UL Spectrum Sharing

To circumvent the challenges discussed above, a new UL/DL decoupling concept was accepted by 3GPP, which will be elaborated on below.

The concept of UL/DL decoupling is to exploit the spare resources in the existing LTE frequency band for 5G-NR operation as a complement of the new 5G wideband spectrum. For example, as shown in Fig. 2.80, the C-band (frequency ranges of 3–5 GHz) TDD carrier can be paired with the UL part of a FDD band overlapped with LTE (e.g., 1.8 GHz). In other words, an UL carrier within the lower frequency FDD band is coupled with a TDD carrier in the higher frequency band for NR users. Then a NR user has two UL carriers and one DL carrier in the same serving cell. By contrast, only one DL carrier and one UL carrier are invoked for a traditional serving cell. With the advent of this concept, the cell-edge NR users can employ either the lower frequency FDD band carrier (UL part) or the higher frequency TDD band carrier to transmit their uplink data. In this case, since the UL propagation loss on the lower frequency band is much lower than that of the higher frequency TDD band, the coverage performance of NR users can be substantially extended and a high UL data-rate is guaranteed even if this user is relatively far from the BS. Moreover, the cell-center users can rely on the higher-frequency TDD band to take advantage of its higher bandwidth.

Usually, it is not necessary to allocate the low-frequency FDD band for the DL of NR since the DL coverage in the C-band is good. Then the low-frequency FDD band is employed in NR only for the UL. In 3GPP, the UL-only carrier frequency is referred to as the supplementary uplink (SUL) frequency from a NR perspective.

NR DL : 3400-3800 MHz
NR UL : LTE low frequency band + 3400-3800 MHz

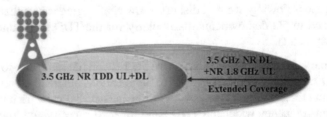

Fig. 2.80   UL/DL decoupling to extend 5G coverage at higher frequencies (e.g., 3.5 GHz)

2.4.1.3   Benefits of 4G/5G Uplink Spectrum Sharing

2.4.1.3.1   Higher Spectrum Utilization Efficiency

UL/DL decoupling is instrumental in striking a compelling trade-off between high spectrum exploitation efficiency and wider DL/UL coverage. For the high-frequency TDD carrier, the DL/UL time-slot (TS) ratio configuration only has to take into account the long-term DL/UL traffic statistics for guaranteeing the DL spectrum exploitation efficiency (usually 4:1). The cell-edge users and IoT devices may opt for the SUL carrier philosophy for their UL transmission. In this case, the high DL/UL Time Slot (TS) ratio on the TDD carrier does not impose any detrimental effects on IoT services. In addition, the lower propagation loss of the lower band is helpful for improving the spectrum efficiency. As a result, given a certain packet size, the requirements imposed on the scheduled bandwidth, or the UE's transmit power are reduced on the lower band compared to that on the higher band.

Let us now observe the UL user throughputs of various UL channel allocations in the 3.5 GHz band, the joint 3.5 and 0.8 GHz bands and joint 3.5 and 1.8 GHz bands seen in Fig. 2.81. The UE's maximum total transmission power across all frequency bands is 23 dBm and the DL/UL TS ratio of the 3.5 GHz TDD system is 4:1. The channel bandwidths of the 3.5, 0.8, and 1.8 GHz scenarios are 100, 10, and 20 MHz, respectively. It can be seen that the UL throughput of the cell-edge UEs relying on the SUL is substantially improved compared to that of an UE operating without SUL, which is a joint benefit of the additional bandwidth, of the lower

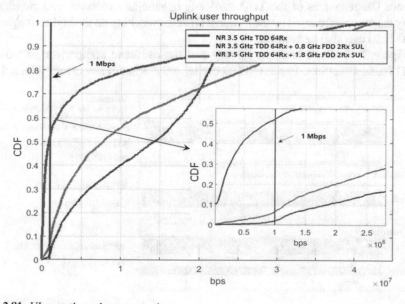

Fig. 2.81 UL user throughput comparison

propagation loss and of the continuous UL resource in the SUL. Additionally, the UL throughput of UEs relying on the SUL at 0.8 GHz is better than that of the UEs with SUL at 1.8 GHz at lower throughput, but it is lower than that of UEs with SUL at 1.8 GHz at higher throughput. The reason for this trend is that when the UL throughput is low, the UEs are usually power-limited and the propagation loss is lower at low frequencies, hence the throughput of the SUL at 0.8 GHz is better than at 1.8 GHz. By contrast, when the throughput is high, the uplink transmission power is not an issue and it is the bandwidth that becomes the bottleneck; thus, the throughput of the SUL at 1.8 GHz within a 20 MHz bandwidth outperforms that at 0.8 GHz with 10 MHz bandwidth. Therefore, with the advent of the UL/DL decoupling concept, the spectrum exploitation efficiency and DL/UL coverage can be beneficially balanced.

2.4.1.3.2 Feedback Latency and Efficiency

From the previous discussions, it is clear that low latency is a critical requirement for URLLC services. In 5G-NR design, a self-contained TDD frame structure [63] is proposed, where in each subframe/slot, both DL and UL can be included. As indicated, frequent DL/UL switching may help reduce the UL latency, but it also introduces a non-negligible overhead, which is inefficient for both of eMBB and URLLC services in a unified system.

Under the UL/DL decoupling concept, the URLLC devices can be scheduled at the SUL carrier for the UL data or control messages, which means that UL resources always exist, whenever an UL message arrives. Thus, the latency due to the discontinuous UL resources of the TDD carrier is beneficially reduced, and simultaneously the overhead caused by the frequent DL/UL switching on the higher-frequency TDD band can also be avoided.

Figure 2.82 shows both the latency and the overhead comparison of various TDD frame structures. For the "TDD carrier only" system associated with a 5 ms

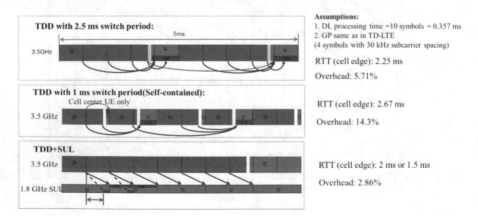

Fig. 2.82 Latency comparison of different TDD frame structures

switch period, the round trip time (RTT) cannot be tolerated by the URLLC services, due to the long feedback latency. If a self-contained TDD frame is applied in the "TDD carrier only" system having a 1 ms switch period, although the RTT is reduced, the overhead increases dramatically due to the frequent DL/UL switch. For the proposed UL/DL decoupling concept, the SUL can provide timely UL feedback without frequent DL/UL switching, which hence beneficially reduces the RTT without any extra overhead. Therefore, the transmission efficiency and latency become well-balanced.

In addition, with this concept, the 5G-NR systems can operate with a combined 5G-NR cell of a TDD carrier in a new frequency band (e.g., C-band or mmWave band) and a SUL carrier in the existing LTE spectrum (lower than 3 GHz). 5G-NR DL traffic is only scheduled on the TDD carrier with dominant DL slots to utilize the large bandwidth of the new frequency band as much as possible, with enhanced DL spectrum efficiency as well as control channel coverage based on massive MIMO and multiple beam scanning.

2.4.1.3.3 Seamless Coverage, Deployment Investment, and Mobility

Seamless coverage is highly desirable for 5G NR in order to provide a uniform user experience. For early 5G-NR deployment, co-site installation with the existing LTE networks would be cost-effective and convenient. Again, it is difficult for 5G NR to achieve seamless coverage in case of co-site deployment with LTE by only using the frequency band above 3 GHz. Due to the higher propagation loss above 3 GHz, one has to introduce denser cells and new sites, which requires higher investment.

With the advent of the UL/DL decoupling, the 5G-NR UL becomes capable of exploiting the precious limited spectrum resources in the lower frequency bands that the operators have been using for LTE. The NR UL coverage can, therefore, be improved to a similar level as that of LTE. This implies that the seamless NR coverage can be supported in co-site NR/LTE deployment.

By doing so, the mobility-related user-experience is also improved. As illustrated in the co-site deployment example of Fig. 2.83a, due to the limited UL coverage, the

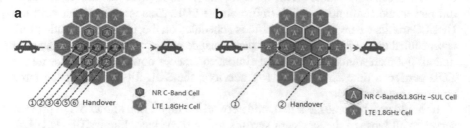

Fig. 2.83 Mobility impact of UL over higher frequency and lower frequency band. (**a**) UL only over higher frequency band. (**b**) UL over both C-band and lower frequency band

radius of 5G C-band cells is much smaller than that of the LTE 1.8 GHz cells. When a UE roams within the cells, inter-Radio Access Technologies (RAT) handovers will occur. Note that each inter-RAT handover will impose interruptions in excess of 100 ms, which is much higher than that of the intra-RAT handover. With the advent of the LTE/NR spectrum sharing concept, the SUL carrier beneficially extends the coverage of 5G cells. As shown in Fig. 2.83b, with the help of SUL, the coverage range of 5G cells and LTE cells becomes similar. Then inter-RAT handovers are only encountered when the UE goes beyond the boundaries between the areas only containing 5G cells and the areas only containing LTE cells. Thus the probability of the inter-RAT handovers is beneficially reduced. Consequently, seamless coverage can be provided by the LTE/NR spectrum sharing concept.

2.4.1.3.4   Unified Network Configuration for Various Traffic Types

5G is developed to support various traffic types such as eMBB, URLLC and IoT services. Different traffic types may require different network deployment if 5G is deployed only on the new spectrum which is TDD spectrum. For eMBB service, it requires much more downlink resources than uplink resources and less UL-DL switching points for higher spectra efficiency. While URLLC service requires continuous uplink and downlink resources in time domain to transmit the packets immediately at the time it arrives and this may require a self-contained slot structure where in each slots there is available downlink and uplink symbols. For IoT service, it may require more uplink resources due to much more uplink traffic for data collection from the devices than downlink control messages. Using only the new TDD spectrum cannot balance the diverse requirements of various traffic types.

By the employment of the UL/DL decoupling one unified network configuration can fulfill all the requirements mentioned above. In the UL/DL decoupling concept, in one cell, a new TDD spectrum with wide channel bandwidth is paired with one continuous uplink (SUL) resources on a lower frequency. For downlink eMBB traffic, it can be scheduled on the TDD carrier with wide channel bandwidth and larger portion of downlink slots providing high downlink throughput. For the uplink eMBB traffic of cell center UE, it can be scheduled on the TDD carrier for higher throughput with wide channel bandwidth. While for the uplink eMBB traffic of cell edge UE, it can be schedule on the SUL for higher throughput due to limited transmit power and continuous uplink traffic on the SUL. This is also similar with the URLLC traffic, where, downlink traffic is scheduled on the TDD carrier and uplink is scheduled on the SUL. The latency performance will be achieved due to the continuous uplink resources on SUL and almost continuous downlink resources on the TDD carrier in time domain. For IoT services, the SUL fill the gap of the large coverage requirements.

It should now become clear that, one network configuration allows 5G to provide serves to all types of devices with various traffic types including eMBB, URLLC, and IoT services via UL/DL decoupling deployments.

2.4.1.4   Summary of LTE/NR Spectrum Sharing Scenarios

In summary, for NR with both standalone deployment and non-standalone deployment, UL sharing between NR SUL carrier and LTE UL carrier is completely supported by 3GPP Release-15. This mechanism provides an efficient and costly network deployment with a good balance between DL capacity and UL coverage, brings benefit in coverage, latency, throughput, etc. It also potentially improves the spectrum utilization of the LTE UL.

2.4.2   LTE/NR Spectrum Sharing: Network Deployment Scenarios

In UL, when NR is deployed in the LTE band, NR and LTE can share the resource in LTE band. LTE-NR UL sharing focuses on the sharing between NR SUL (supplementary uplink) carrier and UL part of LTE FDD.

Generally, the UL resource can be shared in TDM or FDM manner.

- For TDM sharing, both the NR and LTE share the whole UL carrier in different time slots (or symbols), while each of them can schedule the UL data and control signaling transmission in the frequency resources along the whole bandwidth of the UL carrier.
- For FDM sharing, the NR and LTE occupy the UL resource in different frequency resources, which are orthogonal to each other.

From network coupling point of view, the UL resource sharing can be done statically, semi-statically, or dynamically, as illustrated in Fig. 2.84a, b, and c, respectively.

- Static LTE/NR UL sharing normally adopts FDM manner. As shown in Fig. 2.84a, NR TDD carrier at frequency f0 is combined with an SUL carrier at frequency f3, which is independent from the LTE UL carrier at frequency f2. The NR SUL carrier at f3 and the LTE UL carrier at f2 belong to the same FDD band, i.e., the NR SUL transmission and LTE UL transmission share the same UL band. With the static sharing, the NR SUL carrier and LTE UL carrier have independent scheduling, where NR SUL transmission is scheduled by the NR TDD carrier DL control signaling, and the LTE UL carrier is associated to the LTE DL carrier(s) at frequency f1 in the same FDD band.
- For most countries, one operator owns the FDD spectrum with symmetric DL and UL bandwidths. Consequently, when splitting the total UL spectrum into two parts, NR SUL spectrum and LTE UL carrier, the normal static sharing would result in a narrower LTE UL spectrum bandwidth than that of the LTE DL spectrum bandwidth Figure 2.84a illustrates two options of network operation:

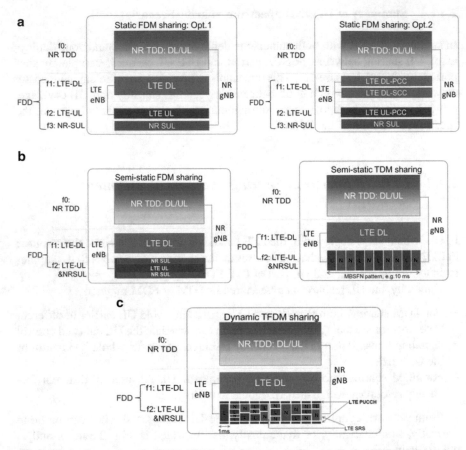

Fig. 2.84 Static, semi-static (TDM, FDM) and dynamic sharing. (**a**) Static LTE/NR UL sharing, by FDM. (**b**) Semi-static LTE/NR UL sharing, by FDM or TDM. (**c**) Dynamic LTE/NR UL sharing, by TFDM (time and frequency domain multiplexing)

- Option1 in the left figure bounds a LTE DL carrier with double bandwidth of that of a LTE UL carrier. However, although 3GPP LTE TS36 series specified the physical-layer signaling of unequal DL and UL bandwidth, but there are no RAN4 test cases specified yet and thus there is no legacy LTE terminals that can support that. To avoid this issue in the legacy LTE system, option1 is not feasible in the practical network deployment.
- Option2 in the right figure uses the equal bandwidth for the LTE DL and UL carrier, while introduces the second component carrier operated with DL CA to make full use of the FDD DL spectrum resources. Among legacy LTE terminals, there are, by the end of 2017, around 20–40% CA capable terminals in the commercial markets in different regions. This percentage will only keep increasing in the future.

   In summary, only option2 of the static LTE/NR UL FDM sharing is feasible in the practical network deployment.

- Semi-static LTE/NR UL sharing can be done in both FDM and TDM manner. Different from the static sharing options, all the semi-static sharing solutions can support the overlapped LTE UL carrier and NR SUL carrier, which can have the same bandwidth, while maintaining the equality of the UL bandwidth with that of the LTE DL carrier as well.

  - The semi-static FDM sharing solution is shown in the left part of Fig. 2.84b. Taken into account that LTE system provides a UE-specific configuration of the frequency-domain-symmetric PUCCH resources, the network can easily perform the FDM sharing by configuring the frequency boundary of the LTE UL control channel PUCCH with a certain distance to the LTE UL carrier boundary assumed by the legacy LTE terminal, which is based on the bandwidth signaled to the UE from network.
  - Moreover, network can schedule all the LTE UL data transmission in the middle part of frequency resources in the carrier, while scheduling the NR SUL data and control signaling transmission at the edge part of the carrier frequency resources. By such a semi-static sharing, LTE system and NR system can do their own scheduling independently, with the only slow or even fixed coordination between the participation of LTE UL and NR SUL frequency resources.
  - The semi-static TDM sharing solution is shown on the right side of Fig. 2.84b right figure. To permits a clean TDM partition between LTE UL and NR SUL transmission, it is important to limit the legacy UE to transmit PUCCH signaling and feedback following the standardized MBSFN pattern. This is because, the legacy LTE terminals would assume those UL subframes that are associated with the DL subframes in the MBSFN pattern are unused and skip that automatically. Based on the MBSFN pattern that is negotiated between LTE and NR network, either fixed or semi-statically, LTE and NR network can do their own scheduling independently.
  - Semi-static solutions are based on RRC signaling to the terminals by the network. It provides the possibilities of proceeding a flexible resource splitting between LTE UL and NR SUL resource, based upon the timely traffic statistics of both LTE and NR systems.

- Dynamic LTE/NR UL sharing is done in TFDM manner (Fig. 2.84c). If NR and LTE are tightly coupled, dynamic sharing, which has no explicit division of UL resource for the two systems, can be used. The NR and LTE can schedule their UE's in any UL resource that is orthogonal to the reserved regions of the legacy LTE PUCCH (at the two edge parts of the carrier frequency) and periodic LTE SRS signals (in the last OFDM symbol of few configured subframes). The interference avoidance and conflict resolution are totally left to the system designer.

From network perspective, there are two kinds of deployment of NR: SA (standalone) and NSA (non-standalone), as illustrated in Sect. 2.4. The differences of UL resource sharing corresponds to these two deployments are introduced in the following subsections.

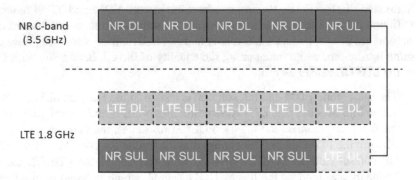

Fig. 2.85 Standalone NR with SUL band combinations, sharing NR-SUL with LTE UL from UE perspective

2.4.2.1   LTE/NR UL Sharing for NR Standalone Deployment

In NR standalone deployment scenario, NR core network and air interface are standalone deployed. The NR UE does not have to access the LTE RAN (radio resource network) or EPC (evolved packet core), but only access NR RAN and NGC (next generation core). In such a standalone scenario, LTE/NR UL sharing is done from network perspective always, i.e., seen from each NR-UE, even though the SUL is deployed in LTE band, the NR UE does not have to know whether an LTE system is deployed in the same band or not. The resource occupied by LTE (if any) is scheduled to other LTE UEs, which are transparent to the NR UE.

An example for the UL sharing in the NR standalone deployment with SUL band combination case is shown in Fig. 2.85. The NR UE does not do concurrent transmission in NR UL and SUL mainly to simplify the implementation complexity at UE side, and to allow concentration of the UE transmit power in a single UL carrier to maximize the UL throughput and coverage in each UL subframe.

2.4.2.2   LTE/NR UL Sharing for Non-standalone NR Deployment, from Network and UE Perspective

In NR non-standalone deployment scenario, the NR UE is expected to have EN-DC capability[6]. In NSA NR EN-DC mode, the NR-UE should access LTE RAN first, then is configured to access an NR cell in the good coverage area of the NR base stations. So, the MCG (master cell group) is using LTE and the SCG (secondary cell group) is using NR. In this case, if NR configures a UE with a SUL band combination, the NR SUL carrier and LTE UL carrier will share the same frequency band. Such sharing can be classified into UE perspective solution and network perspective solution individually.

---

[6]That is the UE is capable of LTE and NR dual connectivity with LTE as the anchor carrier for initial access and mobility.

- LTE/NR UL sharing from UE perspective provides the highest flexibility of sharing the resources. The transmission on SUL carrier scheduled by NR cell and transmission on the UL carrier of LTE FDD scheduled by LTE can be in overlap frequency band. As shown in Fig. 2.86, the UL transmission scheduled by LTE and NR are TDM in the shared carrier. The UE is required to be capable of immediate transmission switching between LTE UL carrier and NR-SUL carrier. It worth noting the premise that the NR UE does not support concurrent transmission on NR UL carrier and NR SUL carrier generally, there exist the case that the UL transmission in a subframe on the NR UL carrier contains only SRS signals, the UL scheduling in the corresponding subframe on the shared carrier is permitted.
- LTE/NR UL sharing from network perspective performs the resource sharing with the relaxed UE implementation complexity, as shown in Fig. 2.87. In such a

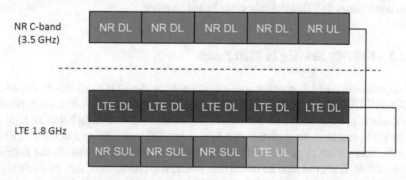

Fig. 2.86 NSA + SUL, UE perspective

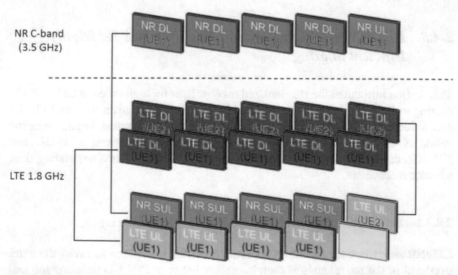

Fig. 2.87 NR non-standalone deployment with SUL band combination, LTE/NR UL sharing from network perspective

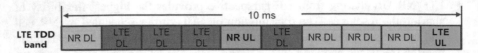

Fig. 2.88 Illustration of UL sharing between LTE and NR in TDD band

case, each UE is scheduled in either LTE-UL transmission or NR-UL transmission at a certain time, i.e., the NR UE does not support concurrent transmission in NR UL or NR SUL. While it works with an NR-SUL transmission mode, the other resources in the same subframe can be allocated to other UEs that may work with LTE-UL transmission mode, and vice versa. Following that solution, the time and frequency domain resources in a network will not be wasted. Note that the LTE UE (UE2 in the figure) is not necessary to be an EN-DC UE. The LTE/NR sharing of the overlapped UL carrier resource among multiple UEs can be done either in TDM manner or in FDM manner

2.4.2.3   LTE/NR Sharing in TDD Band

The aforementioned UL sharing scenarios are all in the LTE FDD bands, the option of LTE/NR UL sharing in TDD bands is available as well. The DL/UL configuration of NR needs to be the same as that of LTE to facilitate UL sharing between LTE and NR in TDM manner. As shown in the Fig. 2.88, NR and LTE can occupy different UL subframes from either network or UE perspective to fully use all the available UL resources. It should also be noted that NR and LTE DL may also be multiplexed in a TDM manner due to the restriction of legacy LTE HARQ timing.

2.4.3   LTE/NR Spectrum Sharing: Requirements for Highly Efficient Sharing

This section introduces the standardized mechanisms for high efficient LTE/NR UL sharing in Release-15, including overhead-avoidance mechanism such as LTE/NR subcarrier alignment and PRB alignment, channel raster alignment to guarantee the validness of NR-SUL shared carrier, synchronization between LTE-UL and NR-SUL carriers for the TDM resource coordination and the corresponding time advance mechanism.

2.4.3.1   Subcarrier Alignment for LTE/NR Spectrum Sharing

LTE/NR subcarrier alignment on the shared carrier is required to avoid the extra overhead of the guardbands at each boundary between LTE UL transmission and NR SUL transmission in the shared spectrum.

As is well known in OFDM numerology design since 4G standardization period, subcarrier misalignment will result in severe performance degradation because of inter-subcarrier interference. To suppress the negative impact caused by inter-subcarrier interference, a hundreds of kHz guard band between any subcarrier-misaligned transmissions should be reserved. It has been evaluated that at least 180 kHz (equals to one LTE PRB) is needed as the guard band for low modulation order, i.e., QPSK. While for higher modulation order including 16QAM and 64QAM, more than 360 kHz (i.e., 2 LTE PRBs) are required for guard band in order to meet to the of out-of-band emission requirement [64]. The guard band will lead to a very inefficient transmission in the LTE/NR shared carrier. Consequently, subcarrier alignment between LTE and NR is a basic requirement in the LTE/NR UL sharing scenario.

To attain the subcarrier alignment between LTE UL and NR SUL carrier, there are two aspects to be taken into account:

- **Alignment of the numerology of subcarrier spacing of the shared carrier:**
  As mentioned in the previous section, UL sharing in the FDM manner has significant benefits compared with TDM. For the case that NR is with subcarrier spacing larger than 15 kHz as in LTE, hundreds of kHz guard band is needed to suppress inter-subcarrier interference between LTE and NR. While for the case that 15 kHz subcarrier spacing is used in NR, inter-subcarrier interference can be avoided if there is subcarrier alignment between LTE and NR and then the guard band is no longer needed.
- **Alignment of the subcarrier shift and DC position of the shared carrier:**
  In LTE, different subcarrier mapping rules are applied to downlink and uplink, where a half-tone shift is adopted to the uplink compared with that in downlink as shown in Fig. 2.89. More concretely, for the LTE downlink, the middle subcarrier is mapping on the carrier frequency and such subcarrier is nulled due to the severe distortion caused by the direct current used for up-conversion. While for the LTE uplink, the middle subcarrier is preferred to be maintained in order to ensure the single carrier property since DFT-S-OFDM waveform is used to obtain a low PAPR. Consequently, for LTE uplink, the half-tone shift can avoid the impact of signal distortion on the middle subcarrier and the distortion can be spread over adjacent multiple subcarriers.

In the early stage of NR discussion, a common subcarrier mapping rule was preferred for both downlink and uplink where the same subcarrier mapping rule as that in LTE downlink was adopted but without nulling the middle subcarrier. Such a common subcarrier mapping rule is beneficial to facilitate cross-link interference mitigation for flexible duplex with DL and UL (and/or backhaul link and D2D link) share the same carrier. Particularly, advanced receiver in baseband is feasible to cancel cross-link interference with this design.

When it comes to the UL sharing scenario between LTE and NR, NR UL subcarriers will be misaligned with LTE UL subcarriers, even when the LTE system is using 15 kHz subcarrier spacing, resulting in inter-subcarrier interference. Such inter-subcarrier interference will highly degrade the performance for both LTE and

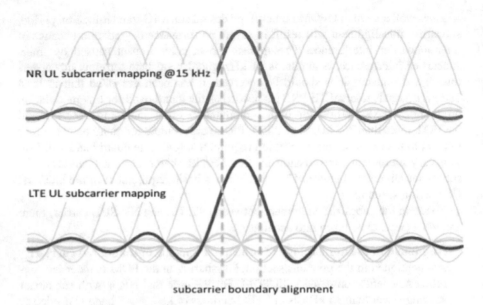

Fig. 2.89 Subcarrier alignment between NR and LTE for LTE-NR UL sharing

NR when they are multiplexed with each other in frequency division. Therefore, NR
introduces flexible configuration between the two types of subcarrier mapping rules
where network will indicate to UE whether to apply a shift of 7.5 kHz for UL transmission, which is different from LTE. The intention is that the UL subcarrier shift
of 7.5 kHz is only supported for the frequency bands that may be shared between
LTE and NR uplink. While for other frequency bands including all TDD bands, a
common subcarrier mapping rule without the 7.5 kHz shift was adopted for both
downlink and uplink.

Theoretically there are multiple implementation solutions at UE side to perform
the half-tone subcarrier spacing. The most used are basedband and RF shift as
shown in Fig. 2.90.

- In baseband shift, the half-tone shift in LTE uplink is achieved during the baseband signal generation. This solution can be straight forward to implement for
  NR to achieve the shift of 7.5 kHz with very low complexity.
- For RF shift, the 7.5 kHz shift is accomplished at the radio frequency (RF) via
  up-converting the baseband signal to the carrier frequency with an addition of
  7.5 kHz offset.

The baseband solution is more suitable than the latter one especially for LTE-NR
dual connectivity (EN-DC) UE. Because EN-DC UE already has a phase locked
loop (PLL) to lock on the carrier frequency for LTE UL. When the 7.5 kHz shift
is applied at the baseband, the EN-DC UE can share the same PLL for LTE UL
without adjusting the locked frequency. While for the RF solution, the UE has to
maintain two different carrier frequencies, and adjust the locked frequency when

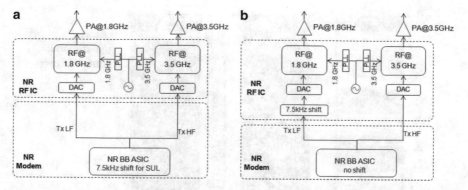

Fig. 2.90 UE implementation solutions of 7.5 kHz subcarrier shift for NR SUL carrier. (**a**) Baseband shift, same as LTE UL. (**b**) RF shift

transmission between LTE and NR UL is switched. Or it may need to implement two PLLs for the shared uplink carrier. Although baseband solution has shown advantage compared with RF solution, but considering it is an implementation issue up to the UE designer, both the options have been captured in the specification.

2.4.3.2   PRB Alignment for LTE/NR Spectrum Sharing

LTE/NR PRB alignment is necessary to reach a non-overhead scheduled resource participation between LTE-UL and NR-SUL transmissions in the shared carrier. Purely aligning subcarrier between LTE and NR, however, can only avoid interference between these two RATs. Scheduling constraint is still needed because the granularity of the frequency-domain resource allocation for NR and LTE are both in a physical resource block (PRB) level, each of which consists of 12 subcarriers. If the PRB grids of NR and LTE are not aligned, it would result in certain unused PRB within the carrier for all static, semi-static and dynamic sharing mechanisms. Furthermore, more number of guard bands are needed when UL sharing between LTE and NR because of more boundaries between LTE UL and NR SUL transmissions, as shown in Fig. 2.91. Such a large number of guard band will highly reduce the spectrum efficiency. PRB alignment between LTE and NR is therefore supported in UL sharing.

There are clear benefits of LTE/NR PRB alignment,

- The LTE/NR UL sharing transmission can be very efficient because additional overhead between LTE and NR in UL is not needed.
- The UL scheduling and network managements in both LTE and NR systems very easy with perfect PRB alignment between LTE UL carrier and NR SUL carrier. It removes the necessity of coordination on the scheduling units and reserved frequency domain resources.

To reach LTE/NR PRB alignment, both the NR scheduling unit of bandwidth part as well as the multi-numerology mechanisms are specified in 3GPP Release-15:

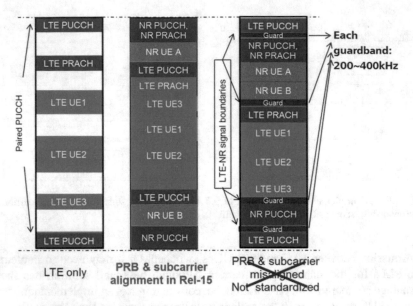

Fig. 2.91 LTE/NR UL sharing: PRB alignment vs. PRB misalignment

- **NR PRB scheduling unit of bandwidth part:**
  In NR, the basic scheduling unit is known as bandwidth part [40]. A bandwidth part is a subset of contiguous common resource blocks. Bandwidth parts can be used to allow the UE to operate with the bandwidths less that of the system bandwidth. Configuration of a bandwidth part is described in clause 12 of [38]. A UE can be configured with up to four different bandwidth parts in both DL and UL with a single downlink bandwidth part being active at a given time. The UE is not expected to receive PDSCH, PDCCH, or CSI-RS (except for RRM) outside an active bandwidth part. If a UE is configured with a supplementary uplink, the UE can in addition be configured with up to four bandwidth parts in the supplementary uplink with a single supplementary uplink bandwidth part being active at a given time. The UE does not transmit PUSCH or PUCCH outside an active bandwidth part. For an active cell, the UE does not transmit SRS outside an active bandwidth part.
- **Multiple Numerologies for a NR cell with a TDD carrier and a SUL carrier**
  As stated before, it is desirable that the same subcarrier spacing is used for the LTE UL carrier and the shared NR SUL carrier. Consequently, multiple OFDM numerologies are supported in Release-15.
  For data channels, subcarrier spacing of 15, 30, and 60 kHz are applicable to frequency range below 6 GHz, while subcarrier spacing 60 and 120 kHz are applicable to frequency range above 6 GHz, (Table 2.39). From that table, each operator can choose their preferred numerologies for both NR TDD carrier and NR SUL carrier in their deployment of the chosen SUL band combinations.

Table 2.39 Channel bandwidths for each NR band [57]

NR band	SCS (kHz)	NR band/SCS/UE Channel bandwidth											
		5 MHz	10 MHz[a,b]	15 MHz[b]	20 MHz[b]	25 MHz[b]	30 MHz	40 MHz	50 MHz	60 MHz	80 MHz	90 MHz	100 MHz
n1	15	Yes	Yes	Yes	Yes								
	30		Yes	Yes	Yes								
	60		Yes	Yes	Yes								
n2	15	Yes	Yes	Yes	Yes								
	30		Yes	Yes	Yes								
	60		Yes	Yes	Yes								
n3	15	Yes	Yes	Yes	Yes	Yes	Yes						
	30		Yes	Yes	Yes	Yes	Yes						
	60		Yes	Yes	Yes	Yes	Yes						
n5	15	Yes	Yes	Yes	Yes								
	30		Yes	Yes	Yes								
	60												
n7	15	Yes	Yes	Yes	Yes								
	30		Yes	Yes	Yes								
	60		Yes	Yes	Yes								
n8	15	Yes	Yes	Yes	Yes								
	30		Yes	Yes	Yes								
	60												
n12	15	Yes	Yes	Yes									
	30		Yes	Yes									
	60												
n20	15	Yes	Yes	Yes	Yes								
	30		Yes	Yes	Yes								
	60												

(continued)

Table 2.39 (continued)

NR band	SCS (kHz)	NR band/SCS/UE Channel bandwidth											
		5 MHz	10 MHz[a,b]	15 MHz[b]	20 MHz[b]	25 MHz[b]	30 MHz	40 MHz	50 MHz	60 MHz	80 MHz	90 MHz	100 MHz
n25	15	Yes	Yes	Yes	Yes								
	30		Yes	Yes	Yes								
	60		Yes	Yes	Yes								
n28	15	Yes	Yes	Yes	Yes								
	30		Yes	Yes	Yes								
	60												
n34	15	Yes	Yes	Yes									
	30		Yes	Yes									
	60		Yes	Yes									
n38	15	Yes	Yes	Yes	Yes								
	30		Yes	Yes	Yes								
	60		Yes	Yes	Yes								
n39	15	Yes	Yes	Yes	Yes	Yes	Yes	Yes					
	30		Yes	Yes	Yes	Yes	Yes	Yes					
	60		Yes	Yes	Yes	Yes	Yes	Yes					
n40	15	Yes	Yes	Yes	Yes	Yes	Yes	Yes	Yes				
	30		Yes	Yes	Yes	Yes	Yes	Yes	Yes	Yes	Yes		
	60		Yes	Yes	Yes	Yes	Yes	Yes	Yes	Yes	Yes		
n41	15		Yes	Yes	Yes			Yes	Yes				
	30		Yes	Yes	Yes			Yes	Yes	Yes	Yes	Yes	Yes
	60		Yes	Yes	Yes			Yes	Yes	Yes	Yes	Yes	Yes
n51	15	Yes											
	30												
	60												

Band	SCS (kHz)													
n66	15	Yes	Yes	Yes	Yes	Yes	Yes	Yes						
	30		Yes	Yes	Yes	Yes	Yes	Yes						
	60		Yes	Yes	Yes	Yes	Yes	Yes						
n70	15	Yes	Yes	Yes	Yes	Yes^c								
	30		Yes	Yes	Yes	Yes^c								
	60		Yes	Yes	Yes	Yes^c								
n71	15	Yes	Yes	Yes	Yes^c									
	30		Yes	Yes	Yes^c									
	60		Yes	Yes	Yes^c									
n75	15	Yes	Yes	Yes	Yes									
	30		Yes	Yes	Yes									
	60		Yes	Yes	Yes									
n76	15	Yes	Yes	Yes										
	30													
	60													
n77	15		Yes	Yes	Yes	Yes	Yes	Yes	Yes					
	30		Yes	Yes	Yes	Yes	Yes	Yes	Yes	Yes	Yes	Yes	Yes	Yes
	60		Yes	Yes	Yes	Yes	Yes	Yes	Yes	Yes	Yes	Yes	Yes	Yes
n78	15		Yes	Yes	Yes	Yes	Yes	Yes	Yes					
	30		Yes	Yes	Yes	Yes	Yes	Yes	Yes	Yes	Yes	Yes	Yes	Yes
	60		Yes	Yes	Yes	Yes	Yes	Yes	Yes					
n79	15							Yes	Yes					
	30							Yes	Yes	Yes		Yes		Yes
	60							Yes	Yes	Yes		Yes		Yes
n80	15	Yes	Yes	Yes	Yes	Yes	Yes							
	30		Yes	Yes	Yes	Yes	Yes							
	60		Yes	Yes	Yes	Yes	Yes							

(continued)

Table 2.39 (continued)

NR band	SCS (kHz)	NR band/SCS/UE Channel bandwidth											
		5 MHz	10 MHz[a,b]	15 MHz[b]	20 MHz[b]	25 MHz[b]	30 MHz	40 MHz	50 MHz	60 MHz	80 MHz	90 MHz	100 MHz
n81	15	Yes	Yes	Yes	Yes								
	30		Yes	Yes	Yes								
	60												
n82	15	Yes	Yes	Yes	Yes								
	30		Yes	Yes	Yes								
	60												
n83	15	Yes	Yes	Yes	Yes								
	30		Yes	Yes	Yes								
	60												
n84	15	Yes	Yes	Yes	Yes								
	30		Yes	Yes	Yes								
	60		Yes	Yes	Yes								
n86	15	Yes	Yes	Yes	Yes			Yes					
	30		Yes	Yes	Yes			Yes					
	60		Yes	Yes	Yes			Yes					

[a] 90% spectrum utilization may not be achieved for 30 kHz SCS
[b] 90% spectrum utilization may not be achieved for 60 kHz SCS
[c] This UE channel bandwidth is applicable only to downlink

Table 2.40 UL/SUL indicator

Value of UL/SUL indicator	Uplink
0	The non-supplementary uplink
1	The supplementary uplink

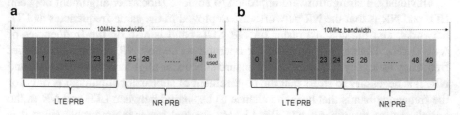

Fig. 2.92 PRB alignment between NR and LTE for LTE-NR UL sharing

2.4.3.3   Channel Raster for the NR-SUL Band

PRB alignment also requires the alignment of the carrier frequency (also refers to RF reference frequency in NR), which is supposed to be placed on the channel raster. In LTE, the channel raster is on the frequencies of $N*100$ kHz where N is an integer. While for NR, especially for the operating bands that can be coexisted with LTE, the corresponding channel raster is also defined on the frequencies of $N*100$ kHz. Thus, it is possible for the operator to deploy both LTE and NR on the same carrier frequency to ensure PRB alignment between LTE and NR. As an example in the Fig. 2.92, the frequency domain gap between LTE PRB and NR PRB is unavoidable when NR PRB is not aligned with LTE PRB, which also results in less available PRB for NR compared with the case of PRB alignment due to the guard band requirement for the given bandwidth.

The channel raster defines a set of RF reference frequencies which is used to identify the position of carriers. In other words, only the frequencies defined by channel raster are valid to deploy carriers where the corresponding carrier frequency must be on the channel raster. In LTE, the channel raster is on the frequencies of $N*100$ kHz where N is a positive integer. While during the discussion of channel raster for NR, companies think the channel raster defined in LTE is rather sparse which may cause severe restriction for operators to deploy networks especially for the operators having few spectrum. Hence, it was proposed to apply the most flexible design where the channel raster is on the frequencies of $N*15$ kHz for all frequency ranges, thus operators can deploy the network more flexibly.

But from the perspective of UL sharing between LTE and NR, such channel raster design for NR will seldom achieve subcarrier alignment between LTE and NR though a 7.5 kHz shift is applied for NR UL. More concretely, subcarrier alignment can only be obtained for the cases that the carrier frequency of LTE uplink is on multiple of 300 kHz. That is to say only one-third of frequencies within LTE channel raster can be used for deployment of UL sharing, which is a tremendous restriction on the usage of UL sharing. Besides, considering that LTE networks have already been deployed in certain frequencies, it is not expected to affect current LTE network when NR shares uplink frequencies with LTE. Otherwise, operator may have to re-deploy current LTE networks whose carrier frequency is not on multiple of 300 kHz.

Obviously, a straightforward approach to achieve subcarrier alignment between LTE and NR is that the NR networks are deployed in the same frequencies as LTE networks, i.e., the channel raster for NR SUL band is also on the frequencies of $N*100$ kHz. Hence, it is possible for operator to deploy NR on the same carrier frequency as legacy LTE network to ensure subcarrier alignment between LTE and NR. It is necessary to stress that the requirement of subcarrier alignment is only for the frequency bands that have the chance to be shared between LTE and NR in the uplink. As for the other bands, the 15 kHz channel raster is preferable since it is more flexible. Owing to the aforementioned reasons, both 100 and 15 kHz channel raster are supported in NR but for different operating bands. Particularly, the 100 kHz channel raster is supported for most operating bands below 3 GHz.

2.4.3.4   Synchronization and Timing for LTE/NR UL Sharing

Synchronization is crucial for both NR SUL band combination and LTE/NR spectrum sharing. For both LTE/NR TDM and FDM semi-static and dynamic sharing, symbol synchronization is required at the BS receiver for the shared carrier in both LTE and NR system, to avoid inter-subcarrier interference and thus avoid the potential guardband between the resources allocated to LTE UL transmission and NR SUL transmission, which is independent of the network perspective sharing or UE perspective sharing. In addition, subframe/slot level synchronization is also necessary for the sharing coordination on the per-subframe resource allocation between the UL carrier in the LTE cell and the shared SUL carrier in the NR cell. Consequently, Release-15 specified the corresponding time advance mechanism for NR SUL band combinations, as well as the enhancement on LTE HARQ timing configuration to guarantee the efficient LTE DL transmission that is associated to the shared carrier.

2.4.3.4.1   Synchronization Requirements Between LTE UL and NR SUL Cells

For NR TDD network, synchronization among different gNBs should be maintained in order to avoid co-channel interference. Cell-level synchronization accuracy for TDD is defined as the maximum absolute deviation in frame start timing between any pair of cells on the same frequency that have overlapping coverage areas. The cell-level synchronization accuracy measured at BS antenna connectors needs to be better than 3 μs, which is the minimum requirement defined in the 3GPP specification.

As for the scenario of LTE and NR UL sharing, synchronization is required at the network side for the LTE UL cell and NR UL cell that serve at least one NR UE with

SUL transmission, to maintain high spectrum efficiency and reduce implementation complexity at gNB side. There are two levels of synchronization requirements:

- **Symbol-level synchronization** is required with both FDM-based and TDM-based LTE/NR spectrum sharing: Taking FDM-based resource sharing as an illustrative example, gNB only needs to perform one reception, e.g., one FFT operation is needed to obtain baseband signal when LTE UL and NR UL are both time and frequency synchronized. Otherwise, the gNB will need to perform separate signal reception for both LTE and NR. While for TDM-based resource sharing, if LTE UL and NR UL are not synchronized, symbols between adjacent LTE and NR subframes would overlapped with each other which cause inter-symbol interference between the two RATs. Guard period with puncturing the corresponding symbols may be reserved to suppress the inter-symbol interference, leading to time-domain resource reduction. LTE and NR is, therefore, required to be synchronized with each other at the symbol level in the shared UL frequency.

- **Subframe/Slot-level synchronization** is mandated for TD-based LTE/NR spectrum sharing, which is also beneficial to FDM-based LTE/NR spectrum sharing because it simplifies scheduling and receiver implementations. As addressed in the previous section, LTE/NR TDM sharing schedules different subframes/slots to different users, and also enables an NR UE to switch between LTE UL transmission mode and NR SUL transmission mode from slot to slot. If the LTE UL cell and the NR SUL cell are asynchronous in the subframe/slot boundary, while both of them are the serving cells for of a certain UE, it'd be hard to schedule the subframe/slot after switching point in either systems for the UL transmission from that UE. Such a restriction on scheduling of the subframe/slot at the boundary would result in a significant overhead loss. To resolve this issue, Release-15 defines a clear guidance of the subframe/slot-level synchronization requirements to align the boundary of the subframes of LTE UL carrier and NR SUL carrier for the LTE cell and NR cell, which are usually provided in collocated base station, when they are sharing the same UL spectrum.

2.4.3.4.2 Timing Advance Mechanisms for LTE/NR UL Sharing

Since time synchronization between LTE UL and NR SUL, and between two ULs for NR are both achieved at network side, timing synchronization for LTE UL and the two ULs of NR at UE side is also needed. Particularly for the two ULs of NR, time synchronization can be obtained by using a single timing adjustment command and defining a same value for TA offset, as described in [38, 65]. As for time synchronization between the LTE UL and NR UL on the shared frequency, it can be achieved through network implementation.

As shown in Fig. 2.93, the same time advance is used for NR TDD UL carrier and NR SUL carrier for a single NR UE, whose transmit time at UE side is $TA_{UE,NR-TDD}$

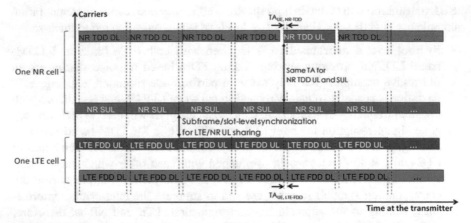

Fig. 2.93 Subframe/Slot-level synchronization for LTE/NR UL sharing

time ahead of the slot boundary at the NR base station transmitter. At the same time the LTE/NR sharing requires the that the boundary of LTE UL subframes and NR UL slots for the TDM sharing are synchronized. To achieve this, the UE at EN-DC mode would transmit the LTE FDD UL signals at the time raster that is multiples of milli-seconds time shift of its transmission of NR-SUL signals. The LTE system has an individual time advance command $TA_{UE,LTE-FDD}$ for the same UE who is working at EN-DC mode. Generally for a co-located deployment of LTE and NR EN-DC, $TA_{UE,NR-TDD}$ and $TA_{UE,LTE-FDD}$ are most likely similar to each other. Thus the LTE FDD DL cell and NR TDD DL cell working jointly to serve the same group of UEs are synchronized in subframe/slot level for the co-site LTE/NR deployment base stations.

It is necessary to note that the time synchronization among different cells or sites in LTE FDD system is not required, and thus different LTE sites may have different timings on the shared UL frequency. Since LTE UL and NR UL are required to be time synchronized on the shared UL frequency, any asynchrony would result in asynchrony among the NR ULs on shared frequency from different sites, i.e., time-domain mismatching (i.e., non-synchronization) between signal and inter-cell interference . Release-15 focused on the scenarios that the LTE-UL cell and the shared NR-SUL cell are collocated, where single time advance command is enough for a NR UE. Release-16 will enhance the LTE/NR sharing to support the non-collocated scenario by providing multiple time advance commands for NR TDD UL carrier and NR SUL carrier, respectively. Note that, for the case that two UL signals at a NR base station receiver are unsynchronized, the gNB is also required to perform separate reception for the UL on TDD frequency and the UL on the shared frequency. While only one time of reception is needed for synchronization case. It now become clear that synchronization between UL on TDD frequency and UL on shared frequency for NR is beneficial to simplify the gNB implementation.

2.4.3.4.3 NSA LTE/NR UL Sharing TDM Configuration for HARQ Timing

Particularly for NSA deployment from the UE perspective, the UE can access both the LTE and NR networks which works in E-UTRA and NR dual connectivity (EN-DC) mode. Focusing on the resource utilization on the shared UL frequency, both FDM-based and TDM-based schemes can be used. For FDM-based resource sharing, the UE can assume that both the LTE UL and NR UL could fully occupy all the UL resource within the shared UL frequency and the network can avoid resource overlapping via scheduling. As for TDM-based resource sharing, the UE will transmit UL using different radio access technology, either LTE or NR, in different UL subframes. Given that the HARQ timing in LTE is not as flexible as that in NR, semi-static TDM-based resource sharing solution is much better than dynamic approach. Moreover, purely adopting subframe separation between LTE and NR will further degrade the DL data rate at the UE side due to the fact that couples of DL subframes cannot be used for DL data transmission because the associated UL subframes used for HARQ transmission have been assigned for NR transmission. To avoid such performance degradation, case 1 HARQ timing has been introduced for EN-DC UE to facilitate the usage of all DL subframes at the LTE side for DL data transmission, as in Fig. 2.94.

In summary, to achieve high spectrum efficiency in LTE/NR sharing scenario, the following mechanisms are required for the shared LTE UL carrier and NR-SUL carrier:

- On the frequency domain both subcarrier and PRB should be aligned between LTE and NR so that all the frequency-domain resource within the carrier can be fully utilized, without additional interference. In addition, the channel raster of NR-SUL carrier should be able to match with the LTE UL carriers in currently deployed commercial networks in different countries.
- On the time domain, network synchronization at both symbol-level and sub-frame/slot-level between LTE UL carrier and NR SUL need to be supported for TDM sharing of the radio resources. The aforementioned requirements are needed for LTE/NR sharing from both network perspective and UE perspective.

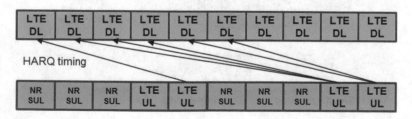

Fig. 2.94 Example of case 1 HARQ timing

2.4.4   NR SUL Band Combinations: Uplink Carrier Selection and Switching

2.4.4.1   Single-Cell Concept

5G-NR introduces a new band type of SUL, at the same time also introduces a new cell structure correspondingly. Cells are the basic elements of a mobile network. The basic logical element of UL/DL decoupling is one cell configured with two uplinks and one downlink. As shown in Fig. 2.95, a typical deployment for a SUL band combination is that every cell has one radio DL carrier associated with two collocated radio UL carriers, one in the normal TDD or FDD band and another in SUL band.

The single cell structure accommodates dynamic scheduling and carrier switching between SUL and the normal UL without any interruption time. It facilitates SUL to be very compatible to carrier aggregation and to maximally reuse L2/L3 designs of single cell. The UE is configured with two ULs for one DL of the same cell, and uplink transmissions on those two ULs are controlled by the network to avoid overlapping PUSCH/PUCCH transmissions in time for the best uplink power concentration. Overlapping transmissions on PUSCH are avoided through scheduling while overlapping transmissions on PUCCH are avoided through configuration, i.e., PUCCH can only be configured for only one of the two ULs of the cell [33, 37].

2.4.4.2   UL Carrier Selection and Switch

As discussed in Sect. 2.4.1, UL/DL decoupling enlarges cell coverage by introducing additional lower uplink frequency to alleviate the coverage bottleneck. Considering the uplink bandwidth of lower frequency is usually smaller than that of

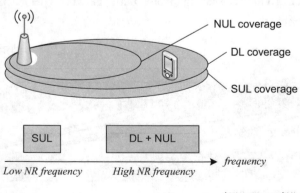

Fig. 2.95 Single cell with two uplinks of two different frequencies. *NUL* normal UL

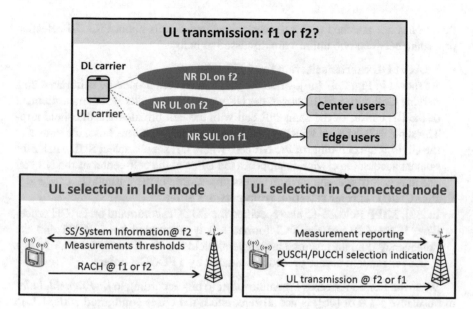

Fig. 2.96 Motivations for uplink selection

higher frequency, e.g., 20 MHz at 1.8 GHz vs. 100 MHz at 3.5 GHz, better uplink throughput can be provided by higher uplink frequency to those UEs who are close enough to a base station or not suffering high large-scale propagation loss on high-frequency uplink. Therefore, UL/DL decoupling should enable a flexible network operation that UEs can access to network on lower frequency f1 for uplink when they are at the cell-edge while on the higher frequency f2 for uplink when they are at the cell-center, as shown in Fig. 2.96. UEs will select the best uplink of the cell based on downlink measurement in both idle mode and connected mode.

Additionally, even for the cell-edge UEs having data uplink transmission on the lower frequency f1, it is necessary to occasionally transmit sounding reference signal (SRS) on the higher frequency f2 to have the reciprocal channel information for DL massive MIMO. This also requires uplink selection and switching. This section discusses the mechanism of UE uplink selection and uplink switching.

2.4.4.2.1   Idle Mode UL Selection: Initial Access with PRACH

To ensure the coverage for initial access, a UE may be configured with PRACH resources on both SUL and NUL (normal uplink). In connected mode, the base station can indicate the selected uplink by DCI to UE for PRACH transmission. However, in idle mode, the uplink for PRACH has to be selected by the UE because the UE has not attached to the network yet. More details about PRACH selection procedure between two uplinks are discussed as below.

- **PRACH UL carrier selection from UE side:**
  As shown in Fig. 2.97 [66], when the Random Access procedure is initiated on a Serving Cell with a SUL carrier, the UE would compare its RSRP measurement of the DL carrier of the same NR cell with the cell broadcasted threshold rsrp-ThresholdSSB-SUL. If RSRP measurement is below the threshold, it means that the UE is at the cell-edge of the NR cell, where UE should select SUL carrier for random access. Otherwise, it supposes the UE is at the cell-center of the NR cell and thus will trigger the random access procedure via the normal UL carrier.
- **PRACH UL carrier selection from network side:**
  In [39], 3GPP Release-15 also specified the PDCCH command on RACH procedure. If the CRC of the DCI format 1_0 is scrambled by C-RNTI and the "Frequency domain resource assignment" field are of all ones, the DCI format 1_0 is for random access procedure initiated by a PDCCH order.

When the Random Access Preamble index (6 bits according to *ra-PreambleIndex* in Subclause 5.1.2 of [66]) is not all zero, and if the UE is configured with SUL in the cell, the following indicators are valid:

- UL/SUL indicator (1 bit) indicates which UL carrier in the cell to transmit the PRACH;
- SS/PBCH index (6 bits) indicates the SS/PBCH that shall be used to determine the RACH occasion for the PRACH transmission;
- PRACH Mask index (4 bits) indicates the RACH occasion associated with the SS/PBCH indicated by "SS/PBCH index" for the PRACH transmission, according to [66];

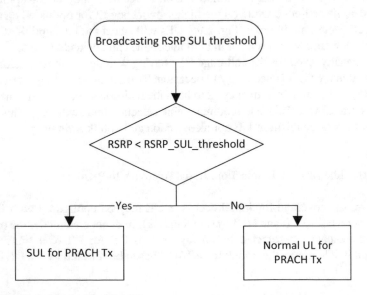

Fig. 2.97 Uplink selection for PRACH

2.4.4.2.2   Connected Mode UL Selection: PUSCH/PUCCH Scheduling

Figure 2.98 shows that a UE can be served by a SUL capable cell with two uplink configurations.

If both normal uplink and SUL are configured within a cell, which uplink to transmit is determined by UL/SUL indication (Fig. 2.99) in the DCI [39] with either Fallback DCI 0-0 or Non-fallback DCI 0-1. The fallback DCI is designed to be compatible to both SUL capable UEs and SUL incapable UEs so that a SUL capable cell can serve SUL incapable UEs who may be roaming from other countries or other carrier network.

The mobility functions including measurement, cell selection, reselection and handover for the SUL capable cell are the same as those for a normal cell. The main difference is only that uplink selection is performed after the cell selection process is completed.

- **UL Data transmission: PUSCH scheduling mechanisms**
  For the UL data transmission, dynamic PUSCH selection is based on DCI 0_0 and DCI 0_1, with the switching time between SUL and non-SUL 0 or 140 us, as specified in Table 5.2C-1 of [57], Table 5.2B.4.2-1 of [60]. Uplink transmission faces many aspects of challenges, such as uplink power control, timing advance, HARQ feedback, and resource multiplexing, which are discussed below.
  In the uplink, the gNB can dynamically allocate resources to UEs via the C-RNTI on PDCCH(s). A UE always monitors the PDCCH(s) in order to find possible grants for uplink transmission when its downlink reception is enabled (activity

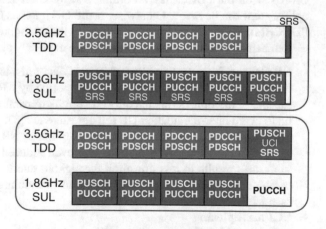

Fig. 2.98 Exemplary Uplink configurations for a cell with SUL

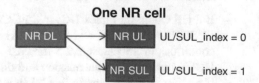

Fig. 2.99 UL/SUL indication for the cell configured with two uplinks

governed by DRX when configured). When CA is configured, the same C-RNTI applies to all serving cells.

- **Dynamic UL/SUL selection:** DCI format 0_0 is used for the dynamic scheduling of PUSCH in one cell [39] based on UL/SUL indicator, where "0" means the selection of normal UL carrier and "1" means the selection of the SUL carrier.
- **Semi-static UL/SUL selection:** In case of the "UL/SUL indicator" is not present, i.e., the number of bits for DCI format 1_0 before padding is not larger than the number of bits for DCI format 0_0 before padding, high-layer RRC configuration on UL/SUL selection for PUSCH transmission is done following the high layer parameter *pucch-Config*.

In addition, with Configured Grants, the gNB can allocate uplink resources for the initial HARQ transmissions to UEs. Two types of configured uplink grants are defined:

- With Type 1, RRC directly provides the configured uplink grant (including the periodicity).
- With Type 2, RRC defines the periodicity of the configured uplink grant while PDCCH addressed to CS-RNTI can either signal and activate the configured uplink grant, or deactivate it; i.e., a PDCCH addressed to CS-RNTI indicates that the uplink grant can be implicitly reused according to the periodicity defined by RRC, until deactivated.

When a configured uplink grant is active, if the UE cannot find its C-RNTI/ CS-RNTI on the PDCCH(s), an uplink transmission according to the configured uplink grant can be made. Otherwise, if the UE finds its C-RNTI/CS-RNTI on the PDCCH(s), the PDCCH allocation overrides the configured uplink grant.

Retransmissions other than repetitions are explicitly allocated via PDCCH(s).

- **UL Control Information Transmission Mechanisms** [12, 20]
  Single uplink operation for EN-DC is supported by NR Release-15 where UE is capable of multiple uplink but only one uplink transmission at one time. It is useful to avoid inter-modulation UE self-interference [37, 60], and to achieve maximum transmission power for either LTE uplink or NR uplink when a UE is not capable of dynamic power sharing. However, reduced number of LTE PUCCH subframes results in less downlink throughput due to lack of HARQ feedback occasion for synchronous HARQ downlink scheduling. Therefore, LTE HARQ case 1 is introduced.
- **UCI multiplexing**
  In Release-15, UE must support UCI multiplexing into PUSCH [38].

  - If a UE would multiplex UCI in a PUCCH transmission that overlaps with a PUSCH transmission, and the PUSCH and PUCCH transmissions fulfill the conditions in Subclause 9.2.5 for UCI multiplexing, the UE multiplexes the UCI in the PUSCH transmission and does not transmit the PUCCH.
  - If a UE multiplexes aperiodic CSI in a PUSCH and the UE would multiplex UCI in a PUCCH that overlaps with the PUSCH, the UE multiplexes the UCI in the PUSCH.

- If a UE transmits multiple PUSCHs in a slot on respective serving cells that include first PUSCHs that are in response to detection by the UE of DCI format(s) 0_0 or DCI format(s) 0_1 and second PUSCHs configured by respective higher layer parameters *ConfiguredGrantConfig*, and the UE would multiplex UCI in one of the multiple PUSCHs, and the multiple PUSCHs fulfill the conditions in Subclause 9.2.5 for UCI multiplexing, the UE multiplexes the UCI in a PUSCH from the first PUSCHs.
- If a UE transmits multiple PUSCHs in a slot in response to detection by the UE of DCI format(s) 0_0 or DCI format(s) 0_1 on respective serving cells and the UE would multiplex UCI in one of the multiple PUSCHs and the UE does not multiplex aperiodic CSI in any of the multiple PUSCHs, the UE multiplexes the UCI in a PUSCH of the serving cell with the smallest *ServCellIndex* subject to the conditions in Subclause 9.2.5 for UCI multiplexing being fulfilled. If the UE transmits more than one PUSCHs in the slot on the serving cell with the smallest *ServCellIndex* that fulfill the conditions in Subclause 9.2.5 for UCI multiplexing, the UE multiplexes the UCI in the PUSCH that the UE transmits first in the slot.

2.4.4.3 SRS Switching

It has been observed that SRS on the TDD uplink can to a large extent enhance its corresponding downlink massive MIMO performance by enabling accurate reciprocal channel information. For the existing UE categories, the typical CA capable UEs only support one or two uplink carriers. Increasing supported number of uplink carriers means higher UE costs, bigger size, and more power consumption. Therefore, an economic practical solution is SRS switching where the same set of UE uplink hardware are shared by different multiple uplink carriers.

- **SRS switching between SUL and normal uplink:**
  Since hardware sharing between different uplink transmissions, switching time may not be zero, which is determined case by case. For a single serving cell configured with SUL, UE is capable of zero us uplink carrier switching time for SRS switching between SUL and NUL, as shown in Table 5.2C-1 of [57]. If the serving cell with SUL is configured with EN-DC and both the SUL and the LTE uplink carrier frequencies are the same, UE is also capable of zero us switching time between SUL and NUL. In case of EN-DC with different carrier frequencies between SUL and LTE uplink, 140 μs switching time is applied to allow RF retuning, as shown in Table 5.2B.4.2-1 of [60].
  For those cases with zero switching time, none performance loss is expected. However, for the case with non-zero switching time, the scheduler at base station has to take it into account and some uplink throughput loss may be inevitable.
  To trigger the SRS switching between SUL and NUL, an UL/SUL indicator field is contained in the SRS request field for both DCI downlink grant and DCI uplink grant [39].

- **SRS switching between LTE uplink and SUL**

- **Switching time ~0 & <20 μs:** S6.3B.1.1 of [60].

- **Carrier aggregation with SRS carrier switching**
  In mobile networks, there are many kinds of downlink heavier traffic, which leads to more number of aggregated downlink component carriers (CC) than the number of (aggregated) uplink CCs. UE generally has the capability of aggregating larger number of DL carriers than that in the UL. As a result, some of TDD carriers with DL transmission for the UE will have no UL transmission including SRS, and channel reciprocity cannot be utilized for these carriers. Allowing fast carrier switching to and between TDD UL carriers becomes a solution to allow SRS transmission on these TDD carriers.
  A UE can be configured with SRS resource(s) on a carrier c_1 with slot formats comprised of DL and UL symbols and not configured for PUSCH/PUCCH transmission. For carrier c_1, the UE is configured with higher layer parameter *srs-SwitchFromServCellIndex* and *srs-SwitchFromCarrier* the switching from carrier c_2 which is configured for PUSCH/PUCCH transmission. During SRS transmission on carrier c_1 (including any interruption due to uplink or downlink RF retuning time [65] as defined by higher layer parameters *rf-RetuningTimeUL* and *rf-RetuningTimeDL*), the UE temporarily suspends the uplink transmission on carrier c_2 [44].

2.4.4.4 Power Control

Ideally, there should be no inter-cell nor inter-user interference in cellular mobile network. However, in practical network there are. For example, the uplink signals of one UE camping in one cell interferes other UEs's uplink reception in the neighboring cell. The higher transmit power of the uplink signal usually means better receiving quality but also may saturate the receiver and results in higher inter-cell interference. To combat this challenge, power control mechanism provides adaptation of dynamic transmit power to the UE environment by measuring the UE receiving SINR and signaling the UE appropriate transmit power.

- **PC for EN-DC**
  If a UE is configured with a MCG using E-UTRA radio access and with a SCG using NR radio access, the UE is configured a maximum power P_{LTE} for transmissions on the MCG by higher layer parameter *p-MaxEUTRA* and a maximum power P_{NR} for transmissions on the SCG by higher layer parameter *p-NR*. The UE determines a transmission power for the MCG as described in [38] using P_{LTE} as the maximum transmission power. The UE determines transmission power for the SCG as described Subclauses 7.1 through 7.5 of [38] using P_{NR} as the maximum transmission power for $P_{CMAX} \leq P_{NR}$.
  If a UE is configured with $\hat{P}_{LTE} + \hat{P}_{NR} > \hat{P}_{Total}^{EN\text{-}DC}$, where $\hat{P}_{LTE}$ is the linear value of

P_{LTE}, $\hat{P}_{\text{NR}}$ is the linear value of P_{NR}, and $\hat{P}_{\text{Total}}^{\text{EN\_DC}}$ is the linear value of a configured maximum transmission power for EN-DC operation as defined in [60] for frequency range 1, the UE determines a transmission power on the SCG as follows.

– If the UE is configured with reference TDD configuration for EUTRA (by higher layer parameter *tdm-PatternConfig-r15* in [38]).
– If the UE does not indicate a capability for dynamic power sharing between EUTRA and NR, the UE is not expected to transmit in a slot on the SCG when a corresponding subframe on the MCG is an UL subframe in the reference TDD configuration; at the same time, the UE is expected to be configured with reference TDD configuration for EUTRA (by higher layer parameter tdm-PatternConfig-r15 in [12].
– If the UE indicates a capability for dynamic power sharing between EUTRA and NR and if the UE is not configured for operation with shortened TTI and processing time on the MCG [12], and if the UE transmission(s) in subframe i_1 of the MCG overlap in time with UE transmission(s) in slot i_2 of the SCG, and if $\hat{P}_{\text{MCG}}(i_1) + \hat{P}_{\text{SCG}}(i_2) > \hat{P}_{\text{Total}}^{\text{EN\_DC}}$ in any portion of slot i_2 of the SCG, the UE reduces transmission power in any portion of slot i_2 of the SCG so that $\hat{P}_{\text{MCG}}(i_1) + \hat{P}_{\text{SCG}}(i_2) \leq \hat{P}_{\text{Total}}^{\text{EN\_DC}}$ in any portion of slot i_2, where $\hat{P}_{\text{MCG}}(i_1)$ and $\hat{P}_{\text{SCG}}(i_2)$ are the linear values of the total UE transmission powers in subframe i_1 of the MCG and in slot i_2 of the SCG, respectively.
– **PC for Physical uplink shared channel** [38]
  If a UE transmits a PUSCH on UL BWP b of carrier f of serving cell c using parameter set configuration with index j and PUSCH power control adjustment state with index l, the UE determines the PUSCH transmission power $P_{\text{PUSCH},b,f,c}(i,j,q_d,l)$ in PUSCH transmission occasion i as

$$P_{\text{PUSCH},b,f,c}(i,j,q_d,l) = \min \begin{cases} P_{\text{CMAX},f,c}(i), \\ P_{\text{O\_PUSCH},b,f,c}(j) + 10\log_{10}\left(2^{\mu} \cdot M_{\text{RB},b,f,c}^{\text{PUSCH}}(i)\right) \\ +\alpha_{b,f,c}(j) \cdot PL_{b,f,c}(q_d) - \Delta_{\text{TF},b,f,c}(i) + f_{b,f,c}(i,l) \end{cases} [\text{dbm}]$$

• **PC for Physical uplink control channel**
  If a UE transmits a PUCCH on active UL BWP b of carrier f in the primary cell c using PUCCH power control adjustment state with index l, the UE determines the PUCCH transmission power $P_{\text{PUCCH},b,f,c}(i,q_u,q_d,l)$ in PUCCH transmission occasion i as

$$\begin{aligned} &P_{\text{PUCCH},b,f,c}(i,q_u,q_d,l) \\ &= \min \begin{cases} P_{\text{CMAX},f,c}(i), \\ P_{\text{O\_PUCCH},b,f,c}(q_u) + 10\log_{10}\left(2^{\mu} \cdot M_{\text{RB},b,f,c}^{\text{PUCCH}}(i)\right) \\ +PL_{b,f,c}(q_d) + \Delta_{\text{F\_PUCCH}}(F) + \Delta_{\text{TF},b,f,c}(i) + g_{b,f,c}(i,l) \end{cases} [\text{bdm}] \end{aligned}$$

- **TPC for SRS**
  TPC for SRS are indicated by three kinds of DCI, DCI UL grant, DCI format 2_2 and format 2_3 [38, 39].

2.4.5   4G/5G DL Spectrum Sharing Design

NR and LTE can share the spectrum resource not only in UL but also in DL. Both FDM and TDM can be used. One of the most important issues is ICI (inter-carrier interference) avoidance between the two systems.

Note that the same waveform with LTE DL, i.e., CP-OFDM, is applied in NR DL. If deployed in LTE band, NR can use the same SCS (subcarrier space) and CP (cyclic prefix) with LTE. Also, NR and LTE have the same subframe length (1 ms) and the same number of OFDM symbols (14) within a subframe. These allow ICI avoidance between NR and LTE, since the subcarriers and symbol boundaries of the two system can be aligned.

Moreover, downlink resource sharing between NR and LTE can benefit from the mechanisms below.

2.4.5.1   Rate Matching Around CRS

NR supports rate matching around LTE CRS (cell-specific reference signal). In LTE, CRS will be sent in every effective DL subframes and DwPTS within the effective bandwidth. Such "always-ON" reference signal will be applied to LTE UEs for CRS-based data demodulation and channel estimation, etc. So, for NR, when coexists with LTE in DL spectrum, it is necessary to support not mapping DL data in the REs (resource elements) that carrying LTE CRS, i.e., rate matching around CRS.

The following IEs (information elements) will be informed to the NR UE to determine the location of CRS:

(a) Central frequency of LTE carrier
(b) DL bandwidth of LTE carrier
(c) Number of LTE CRS port
(d) v-shift parameter of LTE CRS
(e) MBSFN subframe configuration of LTE (optional)

Configured with these IEs, NR UE is aware of the REs that carrying CRS, which is sent by LTE. Then, the NR UE will not receive DL data from such REs. An example is shown in Fig. 2.100.

Note that rate matching around LTE CRS is only supported for NR PDSCH. For other channels, e.g., PDCCH, SSB (synchronization signal block), rate matching around CRS is not supported.

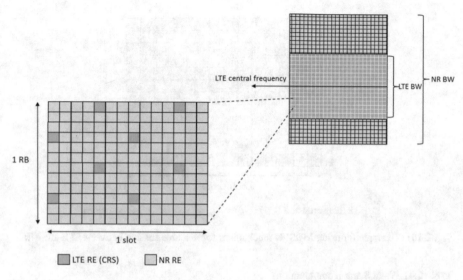

Fig. 2.100 Rate matching around LTE CRS (e.g., 1 CRS port, no MBSFN subframes)

2.4.5.2   MBSFN Type Sharing

MBSFN (Multimedia Broadcast multicast service Single Frequency Network) related mechanism in LTE can be applied to the DL sharing scenario. In LTE, if a subframe is configured as MBSFN subframe, the subframe is divided into two regions: non-MBSFN region and MBSFN region. The MBSFN region spans the first one or two OFDM symbols in an MBSFN subframe. The LTE UE is not expected to receive PDSCH or PDCCH in MBSFN region.

Originally, MBSFN region in the MBSFN subframes is used for PMCH transmission. When NR and LTE coexist, MBSFN mechanism and configuration can be reused for resource sharing. The MBSFN region can be utilized for NR DL transmission, where no LTE PDCCH, LTE PDSCH, LTE CRS will be transmitted, as shown in Fig. 2.101.

Note that the MBSFN configuration in LTE can be transparent to NR UEs. The NR base station can avoid scheduling NR UE in the non-MBSFN region totally up to implementation. This is not the same with the case of rate matching around LTE CRS.

2.4.5.3   Mini-Slot Scheduling

In NR, non-slot-based scheduling is supported both in DL and UL, where only several OFDM symbols within a slot rather than a whole slot will be scheduled. This is also known as mini-slot scheduling. The valid starting symbol and symbol length is given in Table 2.41 [44], as copied below.

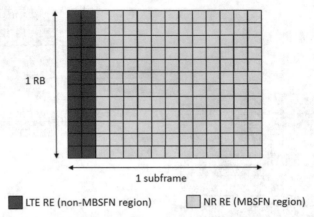

1 subframe

■ LTE RE (non-MBSFN region)   □ NR RE (MBSFN region)

Fig. 2.101 Example for using MBSFN mechanism for resource sharing between LTE and NR

Table 2.41 Valid S and L combinations

PDSCH mapping type	Normal cyclic prefix			Extended cyclic prefix		
	S	L	$S + L$	S	L	$S + L$
Type A	$\{0,1,2,3\}^a$	$\{3,\ldots,14\}$	$\{3,\ldots,14\}$	$\{0,1,2,3\}^a$	$\{3,\ldots,12\}$	$\{3,\ldots,12\}$
Type B	$\{0,\ldots,12\}$	$\{2,4,7\}$	$\{2,\ldots,14\}$	$\{0,\ldots,10\}$	$\{2,4,6\}$	$\{2,\ldots,12\}$

$^aS = 3$ is applicable only if *DM-RS-TypeA-Posiition* = 3

In the above table, S denotes the index of starting symbol, and L denotes the scheduled length of symbols. Mini-slot scheduling can be used in the LTE-NR coexistence scenario. For example, the NR UE can be scheduled in the OFDM symbols that not overlapped with LTE CRS. For another example, the NR UE can be scheduled in the OFDM symbols that within the MBSFN region from LTE's point of view.

2.4.5.4   SS SCS Definition for Coexisting Bands

In NR, SSB includes PSS, SSS, and PBCH. An SSB occupies four consecutive OFDM symbols in time domain and 20 RBs in frequency domain. The NR UE can obtain key system information from SSB, e.g., PCID (physical cell ID) and SFN (system frame number).

SSB is critical for initial access and DL channel measurement. It is important to avoid interference from LTE to NR SSB. In LTE, the PSS, SSS, and PBCH are transmitted in the central 6 RBs of the system bandwidth. In NR, the frequency location of SSB is more flexible and is not necessary to be located in the central bandwidth. When coexisting with LTE in DL, NR SSB can avoid interfering with LTE PSS, SSS, and PBCH by FDM manner. The only thing that should be considered is the interference from LTE CRS, since CRS is always on in non-MBSFN region.

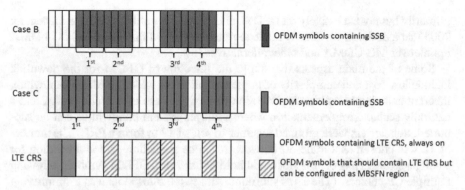

Fig. 2.102 SSB candidates for SCS = 30 kHz

When coexist with LTE in DL (below 3 GHz), NR may use SSB with 30 kHz SCS. NR supports two different SSB candidate patterns when SCS = 30 kHz, i.e., Case B and Case C. In Case B, the first symbols of the candidate SS/PBCH blocks have indexes {4, 8, 16, 20} + 28∗n within a half flame, where n = 0 for carrier frequencies smaller than or equal to 3 GHz (major LTE-NR coexistent band). In Case C, the first symbols of the candidate SS/PBCH blocks have indexes {2, 8} + 14∗n within a half frame, where n = 0, 1 for carrier frequencies smaller than or equal to 3 GHz. The following figure shows the SSB candidates of these two cases, where LTE CRS (four ports) is also illustrated for comparison (Fig. 2.102).

As can be seen from the above figure, even in the worst case (four CRS ports, no MBSFN configuration), Case B still have one SSB candidate (the first one) that not overlapped with any OFDM symbols containing CRS. This guarantees that at least one SSB can be delivered without interfering with LTE CRS. On the other hand, if MBSFN is configured to LTE, all SSB candidates in Case B can be transmitted without interfering with LTE CRS. This provides maximum flexibility of SSB transmission.

2.5   5G-NR New Physical Layer Technologies

2.5.1   Waveform and Multiple Access

OFDM has become a proven success in LTE, but it took some years to be considered mature enough to be selected. OFDM in the ETSI concept group gamma lost out to WCDMA in June 1998 for UMTS. Interest in OFDM did not diminish, however, as seen in various field tests (e.g., [67]) and further study in 3GPP from 2002 to 2004 [68]. The primary track in 3GPP after WCDMA was HSPA, with study approved in March 2000 [69][70], leaving OFDM to be developed in IEEE802.16d/e (WiMAX). For EUTRA, the LTE study approved in RAN#26 in December 2004 focused primarily but not exclusively on OFDM, and it was not until RAN#30 in December 2005 that a decision was firmly made to purse an OFDM downlink and SC-FDMA uplink over MC-CDMA and other alternatives [8].

Some of the main aspects that led to the selection of OFDM for the downlink include low cost and complexity of implementation for wide (20 MHz) bandwidth, inherent and simple multipath protection through the use of a cyclic prefix, and a naturally scalable implementation where multiple system bandwidths can be supported. Subcarriers were grouped together in sets of 12 to form a PRB. The number of 12 was chosen as a trade-off between resource allocation size and padding for small packets [71]. A 0.5 ms slot could hold either 7 or 6 OFDM symbols, allowing multiple CP choices. (The 1 ms subframe with two 0.5 ms slots was a simplifying compromise on proposals ranging (or indeed adapting between) 0.5–2 ms.) Many different SCS were investigated under the basic assumption of keeping a 10 ms radio frame and providing two reasonable sized CP for the expected deployments. 15 kHz offered 4.67 and 16.67 μs CP. Occupancy was targeted at roughly 90%, again for implementation purposes.

For the LTE uplink, a primary concern was an efficient 20 MHz implementation. A 20 MHz uplink represented a big departure from what was commercially used in cellular systems at that time, and there was much worry that the PA and other components would drive the cost too high for widespread adoption. There was much discussion on whether a smaller minimum UE bandwidth could be supported, such as 10 MHz [72]; this idea was eventually discarded because the added system complexity for initial access for a UE bandwidth without a fixed relation to the synchronization signals and the system bandwidth [73]. (More than 10 years later NR repeated the discussion, and given the very large range of carrier bandwidths, decided to support minimum UE bandwidth much smaller than the maximum carrier bandwidth (see Sect. 2.2.1) and locations of synchronization signals not fixed within the UE bandwidth or system bandwidth.) Peak-to-average power ratio (PAPR) was a hot discussion, soon to be replaced with the more accurate but not-as-straightforward to calculate cubic metric (CM) [7]. The selection of a single carrier waveform allowed for a substantially lower CM for the uplink, thus facilitating lower-cost lower-power consumption UE. The SC-FDMA waveform used is "DFT"-spread OFDM (DFT-S-OFDM), using with an extra DFT block making the net signal single carrier. Trade-offs for this waveform included a design that did not mix data and reference symbols in the same SC-FDMA symbol, and restrictions on allocations to only have factors of 2,3,5 for ease of DFT implementation. Multi-cluster UL transmission (with higher CM) or simultaneous PUSCH and PUCCH were not specified until Release-10.

For NR, there were a number of proposals again, but the time for waveform selection was significantly reduced compared to LTE. The NR study item began in April 2016, and the basic waveform for up to 52.6 GHz (the upper limit of a range of bands identified for WRC-19) was selected to be a cyclic-prefix (CP) OFDM spectrally confined waveform. Several different variants of spectrally confined waveforms were studied, including "windowed" and "filtered" OFDM. In the end it was agreed that it would be possible to assign much larger (for example, 106 PRBs

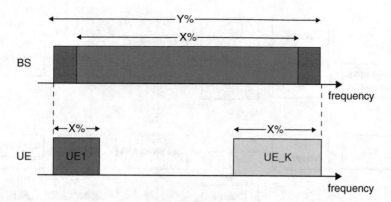

Fig. 2.103 An exemplary illustration of the frequencies used by the system and that by particular UEs

in 20 MHz instead of 100 PRBs as in LTE, see 38.101-1 Sect. 5.3.2) numbers of resources within the carrier bandwidth, higher than the ~90% for LTE [74, 75]. However, the filtering operations necessary to meet those requirements were agreed to be not specified and left up to implementation. As with LTE, the specifications developed in RAN1 still allows the base station to have higher spectrum utilization ($Y\%$) than the minimum requirements defined in RAN4 ($X\%$) [75, 76], # of PRBs [77] (Fig. 2.103).

The CP-OFDM waveform was selected to be the basic waveform for both the downlink and the uplink. DFT-S-OFDM was not selected as the basic waveform for the uplink as part of the initial decision on the uplink waveform. Part of the reason is an increased focus in NR on unpaired spectrum, and symmetry of design including DM-RS was seen to be appealing particularly when considering cross-link interference measurements. DFT-S-OFDM was also selected as a complementary waveform because of its PAPR advantage. All NR UEs were required to provide mandatory support for both CP-OFDM and DFT-S-OFDM [78, 79]. DFT-S-OFDM is intended only for single layer transmission, targeting limited coverage scenarios. The UL waveform is indicated by RMSI for Msg 3 of the RACH procedure and later by UE-specific RRC for non-fallback DCI (fallback DCI uses the same waveform as for Msg 3).

The modulation supported in NR is very similar to that in LTE. QPSK to 256QAM are supported, with 256 QAM mandatory for DL in FR1 and having optional feature signaling for UL and FR2. Higher constellation, such as 1024QAM is not supported in Release-15 NR, though it is supported for the LTE DL intended for high capacity stationary terminals. Release-16 NR will study the usage of 1024QAM as part of the integrated access and backhaul study item.

One difference between NR and LTE is that NR will support $\pi/2$ BPSK for DFT-S-OFDM [80, 81], which with frequency domain spectral shaping (FDSS) in Fig. 2.104 offers lower CM operation. The FDSS filter is up to UE implementation (i.e., the Tx FDSS filter is transparent to the Rx) [82, 83].

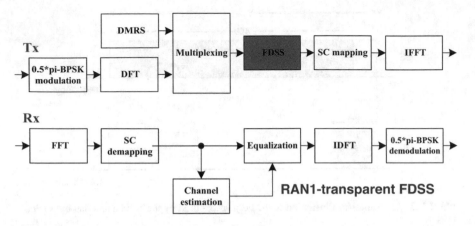

Fig. 2.104 Uplink transmission and reception with π/2 BPSK and FDSS

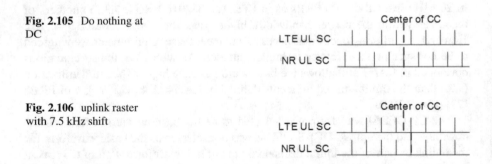

Fig. 2.105 Do nothing at DC

Fig. 2.106 uplink raster with 7.5 kHz shift

The handling of the "DC subcarrier" is different in LTE and NR. For LTE, a DC subcarrier was reserved and not used for data transmission on the DL, and a 7.5 kHz shift was introduced on the uplink to avoid having DC fully aligned with a single subcarrier. For NR, the DL and UL are symmetric with no explicit DC subcarrier reserved. The transmit DC subcarrier is modulated (at either the gNB or UE), and not rate-matched or punctured. The receiver may know if the transmit DC subcarrier is present or not within the receiver BW, and if present the location. At the receive side, no special handling of DC subcarrier(s) is specified in RAN1. One reason for a difference in handling of transmitter and receiver side DC in NR than from LTE is that there may be different minimum bandwidths for different UE, with UE receive bandwidth not necessarily aligned in the center of the gNB carrier bandwidth.

One important issue with NR is coexistence with LTE (see Sect. 2.4). In case NR and LTE are both deployed in the same band, it is desirable to align the uplink subcarriers to minimize the inter-subcarrier interference. Therefore NR has introduced the possibility to shift the uplink raster by 7.5 kHz. In this case, the uplink LTE and NR subcarriers will be aligned (Fig. 2.105 and 2.106).

2.5.2   Channel Coding

Channel coding at peak data rates is a hardware challenge for each new generation; 5G NR is no different. The top LTE UE category is very high (UE cat 26, ~3.5 Gbps), and theoretically LTE can even meet the ITU requirements exceeding 20 Gbps (32 20 MHz carriers at 256QAM/1024QAM). 5G NR will support 20 Gbps or more for eMBB, having similar or greater throughputs as LTE initially and much higher as time goes on. At long block sizes and high data rates, the challenge is keeping good (near Shannon limit) performance with reasonable complexity and latency. At small block lengths (less than a few hundred bits) there is an opportunity for improvement (1–2 dB) over TBCC (tailbiting convolutional codes). In addition, NR must maintain the excellent flexibility and HARQ support as provided in LTE. This is essential for "real world" scheduling and adaptive modulation and coding (AMC).

The focus on short block lengths in addition to long block lengths is one difference of NR compared to LTE. Long blocks may dominate "worst case" implementation complexity and latency seen when considering the peak data rate performance of the system. However, short blocks may dominate overall system performance. A significant percentage (~50–90%) of packets less than 200 bytes for background traffic, instant messaging, gaming, etc. was shown in TR36.822, Enhancements for Diverse Data Applications. System information, RRC signaling, L1 DL/UL control are all typically less than 200 bytes. The performance of small packets and control information may limit the system link budget and system coverage. Coverage for control may be especially important when data is beamformed and has naturally greater coverage than control.

As in LTE, the channel coding discussion for NR occurred in two phases: code down-selection and code design.

- **LTE down-selection**: April 2005 to Aug 2006, WCDMA turbo (legacy) vs. contention free turbo vs. LDPC.
- **LTE design**: Nov 2006 to Feb 2008, contention-free turbo design. (Not as much attention on control channel coding, convolutional code design selected as TBCC March 2007.)
- **NR down-selection**: April 2016 to Oct/Nov 2016, contention-free turbo (legacy) vs. LDPC vs. Polar. (Control channel also considers TBCC.)
- **NR design**: Nov 2016/Jan 2017 to Dec 2017. LDPC and Polar codes.

The discussion of codes for NR focused primarily on turbo codes (as used in LTE), LDPC codes (as discussed for LTE and used in WiFi) and Polar codes (new to standards). At the time of LTE down-selection, LDPC codes [84] although an older concept, were seen as inflexible compared to turbo codes, with implementations that were not as mature compared to turbo codes. Turbo codes were therefore selected for LTE, suitably improved from the version used in HSPA [2]. In the decade since, LDPC codes have been better understood and commercially developed for 802.11n and 802.11ac [85]. Polar codes had few (if any) published implementations, so were a "less well established" candidate. They were however

Table 2.42 Properties of the codes consider for NR

Factor	LTE (turbo)	LDPC	Polar
Maturity	Established (widely implemented)	Established (widely implemented)	Less well established (no commercial implementation at time of NR study item)
Performance	OK for large	OK for large	OK for large Better for small
Complexity	Throughput and area efficiency not sufficient for NR	Best throughput and area efficiency for NR	Implementable with smaller ($L = 8$) list and block sizes
Latency at high data rates	Parallel, but enough?	Best (Highly parallel)	Acceptable
Flexibility	Acceptable	Acceptable	Acceptable
Other	Chase and IR	Chase and IR	Chase and IR

considered promising for low-complexity simple decoding at highest data rates as well as performance at small block sizes [86]. Turbo codes were well-known, but the following was observed for their application to NR: "For turbo codes, there are concerns that implementation with attractive area and energy efficiency is challenging when targeting the higher throughput requirements of NR"[79]. Table 2.42 highlights the selection phase summary of the coding options.

2.5.2.1   LDPC

LDPC codes are used for data transmission in NR for both the DL and UL. There are two base graph designs, targeting small and large block sizes.

An LDPC code takes K information bits and creates N parity bits using an $(N-K) \times N$ bit parity check matrix. The code is represented by a Base Graph whose entries are replaced by sub-matrices of size Z. NR has two Base Graphs, BG1 and BG2. The first Base Graph $= 46 \times 68$, Z is one of 51 values in $\{2,3,4,\ldots,320,352,384\}$. The Information block sizes of $K = 22*Z$ are directly supported, with a maximum $K = 8448$ for $Z = 384$. There is a second Base Graph targeting smaller block sizes, with Base Graph $= 42*52$, the same Z choices, maximum $K = 3840 = 10*Z$ with $N = 50/10*K$, and other small differences. The sub-matrices are all zero, identity, or circular shifted identity matrices.

The block diagram for the downlink is shown in Fig. 2.107. Similar to LTE, rate matching and segmentation/concatenation are defined. The larger information blocks are segmented into multiple code blocks, with a CRC per block, as in LTE. Rate matching supports other information sizes, code rates, and IR-HARQ. The first $2*Z$ bits are not included in the circular buffer, so the mother code rate for BG1 is $22/66 = 1/3$, with $N = 66/22*K$. Four redundancy versions (RVs), as in LTE, are defined. Limited buffer rate-matching (LBRM) may be configured. A bit interleaver for scrambling is applied after rate matching to distribute bits into higher order modulation symbols for a small gain.

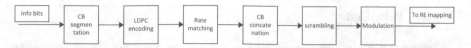

Fig. 2.107 Block diagram of NR DL

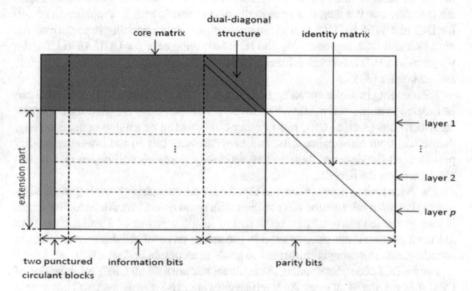

Fig. 2.108 Dual diagonal structure of LPDC codes

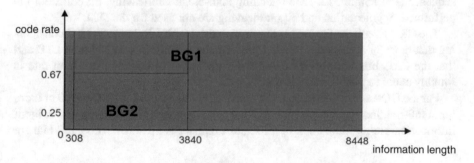

Fig. 2.109 Usage of the two LDPC base graphs in NR

The selected LDPC codes have good properties for implementation (e.g., Quasi-Row Orthogonal (QRO), dual diagonal, etc. Fig. 2.108).

The same base graph is used for the initial transmission and retransmissions of the same code block. Figure 2.109 illustrates the usage of BG1 and BG2 with respect to the code rate (up to 0.95) and the number of information bits (up to 8448 bits). A UE can skip decoding when the effective code rate is >0.95 (some CB sizes may not be decodable above code rate of 0.92).

2.5.2.2 Polar Code

Polar codes are used for control information on the downlink (DCI) and uplink (UCI), as well as for the PBCH, with slight differences in the code construction and design for each of these cases. The information block sizes for Polar codes in NR are therefore smaller than that seen in the data channel, with K a maximum of 140 for DCI and 1706 for UCI (before segmentation). Rate matching is performed for both UCI and DCI, segmentation and bit interleaving only for UCI. As in LTE, the very smallest UCI information block sizes use different coding, repetition, simplex, and RM codes [39].

Polar coding involves encoding in such a way that positions in the codeword can be ordered in terms of reliability. The most reliable positions are used for important "assistant" bits (such as CRC bits or other bits) that can be used to assist decoding, then other input information bits, and then "frozen" bits in the lowest reliability positions. At the decoder, these lowest reliability positions will be set to a known value (zero) in the decoder.

The Polar coding flow is shown in Fig. 2.110. An encoding vector is generated by placing the input information bits together with the assistant bits. An encoding vector is then generated using a Polar sequence of length a power of 2 and greater than the input plus assistant size, ordered in ascending order of reliability. After Arikan encoding, rate matching is performed to produce an output codeword.

For the DCI code construction, the assistant bits are a 24 bit CRC. Initialization of CRC shift register is all ones. An interleaver of size 164 distributes CRC bits within the encoding vector to help provide early termination gains at the decoder. The Polar sequence is of length 512. Rate matching (sub-block interleaving, bit collection) is performed. Segmentation and bit interleaving are not used for the DCI.

The CRC length for the DCI is larger than the 16-bit CRC used in LTE considering that more blind decodes may be performed per unit time in NR than LTE and that the CRC bits are also used for Polar coding. The target false alarm rate is roughly equal to that of a 21 bit CRC.

For the UCI code construction, the assistant bits are an 11 bit CRC for 20 or more input information bits, or a 6 bit CRC and 3 parity check (PC) bits for fewer input information bits. Initialization of CRC shift register is all zeros. The CRC bits are

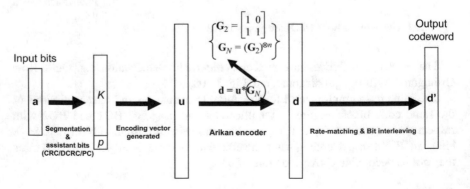

Fig. 2.110 Polar coding flow in NR

appended, not distributed by an interleaver. PC-bits positions are chosen among the highest reliable bits set reserved for the information bits, and are calculated from the information bits. Positions and calculation is known to the decoder. The UCI field order is CRI → RI → Padding bits (if present) → PMI → CQI. The Polar sequence is length 1024. Rate matching also includes bit interleaving. Segmentation and code block concatenation are used when ($K \geq 360$ and $M \geq 1088$ where M is coded bits for UCI), with two equal sized blocks each with CRC based on that segment.

The Polar coding for NR-PBCH is the same as for DCI: polar code with distributed CRC (32 information bits plus 24-bit D-CRC, $N_{max} = 512$. Coding steps for PBCH include payload generation, scrambling, CRC addition, channel coding, and rate matching. Payload generation is unique for PBCH in that the PBCH fields that may have known bit values (e.g., SS block time index, SFN bits in handover cases) are placed in a specific order to enable improved PBCH decoder performance/latency.

Channel coding schemes for control channels are summarized in Table 2.43 below [87].

Table 2.43 Channel coding schemes for NR control channels

Payload type	UCI					BCH	DCI
Payload size (excluding CRC if applicable)	$K = 1$	$K = 2$	$2 < K \leq 11$	$11 < K \leq 19$	$19 < K$	$K = 32$	$11 < K$ (padding is used to reach $K = 12$)
Coding and sub-clause	Repetition 5.3.3.1	Simplex 5.3.3.2	Block code 5.3.3.3	Polar 5.3.1.2	Polar 5.3.1.2	Polar 5.3.1.2	Polar 5.3.1.2
n_{max}	N/A	N/A	N/A	10	10	9	9
CRC polynomial	N/A	N/A	N/A	$g_{CRC6}(D)$	$g_{CRC11}(D)$	$g_{CRC24C}(D)$	$g_{CRC24C}(D)$
Initial CRC remainder	N/A	N/A	N/A	0x00	0x000	0x000000	0xFFFFFF
Maximum number of segments	N/A	N/A	N/A	1	2	1	1
Payload interleaving (I_{IL})	N/A	N/A	N/A	No (0)	No (0)	Yes (1)	Yes (1)
Total number of parity bits (n_{PC})	N/A	N/A	N/A	3	N/A	N/A	N/A
Number of parity bits in minimum weight locations (n_{PC}^{wm})	N/A	N/A	N/A	0 or 1	N/A	N/A	N/A
Coded-bit interleaving (I_{BIL})	N/A	N/A	N/A	Yes (1)	Yes (1)	No (0)	No (0)

2.5.3   MIMO Design

MIMO operation in NR is based very closely on that developed in Release-14 LTE, with a few main differences summarized below and discussion in details in following sections and an overall summary is given in Table 2.44.

As discussed in Sect. 2.2.4 on reference signals, cell-common reference signal as in LTE no longer exists in NR standards and the whole framework of reference signals designs changes. As a consequence, MIMO transmission in NR is based only on DM-RS for demodulation with assistance of TRS for time and frequency tracking and of PT-RS for phase tracking in FR2. Note that there is no uplink or downlink transmit diversity, however, which means NR could be less robust in certain environments (such as high speeds).

Table 2.44 An overall summary of the MIMO feature in 5G

Aspect	Description
Reference signal framework	Removed LTE-like CRS, introduced PT-RS for phase noise compensation, and use various CSI-RS configurations for multiple functionalities such as fine T/F synchronization and QCL assumption, RRM, CSI acquisition, and beam management
DMRS	Extended and more flexible than LTE. Introduced Type-1 and Type-2 configurations that occupy configurable 1 or 2 symbols. Support up to 12 orthogonal layers with Type-2 and 2 symbol configuration. Support front loaded only (for low latency) and front loaded plus additional DMRS location(s) for mobility
CSI-RS	Expanded use from LTE with more options of T/F patterns and densities. Introduced different configurations to support, in addition to CSI acquisition, BM, tracking (TRS), NZP IM for probing, RRM (optionally), and RLM
SRS	Similar as Rel-15 LTE, enhanced to support up to four ports, antenna and carrier switching, as well as non-codebook-based UL MIMO
Configurable scrambling/initialization	Available for UL and DL data, PUCCH, UL and DL RS. Based on cell ID may be default, but enables the possibility of UE-centric operation
Beam management	New for NR, needed for >6GHz, can be used below 6GHz. "Beam" is used in RRC, and "Transmission Configuration Indication" (TCI) with Type-D QCL is used in PHY to indicate beam
Beam failure recovery	New for NR. UE measures downlink signal (CSI-RS/SSB) to detect beam failure and to find new candidate beam(s), sends PRACH to report, and then looks on corresponding CORESET for gNB response
DL MIMO	Type I codebook is very similar to that of LTE Rel-14. Type II codebook introduced with higher accuracy. CSI reporting framework is extended in terms of flexibility, and supporting of beam management and interference measurements. Non-transparent MU-MIMO and simpler codeword to layer mapping
UL MIMO	Similar as LTE: Codebook based. Framework is LTE (codebooks updated) New for NR: non-codebook based. Better resolution than codebook. Needs extra signaling and improved for using TDD reciprocity-based operation

A large number of antennas is envisioned in NR networks while a relatively small number of antennas are at the handsets; therefore multi-user (MU) MIMO is critical to really take advantages of the spatial dimension of the multi-antenna system for high spectrum efficiency. To achieve good trade-off between the MU-MIMO performance and overhead associated with the potentially large number of layers for the UEs, explicit indication of DM-RS antennas ports utilized for multiple UEs is supported which is not the case for LTE. Codeword to layer mapping and RE mapping are now also different in NR.

CSI report framework is expanded and modified in NR in order to accommodate features like beam management, new non-zero power (NZP) CSI-RS interference measurement resource, etc. Feedback codebook design is also expanded to cover multiple panels, and to have higher accuracy with larger overhead.

For UL MIMO, in addition to codebook-based MIMO scheme, non-codebook-based MIMO scheme is introduced.

To support FR2 in higher frequency, beam management procedures are introduced for beam sweeping/selection, beam failure recovery, etc.

The fact that NR does not have a number of transmission modes like LTE does not mean that NR is simpler than LTE. In actuality, this "single mode" is extremely configurable, including the new beam management functionality, RS, and other enhancements. NR has upwards of about 50 related UE capabilities and about 200 RRC parameters, a significant increase over the ~15 and ~50 seen in LTE Release-13/14 MIMO. It is likely that significant efforts will be needed to ensure that gNB and devices can interoperate beyond the most basic functionality and defaults to reach the full potential of NR MIMO.

2.5.3.1   DM-RS-Based MIMO Transmission

In NR, both uplink and downlink data transmissions (PUSCH and PDSCH) are based on DM-RS for demodulation. For DM-RS design, please refer to Sect. 2.2.4.2. For SU-MIMO, NR supports maximum eight orthogonal DL DM-RS ports and maximum four orthogonal UL DM-RS ports conditioning on the related UE capability. For MU-MIMO, NR supports maximum 12 orthogonal DM-RS ports for both downlink and uplink. For DFT-s-OFDM waveform, only single layer (i.e., rank 1) per UE is supported. For CP-OFDM waveform, the actually supported maximum ports per UE depends on the configurations of DM-RS types and number of occupied OFDM symbols as shown in Table 2.45.

In the following, more details are provided for codeword to MIMO layer mapping, PRB bundling for precoding, and DCI indication for DM-RS ports.

2.5.3.1.1   Codeword to Layer Mapping

Comparing to LTE where generally two codewords are mapped once there are more than one layer for PDSCH/PUSCH transmission, in NR up to four layers are mapped to a single codeword and five or more layers are mapped to two codewords. Rationale behind this design includes signaling overhead, robustness of link adaptation,

Table 2.45 Maximum number of ports per UE for CP-OFDM

DM-RS configuration	Max ports per UE in SU-MIMO (CP-OFDM)		Max ports per UE In MU-MIMO (CP-OFDM)	
	DL	UL	DL	UL
Type 1 with 1-symbol	4	4	2	2
Type 1 with 2-symbol	8	4	4	4
Type 2 with 1-symbol	6	4	4	4
Type 2 with 2-symbol	8	4	4	4

demodulation/decoding latency, etc. Since a single MCS level is assigned for a codeword of multiple MIMO layers whose channel quality can be very different, there may be potential performance degradation while the exact amount of loss depends on many factors.

For five or more layers, first $\lfloor L/2 \rfloor$ layers are mapped to the first codeword (CW0) and the remaining layers to the second codeword (CW1).

When mapping coded modulation symbols, the order is layer first, then frequency (in terms of subcarrier), and last time (in terms of OFDM symbol).

2.5.3.1.2   PRB Bundling

Frequency domain granularity of precoding of DM-RS trades off more precise precoding vs. better channel estimation performance. Similar to LTE, this frequency domain granularity is defined via PRB bundling where UE may assume the same precoding is applied for a set of contiguous (and bundled) PRBs. The PRB bundling size is configurable in NR. The candidate sizes include 2, 4, or wideband. In the case of wideband PRB bundling, the UE is not expected to be scheduled with non-contiguous PRBs and the UE may assume that the same precoding is applied to the allocated resource. In the case of PRB bundling size of 2 or 4, the corresponding bandwidth part (BWP) is partitioned into precoding resource block groups (PRGs) of size 2 or 4 with aligned boundary to facilitate multiple user pairing. PRB bundling size is UE-specifically configured. It can be configured as static with 1 value. It can also be configured for dynamic indication via a 1 bit field in DCI and PDSCH resource allocation to jointly determine the size. Default size 2 is used before RRC configuration and for PDSCH with DCI format 1_0. The procedure to determine PRB size is shown in Fig. 2.111

Note that NR also supports DM-RS bundling in time domain across slots when contiguous slots are assigned for a PDSCH transmission.

2.5.3.1.3   DCI for MU-MIMO

To achieve high spectrum efficiency, MU-MIMO transmission and reception has to adapt dynamically to channel conditions, UE distribution, data traffic, etc. This implies that the number of MIMO layers and the occupied DM-RS ports for

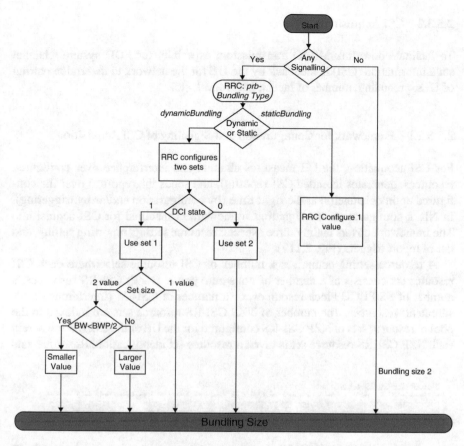

Fig. 2.111 Procedure to determine PRB bundling size

the paired UEs vary with time (from transmission to transmission) and frequency (from RBG to RBG). More transmission layers may provide higher throughput at the cost of DM-RS overhead. In NR, in addition to the DM-RS ports used for data transmission (PDSCH or PUSCH) of the intended UE, DCI also indicate the number of DM-RS CDM group(s) that are without data mapped to their corresponding REs. These DM-RS CDM groups include of course the CDM group(s) of the UE's DM-RS ports, and in addition, it can include CDM group(s) that may be for other UEs' DM-RS ports. Therefore, this signal can be used to indicate MU-MIMO transmission and dynamically adjust the associated overhead. For downlink (and in a sense uplink as well), this falls between transparent MU-MIMO where the UE has no knowledge of paired UE(s) in terms of their used DM-RS ports and the nontransparent MU-MIMO where the UE knows exactly which DM-RS ports are used for other UE(s).

2.5.3.2   CSI Acquisition

To facilitate downlink MIMO transmission, especially for FDD system, channel
state information (CSI) is feedback by the UE for the network to determine pairing
of UEs, precoding, number of layers, MCS level, etc.

2.5.3.2.1   Framework for Configuration and Signaling of CSI Acquisition

For CSI acquisition, the UE measures channel and interference over configured
resources, generates intended CSI report(s), and sends the report(s) over the configured uplink channel(s) at the right time (by configuration and/or by triggering).
In NR, a configuration and signaling framework is specified for CSI acquisition.
The framework covers mainly three aspects: resource setting, reporting setting, and
list of report triggers (Fig. 2.112).

A resource setting comprises a number of CSI resource sets where each CSI
resource set consists of a number of none zero power (NZP) CSI-RS resources, a
number of SS/PBCH block resources, or a number of CSI-IM (interference measurement) resources. The number of NZP CSI-RS resource sets are indexed to the
pool of resource sets of NZP CSI-RS configured for the UE of the CC/BWP wherein
each NZP CSI-RS resource set is given a resource set identification. Resource sets

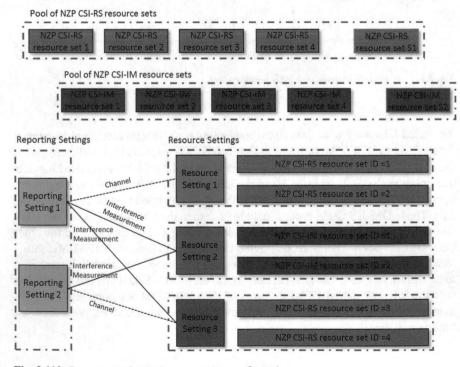

Fig. 2.112 Resource setting and report setting configuration

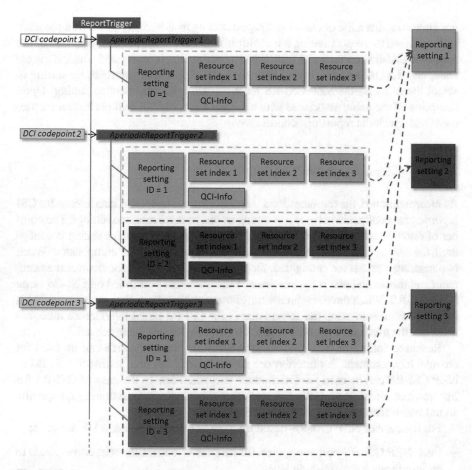

Fig. 2.113 Report trigger list for CSI acquisition

of SS/PBCH block resources and of CSI-IM are configured in the same way. A number of resource setting may be configured for the UE of the CC/BWP.

A number of report setting may be configured for the UE of the CC/BWP. A report setting configures information for the UE to measure and report certain channel state information. These information include measurement resources (as resource setting(s)) where channel measurement is performed and if configured where interference measurement is performed, report type (periodic, semi-persistent on PUCCH or PUSCH, or aperiodic), report quality, report frequency configuration (wideband vs. subband), time restriction for channel or interference measurement (on or off), codebook configuration, group-based beam reporting, CQI table, subband size, non-PMI port indication, etc. (Fig. 2.113).

A list of trigger states can be configured for aperiodic CSI reporting triggered by DCI field of CSI request and another list of trigger states may be configured for semi-persistent CSI reporting on PUSCH. For aperiodic CSI reporting, each code point of the DCI field of CSI request is associated with one trigger state. Each trigger state contains a list of associated report setting as described above. If a resource setting linked to a report setting has multiple aperiodic resource sets, only one of the aperiodic CSI-RS resource sets from the resource setting is associated with the trigger state, and the UE is higher layer configured per trigger state per resource setting to select the one CSI-IM/NZP CSI-RS resource set from the resource setting. Upon reception of the value associated with a trigger state, the UE will perform measurement and aperiodic reporting accordingly.

2.5.3.2.2   Measurement for CSI Acquisition

As aforementioned, the resources for a UE to measure channel and interference for CSI reporting are configured in the report setting as one or more resource settings. The number of resource settings can be one, two, or three. When one resource setting is configured, the resource setting is for channel measurement for L1-RSRP computation. When two resource settings are configured, the first resource setting is for channel measurement and the second one is for interference measurement performed on CSI-IM or on NZP CSI-RS. When three resource settings are configured, the first resource setting is for channel measurement, the second one is for CSI-IM-based interference measurement and the third one is for NZP CSI-RS-based interference measurement.

Resource setting of NZP CSI-RS or SS/PBCH block resource can be used for channel measurement. For interference measurement, resource setting of CSI-IM or NZP CSI-RS can be used for interference measurement. In the case of CSI-IM for interference measurement, the UE measures the interference assuming no specific signal was transmitted over the REs of the CSI-IM resources.

For the case of NZP CSI-RS-based interference measurement, a UE assumes:

- Each NZP CSI-RS port configured for interference measurement corresponds to an interference transmission layer.
- all interference transmission layers on NZP CSI-RS ports for interference measurement take into account the associated EPRE ratios.
- other interference signal on REs of NZP CSI-RS resource for channel measurement, NZP CSI-RS resource for interference measurement, or CSI-IM resource for interference measurement.

The UE accumulates interference measured over these resources to derive CSI report. The UE measurement behavior defined here combining with proper network implementation can realize much more accurate link adaptation and hence improve system performance. For example, precoded CSI-RS may be sent over the NZP CSI-RS resources for interference measurement where each port is mimicking an intended data transmission layer of users paired for MU-MIMO transmissions. UE measures these NZP CSI-RS resources for interference (and for channel) such that the reported CQI reflects the actual MU-MIMO interference condition with much higher accuracy. Similarly, combining CSI-IM-based and NZP CSI-RS-based interference measurements, the network can send proper CSI-RS signals to probe the inter-cell and intra-cell interference condition and obtain CQI with high accuracy and hence better link adaptation performance.

2.5.3.2.3   Feedback Report and Calculation

A list of parameters are supported under NR CSI acquisition procedure. In addition to MIMO transmission related reports such as CQI (channel quality indicator), PMI (precoding matrix indicator), and RI (rank indicator), beam management related reports such as CRI (CSI-RS resource indicator), SSBRI (SSB resource indicator) and L1-RSRP are also supported. Brief description for each of these parameters are given below:

- **CRI** (CSI-RS resource indicator) indicates the selected CSI-RS resource from the multiple CSI-RS resources of the NZP CSI-RS resource set configured for the CSI reporting. The higher layer parameter "repetition" should be set to "off" for the NZP CSI-RS resource set as different transmission beams may be applied on these CSI-RS resources to facilitate beam selection and reporting at the UE.
- **SSBRI** (SSB resource indicator), similar to CRI, indicates the selected SSB resource from the multiple SSB resources configured for reporting to facilitate SSB-based beam selection and reporting at the UE.
- **L1-RSRP** (layer 1 reference signal received power) is measured either over the SSB resource indicated by the reported SSBRI or over the CSI-RS resource indicated by the reported CRI.
- **RI** (rank indicator) reports the rank for PDSCH transmission selected by the UE.
- **PMI** (precoding matrix indicator) reports the precoding matrix selected from the codebook. Details of codebook and PMI are given in the following section.
- **LI** (layer indicator) indicates which column of the precoder matrix of the reported PMI corresponds to the strongest layer of the codeword corresponding to the largest reported wideband CQI. If two wideband CQIs are reported and have equal value, the LI corresponds to strongest layer of the first codeword.
- **CQI** (channel quality indicator) reports the highest CQI index such that a single PDSCH transport block with a combination of modulation scheme, target code rate and transport block size corresponding to the CQI index, and occupying a group of downlink physical resource blocks termed the CSI reference resource, could be received with a transport block error probability not exceeding a target BLER value. Note that in NR there are two different target BLER values, 0.1 (for general traffic) and 0.00001 (for URLLC traffic). Here for CSI reporting, the CSI reference resource is defined in frequency domain as the group of downlink physical resource blocks corresponding to the band to which the derived CSI relates and in time domain as a single downlink slot derived based on the reporting time, downlink and uplink numerology, the measurement CSI resources, UE capabilities, etc. In addition, to calculate and report CQI (and PMI and RI if configured), a set of common assumptions of PDSCH transmission are assumed by the UE, including number of OFDM symbols for control channel, DM-RS and PDSCH, amount of overhead for other channels, redundancy version for channel coding, PRB bundling size, PDSCH antenna ports, etc.

Note that there is dependency between the different CSI parameters. More specifically, RI shall be calculated conditioned on the reported CRI; PMI conditioned on reported RI and CRI; CQI conditioned on reported PMI, RI, and CRI; and LI conditioned on reported CQI, PMI, RI, and CRI.

In the report setting, a list of possible combination of CSI parameters that may be configured for reporting are "none," "cri-RI-PMI-CQI," "cri-RI-i1," "cri-RI-i1-CQI," "cri-RI-CQI," "cri-RSRP," "ssb-Index-RSRP," or "cri-RI-LI-PMI-CQI." When report quantity is set to "none," UE will not report any quantity for the corresponding report setting. When report quantity is set to "cri-RI-i1," the UE expects to be configured with Type I single panel codebook and to report a PMI consisting of a single wideband indication for the entire CSI reporting band. When report quantity is set to "cri-RI-i1-CQI," the UE expects to be configured with Type I single panel codebook and to report a PMI consisting of a single wideband indication (i.e., "i1") for the entire CSI reporting band. The CQI is calculated conditioned on the reported wideband indication assuming PDSCH transmission with a number of precoders (corresponding to the same i1 but different i2), where the UE assumes that one precoder is randomly selected from the set of precoders for each PRG on PDSCH. When report quantity is set to "cri-RI-CQI," the UE calculates the CQI for the reported rank using the ports indicated (via higher layer configuration) for that rank for the selected CSI-RS resource.

CSI reporting is effectively very calculation and memory heavy as it involves channel and interference estimation and measurement, codebook selection, feedback report channel coding and modulation, etc. However, it is not an easy task to quantify the amount of resource needed for different types of CSI reporting and to allowing reuse of computing and memory resource in a handset across different reporting events. In NR specification, UE reports the number of supported simultaneous CSI calculation as number of CSI Processing Unit (CPU) for processing CSI reports across all configured cells. Processing of a CSI report will occupy a number of CPUs for an amount of time (in terms of number of OFDM symbols) where the exact number and time for different report quantity and reporting situation varies. When CSI reports requires more CPUs than the UE supports, the UE is not required to update some of the request CSI reports according to the time that these CSI reports start occupying the CPUs and their priorities.

The priority of a CSI report is given by its reporting type in the order of aperiodic CSI on PUSCH > semi-persistent on PUSCH > semi-persistent on PUCCH > periodic on PUCCH, then CSI report carrying L1-RSRP > others, then by the serving cell index, and then by report setting ID. This priority is also used to determine whether and which CSI report(s) to drop when physical channels carrying the CSI reports overlap in at least one OFDM symbol and are transmitted on the same carrier.

CSI reports can be carried on PUSCH and/or PUCCH as illustrated in Table 2.46.

Table 2.46 CSI reports and uplink channels

	Periodic CSI	Semi-persistent CSI	Aperiodic CSI
Codebook(s) and reporting channel	Type I CSI • Short PUCCH • Long PUCCH	Type I CSI • Short/long PUCCH • PUSCH Type II CSI • Long PUCCH (only part 1) • PUSCH	Type I CSI • Short PUCCH • PUSCH Type II CSI • PUSCH

2.5.3.2.4   Codebooks for PMI Report

NR codebooks for PMI report include Type I and Type II codebooks.

Type I codebook is very similar to that of LTE later releases with some moderate extensions and modifications. In Release-13 and 14, beam selection codebook was introduced into LTE known as Class A codebook. Same design principle is used to design NR Type I codebook with slightly different structure and parameters. Type I codebook is defined for single antenna panel case and multiple (2 or 4) panel cases. In the case of single panel codebook, beam selection is done by first choosing the beam of the first layer and then choose the adjacent orthogonal beams for the other layers which is different from that of LTE Class A codebook. The multi-panel codebook is an extension of single panel codebook by adding inter-panel wideband or subband co-phasing parameter.

Instead of beam selection, combination of multiple beams is used in Type II codebook design to improve the accuracy of CSI feedback at the cost of much higher overhead. Type II codebook includes single panel codebook and port selection codebook. Only rank 1 and 2 codebook is defined for Type II single panel codebook in Release 15. In addition to subband phase reporting, amplitude reporting of wideband or subband are supported with un-even bit allocation for subband where less quantization bits are used for the coefficients with smaller amplitude. Port selection Type II codebook is an extension from the Type II single panel codebook to combine multiple ports based on beamformed CSI-RS.

2.5.3.3   Uplink MIMO

For NR uplink, in addition to the codebook-based MIMO scheme similar to that of LTE, non-codebook-based uplink MIMO scheme is introduced.

2.5.3.3.1   Codebook-Based Uplink MIMO

For codebook-based uplink MIMO scheme, transmission codebooks for both DFT-s-OFDM and CP-OFDM waveforms are designed. The codebook for DFT-s-OFDM is based on the LTE Release 10 design and only rank 1 (i.e., single layer) transmission is supported. For CP-OFDM, DFT-based codebook design is adopted.

Coherent capability of the UE transmit antennas is reported by the UE that whether all the UE antennas are coherent (denoted as "fullyAndPartialAndNon-Coherent"), or there are two groups of UE antennas where antenna within a group are coherent and the antennas within different groups are non-coherent (denoted as "partialAndNonCoherent"), or all UE antennas are non-coherent (denoted as "nonCoherent"). Subsets of codebook are then designed to match the UE antenna coherent capability.

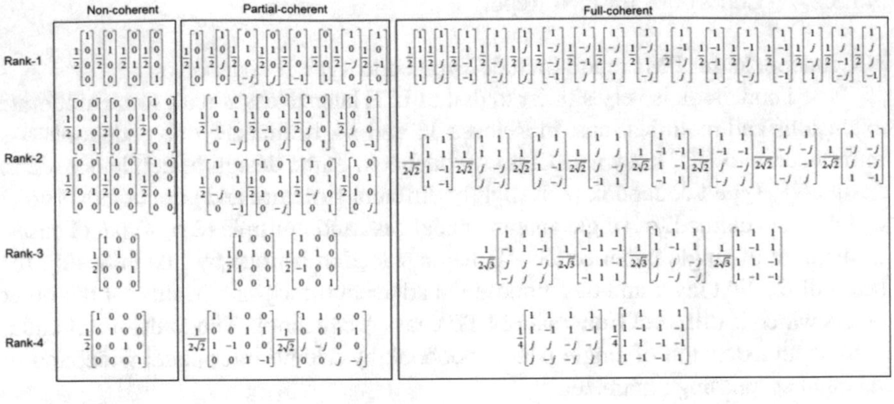

Fig. 2.114 Codebook for uplink MIMO

For two transmit antenna codebook, there are two subsets. One "fullyAndNon-Coherent" subset contains codewords with one non-zero port per layer and codewords with two non-zero ports per layer and this subset can be indicated for transmission by UE with full antenna coherent capability. Another "nonCoherent" subset contains codewords with one non-zero port per layer which can be indicated for transmission by UE with non-coherent capability.

For four transmit antenna codebook, there are three subsets. One "fullyAndNonCoherent" subset contains codewords with one non-zero port per layer and two non-zero ports per layer and four non-zero ports per layer which can be indicated for transmission by UE with full antenna coherent capability. One "partialAndNonCoherent" subset contains codewords with one non-zero port per layer and two non-zero ports per layer which can be indicated for transmission by UE with partial coherent capability. And one "nonCoherent" subset contains codewords with one non-zero port per layer which can be indicated for transmission by UE with non-coherent capability. Note that port grouping is aligned for each codeword.

Transmission PMI is determined based on SRS measurement at the gNB and indicated to the UE by DCI via fields including SRI (SRS resource indicator), TRI (transmit rank indicator), and TPMI (transmit PMI) (Fig. 2.114).

2.5.3.3.2   Non-codebook-Based Uplink MIMO

In non-codebook-based uplink transmission, the precoder applied at the UE side is determined at the gNB with assistance from UE via uplink SRS transmission and no explicit transmission PMI is needed from the gNB.

As illustrated in Fig. 2.115, a UE can be configured with a number of SRS resources based on its capability. The UE determines precoder for each SRS resource based on downlink CSI-RS measurement. The gNB measures these SRS

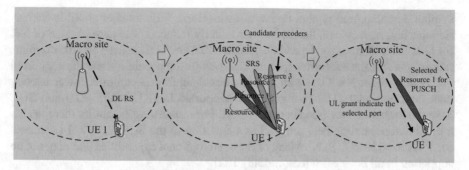

Fig. 2.115 Illustration of non-codebook based uplink MIMO

transmissions to select the proper precoder(s) associated with a subset of the SRS resources and then indicates via SRI (SRS resource indicator) to the UE the selected precoder(s) (and transmission rank) for PUSCH transmission. The SRS resource set and the associated CSI-RS resource can be jointly triggered by SRS request field in DCI.

2.5.4 5G-NR Unified Air Interface Design for eMBB and URLLC

The LTE air interface provides flexible support for a large number of services, including MBB, VoIP, and MBMS. In later releases specific optional enhancements were developed for verticals such as V2X and MTC [NB-IoT(Narrow Band-Internet of Things)/eMTC (enhanced Machine Type Communication)]. The MTC enhancements made extensive use of repetition to reach the coverage enhancement targets. For LTE, some additional functionality targeting low latency and high reliability was introduced in Release-15 [88, 118]. Though these URLLC enhancements came later in the life of LTE, they were actually considered in the very early stages of LTE development. It was conceived that users could be scheduled from 1 to 4 slots depending on characteristics such as latency and reliability [89], but in the end to stabilize the system design a subframe was fixed at two slots [71].

For NR, the system has been designed to accommodate a range of URLLC services from the very beginning. The ITU URLLC targets [90] will generically support URLLC services already in Release-15. These requirements are a "5 nines" (99.999%) reliability and 1 ms latency for single UE for a single 32byte layer 2 PDU. Table 2.47 below summarizes some of the key differences between LTE and NR for the support of URLLC services

The primary latency reduction in LTE Release-15 is to schedule slot or subslot transmissions. NR has similar flexibility, with the ability to schedule as short as two symbol downlink/one symbol uplink transmissions. With a larger possible subcarrier spacing, latency can, thusly, be significantly reduced. The performance of NR for URLLC will strongly depend on whether FDD or TDD is used and the configuration for TDD, the subcarrier spacing, the monitoring capability, and the supported processing times. NR TDD with self-contained subframes can offer lower latency than the traditional TDD configurations supported in LTE, while FDD and SUL have the lowest latency (see Sect. 2.4.1.3.2). For monitoring capability there is also a minimum separation between DCI as a function of the SCS (2, 4, 7, 14 symbols for 15/30/60/120 kHz SCS). Many of these are optional capabilities that may not be supported in all devices, at least initially [92].

For reliability, LTE in Release-15 supports some repetition of DL and UL (SPS) data, as well as multiple decoding occasions for sPDCCH. NR also has built in repetition capability for DL and UL data, and additional enhancements such as higher aggregation level PDCCH, "grant-free" operation (configured UL grant type 1, fully RRC configured), and MCS/CQI tables targeted for high-reliability. URLLC traffic may also be directed to an SUL with superior coverage/reliability (see Sect. 2.4.1.3.3). NR also has considered that eMBB traffic may need to be pre-empted in order to serve URLLC traffic, including a "group common" PDCCH as a pre-emption indicator that helps eMBB UEs avoid corrupting their HARQ buffers. Both LTE and NR can take advantage of higher layer packet duplication, which may be available and useful especially for the downlink carrier aggregation case.

Studies are underway to further expand the supported services for the transport industry, factory automation/IIoT, power/smart grid, and VR/AR [93–96], taking into account requirements for these services (c.f. Table 2.47 taken from 3GPP SA1 TS22.104). A key difference from Release-15 is that supporting more than a single user is considered, which may result in enhancements needed to help increase the number of users supported that can meet the latency and reliability requirements. The studies will investigate PDCCH enhancements (compact DCI, PDCCH repetition, increased PDCCH monitoring capability), HARQ and CSI enhancements, PUSCH repetition enhancements, timeline improvements, intra- and inter-UE multiplexing, enhanced configured grant, and other IIoT improvements for time sensitive networks. Work under MIMO will also specify multi-TRP operation to improve reliability and robustness.

2.5.5   mMTC

Machine type communication (MTC) is a broad class of communication where the cellular network connects devices or applications that are machines. As we have seen in the previous section, factory automation / IIoT is in fact a type of MTC, albeit one that may have very stringent requirements of latency and reliability and so therefore is treated as URLLC. Massive MTC (mMTC) is most often associated with LPWA (low power wide area) communication, a type of MTC that 3GPP spent many years developing in the form of NB-IoT and eMTC. The hallmarks of LPWA include:

Table 2.47 LTE and NR URLLC service details

Features supporting URLLC	LTE	NR
Network slicing	No	Yes (see [91])
Bandwidth parts	No	Yes
SCS	15 kHz	15, 30, 60 kHz in FR1 60, 120 kHz in FR2
Time-domain scheduling	Subframe (1 ms, 14 symbols) Release-15: 2/3 or 7 symbols	14 symbol slot 2/4/7 symbols DL minislot 1–14 symbols UL minislot
PDCCH monitoring	Subframe/slot/subslot	Slot Multiple per slot
FDD/TDD (configuration)	FDD/TDD (subframe-level semi-static configuration)	FDD/TDD (LTE-like, self-contained slot, dynamic per symbol)
SUL	N/A	Yes (for latency and reliability)
Processing times	Normal $(n + 4)$ Release-15: Shortened $(n + 3$ subframe, $n + 4$ slot/subslot)	Normal (set #1) Shortened (set #2)
Pre-emption indication	No	Yes for DL
PDCCH aggregation level	Up to 8	Up to 16
Repetition (lower layers)	PUCCH (up to 6) Release-15: UL SPS, PDSCH (up to 6), PDCCH (no combining)	PUCCH, Up to 8 for data
Repetition (higher layers)	PDCP repetition	PDCP repetition
PDCCH-less transmission	SPS	SPS, Grant-Free
MCS tables	MBB w/ some VoIP entries	eMBB, URLLC
CQI tables	MBB	eMBB, URLLC

– Very low power consumption (>10–15 years with simple batteries)
– Wide area (sometimes in coverage limited locations)
– Low cost
– High connectivity

Even for mMTC there is no uniform use case, but applications do include sensors, monitors, and meters, as well as movable or even mobile devices. The designs are optimized for smaller, more infrequent transmissions. As expected, the supported data rates go down as cost, coverage, power consumption, and connectivity improve [97].

2.5.5.1   NB-IoT

NB-IoT has been widely deployed or launched over the world based on the industry report [98] since Release-13 NB-IoT standard was finalized in June 2016. Before NB-IoT going to Release-13 standardization, there were extensive discussions on the cellular-based technologies supporting IoT in 3GPP. In May 2014, a Study Item

(SI) of the cellular system-based M2M technology supporting low throughput and low complexity MTC was first approved in 3GPP GERAN [99]. One reason doing it in GERAN was that many M2M business at that period were relied on the use of legacy GPRS, and it was motivated to study new solution for M2M with much competitiveness in terms of coverage, complexity, cost, power efficiency, connectivity, etc. compared to the legacy GPRS. Both backward compatible with GPRS and clean slate solution are under the study.

During the study period, a clean slate proposal Narrow Band M2M (NB-M2M) [100] and a backward compatible with GSM proposal Extended Coverage for GSM (EC-GSM) [101] were first discussed, and later on there was another clean slate solution Narrow Band OFDMA (NB-OFDM) [102] proposed. In May 2015, the air interface technologies of NB-M2M and NB-OFDM were merged to use OFDMA in downlink and FDMA in uplink, and the name was changed to NarrowBand Cellular IoT (NB-CIoT)[103]. At the last meeting of the SI in August 2015, an additional proposal Narrow Band LTE (NB-LTE) [104] was submitted. In GERAN#67 meeting, the SI was completed and there was the outcome that NB-CIoT and EC-GSM concluded and shown compliance to all objectives [103], but there is no agreement whether NB-LTE is a clean slate or not in the SI.

According to the decision from 3GPP PCG#34 [105], the normative work of "clean slate" scheme resulting from the study will be undertaken within TSG RAN. Then in RAN#69, the work item on NB-IoT [106] was approved after extensive discussions on the submitted proposals NB-CIoT and NB-LTE, and it is targeted to be finalized in June 2016 during Release-13 to fulfill the requirement of time to market.

Considering the deployment flexibility, the system bandwidth of one NB-IoT carrier is designed to be 180 kHz, which is helpful to gradually refarm GSM carrier with 200 kHz each. It also enables the system to be easily deployed in some scenarios with at least 180 kHz contiguous spectrum for IoT. In these scenarios, NB-IoT is deployed in a standalone manner. In addition, the 180 kHz carrier bandwidth of NB-IoT is compatible with one resource block in LTE, and NB-IoT also supports in-band and guard-band deployments in LTE. One NB-IoT carrier occupies one resource block of LTE carrier for in-band deployment. Multiple NB-IoT carriers can be deployed over different resource blocks of the LTE carrier. In this case, one NB-IoT carrier is the anchor carrier which conveys the basic system information, e.g., synchronization signal and primary broadcast channel. The remaining carriers are non-anchor carriers without these system information transmissions, which are only used for random access, paging, and data transmission with low overhead. For guard-band deployment, NB-IoT is deployed in the unutilized resource block(s) located at the guard band of LTE carrier. The three deployments of NB-IoT are illustrated in Fig. 2.116.

For NB-IoT, the downlink is based on OFDMA with 15 kHz subcarrier spacing which is compatible with LTE numerology. The downlink resource allocation granularity is 180 kHz for data transmission. The uplink supports single tone and multitone (including 3/6/12 tones) transmission as shown in Fig. 2.117. Two subcarrier spacing values are supported as 15 and 3.75 kHz in case of single tone transmission.

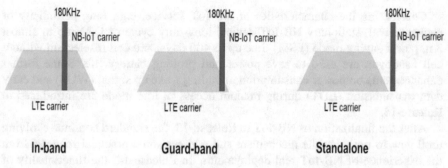

In-band   **Guard-band**   **Standalone**

Fig. 2.116 Three deployment modes of NB-IoT

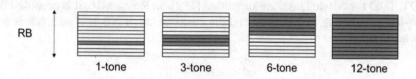

Fig. 2.117 single tone and multi-tone (3/6/12) uplink resource allocation with 15 kHz subcarrier spacing

As most IoT services are from the uplink, i.e., the device sends the message to base station like sensor result reporting, the uplink transmission shall be effective to fulfill the massive connectivity requirement. For NB-IoT, the uplink single tone narrow band transmission results in high power spectral density (PSD), which is effective to improve the number of connections especially in the coverage limited scenario. The advantage of NB-IoT in the aspect of connection density can be seen in the system simulations in Sect. 5.4.4. For multi-tone transmission, only 15 kHz subcarrier spacing is supported and SC-FDMA is applied in this case.

Deep coverage is required for NB-IoT, and there is the requirement of 20 dB coverage enhancement compared to GPRS with 144 dB Maximum Coupling Loss (MCL). Normally, uplink is the coverage bottleneck due to the restriction of UE transmission power. For NB-IoT, the narrow band transmission (e.g., 3.75 kHz) with PSD boost is helpful for coverage extension. As LPWA kind of service is typically small packet and not latency sensitive, repetition is an effective method to improve the coverage, and up to 128 repetitions for NB-IoT uplink transmission is supported. In addition, other methods, e.g., new modulation scheme as $\pi/2$-BPSK and $\pi/4$-QPSK are supported for single-tone transmission with lower PAPR.

Regarding low cost for NB-IoT, from hardware complexity reduction perspective, the technical components include the 180 kHz system bandwidth reducing the RF and base band cost due to the limited bandwidth, single receive RF chain, 20 and 14 dBm low transmit power classes, half-duplex operation, etc. In addition, some further steps are performed, e.g., supporting only 1 or 2 HARQ processes to reduce the soft buffer size, low order modulation, etc. [107].

Considering the characteristics of NB-IoT service, e.g., long periodicity of packet arrival, stationary NB-IoT device, long duty cycle of DRX up to almost 3 h, power saving mode (PSM) time up to 400 days, and cell reselection without cell handover, are used to save power and prolong battery life. Some further enhancements on power consumption including wake up signal (WUS) and early data transmission (EDT) during random access in idle mode are introduced in Release-15.

After the finalization of NB-IoT in Release-13, the standard continue evolving until now to optimize the mechanism and address some practical requests from the experience of NB-IoT real deployments. In Release-14, the functionality of multicast and positioning are supported, and category NB2 with higher peak data rate (>100 kbps) is introduced [108]. In Release-15, as mentioned above, WUS, EDT, TDD for NB-IoT, etc. are supported [109]. In Release-16, it is proposed that NB-IoT connecting to NG Core and the coexistence of NB-IoT with NR is supported [110].

2.5.5.2   eMTC

eMTC is a feature originated from LTE for machine type communication, which is introduced in Release-13. Before Release-13, first there was a study item on low-cost MTC UEs based on LTE in Release-12 [111], which is targeted to study LTE-based MTC solutions with low cost and good coverage. It is called as low-cost MTC. A set of proposals for reducing cost and improving coverage were studied [112]. In Release-12, only low cost part is specified and 15 dB coverage enhancement was removed at the later stage because of standard progress. A new UE category with low cost (i.e., Cat.0) is introduced in Release-12. The cost is reduced to an extent compared to UE Cat.1 by using peak data rate reduction (the maximum transport block size for unicast is restricted to 1000 bits), single receive RF chain and half-duplex [112] . In Release-13, the target is to further reduce the cost and support 15 dB coverage enhancement, and the name is changed to be eMTC which is commonly used in the standard and industry field now. Another UE category (i.e., Cat. M1) is introduced [113]. The standardization of eMTC is completed in June 2016. Currently, the deployment and launch of eMTC is ongoing over the world [98].

As eMTC is based on LTE, it only supports in-band transmission before Release-16, i.e., eMTC is embedded in the LTE network. It is different from NB-IoT as a clean slate proposal with totally new design for these signals and channels, eMTC reuses the signals and physical channels of LTE for initial access including PSS/SSS/PBCH/PRACH. Hence, the minimum system bandwidth of eMTC UE cannot be smaller than six resource blocks, which is much larger than 180 kHz NB-IoT. To reduce the cost of eMTC UE, the bandwidth is limited to 1.4 MHz for both downlink and uplink. In addition, reducing maximum UE transmit power, UE processing relaxing (e.g., number of HARQ process, relaxed TX/RX EVM requirement), etc. are applied.

The scheduling granularity of eMTC for downlink and uplink data transmission is one resource block (i.e., 180 kHz), and up to six resource blocks can be allocated which is suitable for medium data rate IoT service. As mentioned in Sect. 2.5.5.1, narrow band transmission is more effective especially in the coverage limited scenario. Due to the wider uplink transmission, the connection density of eMTC is smaller than NB-IoT (see Sect. 5.4.4). To further increase the connection density of eMTC, it is enhanced to support 2/3/6 tones uplink transmission in Release-15.

There are two coverage enhancement modes defined for eMTC as CE mode A and B which correspond to 0~5 and 5~15 dB coverage enhancement, respectively. Repetition in time domain is a key method for coverage enhancement, and up to 2048 repetitions is supported for data channel. Repetition is applied for common channel, control channel and data channel to improve the coverage. The physical downlink control channel of eMTC (i.e., MPDCCH) has a big change compared to the conventional PDCCH in order to address the requirement of coverage. MPDCCH spans the whole subframe and multiplexed with PDSCH in the frequency domain. In addition, frequency hopping is supported to achieve frequency diversity gain.

Similar to NB-IoT, eMTC continues evolution to optimize the performance and extend the functionality after Release-13. In Release-14, the functionality of positioning and single cell-point to multipoint transmission (SC-PTM)-based multicast is supported. Multicast is an effective method for the software upgrade of eMTC UEs. The coverage enhancement techniques are extended for normal UE and VoLTE. In Release-15, to differentiate NB-IoT and eMTC and avoid the confusion in the market, it is agreed [114]

- eMTC capable only UE category with a maximum supported channel bandwidth is no less than 6 PRB in UL and DL.
- NB-IoT capable only UE category with a maximum supported channel bandwidth is no more than 1 PRB in UL and DL.

Under this guidance, resource element level uplink scheduling (i.e., 2/3/6 tones) is supported for connection density improvement. In Release-16, it is proposed to consider that eMTC connecting to NG Core and the coexistence of eMTC with [110].

2.5.5.3   NR mMTC

3GPP LPWA solutions have been rapidly deployed worldwide. Given the significant effort spent to develop NB-IoT/eMTC, the efficiency of those solutions for LPWA, and the very long lifetime of the deployed devices, there is no hurry to replicate LPWA functionality on NR. The 5G ITU requirements for mMTC are easily met by NB-IoT and eMTC. (Compare results in Sect. 5.4.4 to the requirement in Table 5.3.) The work in Release-15 for NR has instead included mechanisms designs to allow good coexistence with LTE, including NB-IoT and eMTC. In particular, NR supports PDSCH RE-mapping patterns which indicate REs declared as not

Table 2.48 LTE and NR properties for LPWA service

mMTC characteristic	LTE(NB-IoT and eMTC)	NR
Connection density	Meets ITU requirement	Meets ITU requirement (using 5MHz BW, not as good as NB-IoT)
Battery life	Highly optimized design eDRX, PSM, WUS	Small BWP/UE BW Grant-free Release-16 SI: WUS, 1T2R
Wide area (coverage, reliability)	Highly optimized design (normal/small and large) coverage extension)	Repetition (up to 8) 16 CCE
Low cost	Highly optimized design (including low end software DSP NB-IoT implementation)	Small BWP/ UE BW $\pi/2$ BPSK UL Release-16 SI: 1T2R

available for PDSCH, which can include RB and symbol level bitmaps [92]. BWPs may also be defined that avoid resources in use by legacy NB-IoT/eMTC.

In the future there may be mMTC applications that are LPWA-like that are not well served by LTE solutions. One example is high definition video surveillance [115]. As seen in the previous sections, the techniques to improve coverage for NB-IoT/eMTC are similar to those described or considered above for URLLC reliability, namely repetition of control and data channels. Devices built for the scenario without any or with a small amount of coverage enhancement may involve just a small number of repetitions as defined in Release-15 NR. Although lowest cost may not be the focus of NR mMTC, NR in Release-15 supported 5 MHz bandwidth devices that could use single PRB transmission and low PAPR modulation. Such "narrowband" devices may take longer to transmit a packet but these devices can have lower complexity, cost, and operating power than a typical NR device. Narrower band transmissions can also have a higher PSD which can help coverage and connection efficiency. Evaluations show that 5 MHz NR mMTC can meet the 5G ITU mMTC requirements for connectivity [116].

There are some studies ongoing in Release-16 that could be used to improve reliability (URLLC SI), have faster transmissions from idle (NR-U SI), save power (power savings SI), reduce cost by allowing a 1T2R UE (V2X RAN4 SI), or increase connectivity by overloading resources (NOMA SI), even though there is no concerted effort to create an NR-IoT LPWA device. Table 2.48 summarizes LTE and NR with respect to LPWA characteristics.

2.6 Summary

The basic air-interface design of 5G NR was describe previously to give the framework of NR physical layer. As the NR air-interface reuses some design from LTE, the significant parts of LTE design are reviewed first, which help understand the new considerations of NR in these aspects and the corresponding new designs.

For NR, it needs to fulfill the diverse requirements defined in Sect. 1.3. In addition, the deployment spectrum has much wider frequency range than LTE (restricted below 6 GHz), and therefore some new designs for NR are needed. The design of NR in the aspects of numerology, frame structure, physical channels, reference signals, waveform, channel coding, and MIMO is elaborated. Furthermore, because C-band will become a typical spectrum for the first round of NR deployment all over the world, it is identified that coverage is a key issue to be solved for NR. To address this issue, the DL/UL decoupling scheme (i.e., LTE/NR spectrum sharing) is briefly introduced.

For LTE, the initial design was that the transmission scheme is based on CRS being always on in every subframe, which can simply the system operation. During the evolution of LTE, due to the requirement of backward compatibility, the always on CRS limits the flexibility of system design and the compatible with new features. When LTE was evolved to TM9 based on DM-RS, the coexistence of CRS and CSI-RS/DM-RS impacts on the system performance due to the overhead. Taking the lesson from LTE, for NR, *forward compatibility* is required which is a big change compared to LTE. To achieve the forward compatibility, always on CRS is eliminated. However, the functionalities of CRS still have to be implemented, and these functionalities are achieved by a set of NR reference signals such as CSI-RS, DM-RS, and PT-RS.

From the deployment spectrum perspective, NR spans much wider frequency range which includes FR1 and FR2 corresponding to 450 6,000 MHz and 24,500–52,600 MHz, respectively, and *wider carrier bandwidth* is supported. Because of the available wider spectrum, e.g., C-band, Mm-wave bands, the maximum carrier bandwidth is extended to 100 MHz for FR1 and 400 MHz for FR2, respectively, which is very helpful to achieve the peak data rate of 5G. In addition, multiple numerologies in terms of subcarrier spacing and cyclic prefix are supported to adapt different frequency band and service requirement. A set of NR frequency bands are defined and the corresponding subcarrier spacing(s) is given for each band.

Although the available spectrum bandwidth is rich for higher frequency band, there is the high propagation loss which limits the coverage especially for UL due to the limited user power. The new spectrum released for 5G deployments, primarily above 3 GHz, and therefore the coverage of NR is a serious problem to be solved; otherwise, the cost will be increasing more base stations to guarantee the experience which is not desirable for operators. In addition, because it is more and more difficult to find the base station sites, it is motivated to enable NR with C-band and LTE co-located, i.e., they have the same coverage. The *UL/DL decoupling based on LTE/NR spectrum sharing* can enable to achieve the target, which is to accommodate the UL resources in an LTE FDD frequency band as a supplemental UL carrier in addition to the NR operation in the TDD band above 3 GHz.

For duplexing, FDD and TDD are two conventional dominant duplex modes applied for paired spectrum and unpaired spectrum, respectively, which cannot flexibly adapt to the traffic load variation. Due to the more and more serious imbalance between DL and UL traffic load, the symmetric DL/UL bandwidths in FDD results in the ineffective utilization of the UL spectrum. Although TDD can be configured with different DL/UL subframe, due to the requirement of synchronization, the configuration is semi-static and relatively slow. To address this issue, *flexible duplex* in addition to FDD and TDD is introduced in NR which enables the spectrum utilization to adapt the traffic load. Due to the flexible duplexing, there is a very flexible frame structure for NR and each symbol in one slot can be flexibly configured as downlink or uplink transmission in principle.

Regarding the NR new physical layer technologies, the CP-OFDM waveform was selected to be the basic waveform for both the downlink and the uplink. DFT-S-OFDM is also supported in uplink for the sake of coverage which is only applicable for rank-1 transmission. Given the CP-OFDM-based waveform, the spectrum utilization is improved to be larger or equal to 90% which depends on the frequency band and the carrier bandwidth. The channel coding for NR is totally different from LTE, i.e., LDPC and Polar code were selected as the coding scheme for data channel and control channel, respectively. The selection of LDPC for both DL and UL data channel is due to the advantage in the decoding latency, complexity and performance especially for high data rate. The polar code shows the performance advantage when it is used for control channel. Massive MIMO is an important technical component of NR to increase the spectrum efficiency by effectively utilizing MU-MIMO. The DM-RS, interference measurement, and codebook are newly designed to show the performance advantage. In addition, beam management is supported for high-frequency scenario. Finally, NR is designed to have unified air interface to accommodate eMBB, URLLC and mMTC, which enables one carrier to support the multiplexing of different kinds of service with diverse requirements. For mMTC, LPWA is still based on NB-IoT/eMTC and it is not touched in NR Release-15 and 16. As the industry of NB-IoT is becoming mature and there are already large scale of NB-IoT/eMTC commercial deployment over the world, the industry trend is to protect the existing investments by continuously evolving NB-IoT/eMTC for LPWA and it is not urgent to do a new LPWA in NR. For NR, it can address some new MTC services which are complementary with LPWA, e.g., video surveillance and industry sensor.

It should now be clear that there are significant physical layer changes in 5G such that it can satisfy the various requirements that we discussed in Part 1. Due to the aforementioned newly introduced physical layer technologies in this part, the physical layer procedures and RAN protocol have some changes which will be described in some detail next.

References

1. E. Dahlman, S. Parkvall, J. Skold, *4G: LTE/LTE-Advanced for Mobile Broadband* (Elsevier, Amsterdam, 2011)
2. S. Sesia, I. Toufik, M. Baker, *LTE: The UMTS Long Term Evolution, section* (Wiley, New York, 2011)
3. 3GPP TS 36.211, Physical Channels and Modulation (Release 15), June 2018
4. ITU-R M.2370-0, IMT traffic estimates for the years 2020 to 2030, July 2015

5. R.V. Nee, R. Prasad, *OFDM Wireless Multimedia Communications* (Artech House Publishers, Boston, 2000)
6. C.C. Yin, T. Luo, G.X. Yue, *Multi-Carrier Broadband Wireless Communication Technology* (Beijing University of Posts and Telecommunications Press, Beijing, 2004)
7. Motorola, R1-060385, Cubic Metric in 3GPP-LTE, 3GPP TSG-RAN WG1 Meeting #44, Denver, CO, Feb 2006.
8. 3GPP TR 25.814, Physical Layer Aspects for Evolved Universal Terrestrial Radio Access (UTRA) (Release 7), Sept 2006
9. 3GPP TS 36.104, Base Station (BS) Radio Transmission and Reception (Release 8), March 2010
10. 3GPP TR25.913, Requirements for Evolved UTRA (E-UTRA) and Evolved UTRAN (E-UTRAN) (Release 8), Dec 2008
11. 3GPP TS36.331, Radio Resource Control (RRC) Protocol Specification (Release 8), Sept 2009
12. 3GPP TS 36.213, Physical Layer Procedures (Release 15), June 2018
13. 3GPP TS 36.214, Physical Layer-Measurements (Release 15), June 2018
14. Samsung, Nokia Networks, RP-141644, New SID proposal: study on Elevation Beamforming/Full-Dimension (FD) MIMO for LTE, 3GPP TSG RAN Meeting #65, Edinburgh, Scotland, 9–12 Sept 2014
15. Samsung, RP-160623, New WID proposal: enhancements on full-dimension (FD) MIMO for LTE, 3GPP TSG RAN Meeting#71, Göteborg, Sweden, 7–10 March 2016
16. 3GPP TR 36.872, Small Cell Enhancements for E-UTRA and E-UTRAN, Physical Layer Aspects (Release 12), Dec 2013
17. D.C. Chu, Polyphase codes with good periodic correlation properties. IEEE Trans. Inf. Theory **8**, 531–532 (1972)
18. Panasonic, NTT DoCoMo, R1-073626, reference signal generation method for E-TTRA uplink, 3GPP TSG-RAN WG1 Meeting #50, Athens, Greece, 20–24 Aug 2007
19. 3GPP TS 36.212, Multiplexing and Channel Coding (Release 15), June 2018
20. 3GPP TS 36.331, Radio Resource Control (RRC); Protocol Specification (Release 13), Sept 2018
21. Motorola, R1-080072, TBS and MCS Table Generation and Signaling for E-UTRA, 3GPP TSG RAN1 #51bis, Seville, Spain, 14–18 Jan 2008
22. Ericsson, R1-080556, Outcome of Ad Hoc Discussions on TB Size Signaling, 3GPP TSG RAN1 #51bis, Seville, Spain, 14–18 Jan 2008
23. Huawei, HiSilicon, A/N Coding Schemes for Large Payload Using DFT-S-OFDM, 3GPP TSG RAN1 #62bis, Xi'an, China, 11–15 Oct 2010
24. P. Ericsson, Motorola, R1-082091, MCS and TBS Tables for PUSCH, 3GPP TSG RAN1 #53. Kansas City, MO, USA, May **5-9** (2008)
25. Ericsson, etl. R1-073871, Maximum Number of Hybrid ARQ Processes, 3GPP TSG RAN1 #50, Athens, Greece, 20–24 Aug 2007
26. Huawei, etl., R1-081124, Way Forward for TDD HARQ Process, 3GPP TSG RAN1 #52, Sorrento, Italy, 11–15 Feb 2008
27. S. Wicker, *Error Control Systems for Digital Communication and Storage* (Prentice Hall, Englewood Cliffs, 1995)
28. 3GPP TS 36.104, Base Station (BS) Radio Transmission and Reception (Release 15), June 2018
29. 3GPP TR 36.913, Requirements for Further Advancements for Evolved Universal Terrestrial Radio Access (E-UTRA) (LTE-Advanced) (Release 8), March 2009
30. GSA Report, Evolution from LTE to 5G, April 2018
31. Huawei, R1-072321, P-SCH Sequences, 3GPP TSG RAN1 #49, Kobe, Japan, 7–11 May 2007
32. Texas Instruments et al., R1-074143, Way Forward for Secondary SCH Mapping and Scrambling, 3GPP TSG RAN1 #50, Shanghai, China, 8–12 Oct 2007
33. 3GPP TS 36.300, Overall Description, Stage 2 (Release 15), June 2018

34. 3GPP TS 38.104, Base Station (BS) Radio Transmission and Reception (Release 15), Sept 2018
35. IMT-2020 (5G) PG, "5G Vision and Requirement White Paper" (May 2014), http://www.imt-2020.cn/zh/documents/download/1
36. 3GPP TR 38.901, Study on Channel Model for Frequencies from 0.5 to 100 GHz (Release 15), June 2018
37. 3GPP TS 38.331, NR, Radio Resource Control (RRC) Protocol Specification (Release 15), Sept 2018
38. 3GPP TS 38.213, NR, Physical Layer Procedures for Control (Release 15), Sept 2018
39. 3GPP TS 38.212, NR, Multiplexing and Channel Coding (Release 15), Sept 2018
40. 3GPP TS 38.211, NR, Physical Channels and Modulation (Release 15), Sept 2018
41. 3GPP TR 38.913, Study on Scenarios and Requirements for Next Generation Access Technologies (Release 15), June 2018
42. Huawei, HiSilicon, R1-167224, Principles for reference signal design and QCL assumptions for NR, 3GPP TSG RAN WG1 Meeting #86, Gothenburg, Sweden, 22–26 Aug 2016
43. Liu, J., Xiao, W., and Soong, A.C.K., Dense network of small cells, in (A. Anpalagan, M. Bennis and R. Vannithamby eds.), *Design and Deployment of Small Cell Networks*, Cambridge University Press, Cambridge, 2016
44. 3GPP TS 38.214, NR, Physical Layer Procedures for Data (Release 15), Sept 2018
45. NTT DoCoMo, R4-1706982, WF on band specific UE channel bandwidth, 3GPP TSG-RAN WG4-NR Meeting #2, Qingdao, China, 27–29 June 2017
46. GSA White Paper "The Future Development of IMT in 3300-4200 MHz Band" (June 2017), https://gsacom.com/.
47. TD-LTE Industry White Paper, TD Industry Alliance, Jan 2013
48. ECC Report 281, Analysis of the Suitability of the Regulatory Technical Conditions for 5G MFCN Operation in the 3400-3800MHz band, 6 July 2018
49. Ofcom, Award of the 2.3 and 3.4GHz Spectrum Bands, Competition Issues and Auction Regulations (11 July 2017), https://www.ofcom.org.uk/__data/assets/pdf_file/0022/103819/Statement-Award-of-the-2.3-and-3.4-GHz-spectrum-bands-Competition-issues-and-auction-regulations.pdf
50. Ofcom, Award of the 2.3 and 3.4GHz Spectrum Bands, Information Memorandum (11 July 2017), https://www.ofcom.org.uk/spectrum/spectrum-management/spectrum-awards/awards-archive/2-3-and-3-4-ghz-auction
51. L. Wan, M. Zhou, R. Wen, Evolving LTE with flexible duplex, in *2013 IEEE Globecom Workshops (GC Workshops)*, Atlanta, GA (2013), p. 49–54
52. E. Everett, M. Duarte, C. Dick, et al., Empowering full-duplex wireless communication by exploiting directional diversity, in *2011 Conference Record of the Forty Fifth Asilomar Conference on Signals, Systems and Computers (ASILOMAR)*, 6–9 Nov 2011, p. 2002–2006
53. J. Choi, M. Jain, K. Srinivasan, P. Levis, S. Katti, Achieving single channel, full duplex wireless communication, in *Proceedings of the ACM MobiCom Conference*, 2010, p. 1–12
54. 3GPP TR 36.828, Further Enhancements to LTE Time Division Duplex(TDD) for Downlink-Uplink (DL-UL) Interference Management and Traffic Adaptation (Release 11), June 2012
55. K. Doppler, M. Rinne, C. Wijting, et al., Device-to-device communication as an underlay to LTE-advanced networks. IEEE Commun. Mag. **47**(12), 42–49 (2009)
56. O. Sahin, O. Simeone, E. Erkip, Interference channel with an out-of-band relay. IEEE Trans. Inf. Theory **57**(5), 2746–2764 (2011)
57. 3GPP TS 38.101-1, NR, User Equipment Radio Transmission and Reception; Part 1: Range 1 Standalone (Release 15), June 2018
58. Vodafone, RP-172788, Mandatory 4Rx Antenna Performance for NR UE, 3GPP TSG RAN#78, Lisbon Portugal, 18–21 Dec 2017
59. 3GPP TS 38.101-2, NR, User Equipment Radio Transmission and Reception; Part 2: Range 2 Standalone (Release 15), June 2018
60. 3GPP TS 38.101-3, NR, User Equipment Radio Transmission and Reception; Part 3: Range 1 and Range 2 Interworking Operation with Other Ratios (Release 15) June 2018

61. RP-190714 WID Revision: SA NR SUL, NSA NR SUL, NSA NR SUL with UL Sharing from the UE Perspective (ULSUP); Rapporteur, Huawei
62. 3GPP TS 36.101, E-UTRA, User Equipment (UE) Radio Transmission and Reception (Release 15), Sept 2018
63. Wang, Q., Zhao, Z., Guo, Y., Gong, X., Schubert, M., Schellmann, M., Xu, W., Enhancing OFDM by pulse shaping for self-contained TDD transmission in 5G, in *2016 IEEE 83rd Vehicular Technology Conference (VTC Spring)*, Nanjing, 2016, p. 1–5
64. Huawei, HiSilicon, R1-1709980, Subcarrier mapping for LTE-NR coexistence, 3GPP TSG RAN WG1 NR AH Meeting, Qingdao, China, 27–30 June 2017
65. 3GPP TS 38.133, NR, Requirements for Support of Radio Resource Management (Release 15), Sept 2018
66. 3GPP TS 38.321, NR, Medium Access Control (MAC) Protocol Specification (Release 15), Dec 2018
67. M. D. Batariere, J. F. Kepler, T. P. Krauss, et al., An experimental OFDM system for broadband mobile communications, in IEEE VTC Fall 2001, 7–11 Oct 2001
68. 3GPP TR 28.892, Feasibility Study for Orthogonal Frequency Division Multiplexing (OFDM) for UTRAN Enhancement (Release 6), June 2004
69. Motorola, RP-000032, Work item description sheet for high speed downlink packet, 3GPP TSG RAN Meeting #7, Madrid, Spain, 13–15 March 2000
70. Motorola, RP-0000126, Details of high speed downlink packet access, 3GPP TSG RAN Meeting #7, Madrid, Spain, 13–15 March 2000
71. 3GPP MCC, R1-063013, Approved Minutes of 3GPP TSG RAN WG1 #46 Tallinn, Tallinn, Estonia, 28 Aug–1 Sept 2006
72. 3GPP MCC, R1-061099, Approved Report of 3GPP TSG RAN WG1 #44 Meeting, Denver, CO, USA, 13–17 Feb 2006
73. 3GPP MCC, R1-070633, Approved Report of 3GPP TSG RAN WG1 #47, Riga, Latvia, 6–10 Nov 2006
74. Huawei, HiSilicon, et al. R1-167963, Way forward on waveform, 3GPP TSG RAN WG1 #86 Meeting, Gothenburg, Sweden, 22–26 Aug 2016
75. 3GPP MCC, R1-1608562, Final Report of 3GPP TSG RAN WG1#86, Gothenburg, Sweden, 22–26 Aug 2016
76. 3GPP RAN WG1, R1-1715184, LS response on spectrum utilization, 3GPP TSG RAN WG1 #90 Meeting, Prague, Czech Republic, 21–25 Aug 2017
77. Samsung, et al. R4-1709075, Way forward on spectrum utilization, 3GPP TSG RAN WG4 #84 Meeting, Berlin, Germany, 21–25 Aug 2017
78. Qualcomm, et al. R1-1610485, WF on Waveform for NR Uplink, 3GPP TSG RAN WG1 #86bis, Lisbon, Portugal, 10–14 Oct 2016
79. 3GPP MCC, R1-1611081, Final Report of 3GPP TSG RAN WG1#86bis, Lisbon, Portugal, 10–14 Oct 2016
80. IITH, et al. R1-1701482, WF on pi/2 BPSK modulation with frequency domain shaping, 3GPP TSG RAN WG1 AH_NR Meeting, Spokane, USA 16–20 Jan 2017
81. 3GPP MCC, R1-1701553, Final Report of 3GPP TSG RAN WG1#AH1_NR, Spokane, USA, 16–10 Jan 2017
82. 3GPP TSG RAN1, R1-1715312, LS on further considerations on pi/2 BPSK with spectrum shaping, 3GPP TSG RAN WG1 Meeting #90, Prague, CZ, 21–25 Aug 2017
83. 3GPP MCC, R1-1716941, Final Report of 3GPP TSG RAN WG1#90, Prague, Czech Republic, 21–25 Aug 2017
84. R.G. Gallager, *Low-Density Parity-Check Codes* (M.I.T. Press, Cambridge, 1963)
85. H. Park, S. Lee, *The hardware design of LDPC decoder in IEEE 802.11n/ac*, in Proceedings of the 2014 6th International Conference on Electronic, Computer and Artificial Intelligence (ECAI) (2014), pp. 35–38
86. E. Arikan, Channel polarization: a method for constructing capacity-achieving codes for symmetric binary-input memoryless channels. *IEEE Trans. Inf. Theory* **55**(7), 3051–3073 (2009)

87. Qualcomm, R1-1801271, Polar coding, 3GPP TSG RAN WG1 Meeting #AH1801, Vancouver, Canada, 22–26 Jan 2018
88. Ericsson, RP-181870, Summary for WI shortened TTI and processing time for LTE, 3GPP TSG RAN Meeting #81, Gold Coast, Australia, 10–13 Sept 2018
89. Motorola, R1-050246, Downlink multiple access for EUTRA, 3GPP TSG RAN1#40bis Meeting, Beijing, China, 4–8 April 2005
90. Report ITU-R M.2410, Minimum Requirements Related to Technical Performance for IMT-2020 Radio Interface(s), Nov 2017
91. 3GPP TS 38.300, Technical Specification Group Radio Access Network; NR; NR and NG-RAN Overall Description; Stage 2
92. R1-1812064 3GPP RAN1 NR UE Features
93. 3GPP TR 22.804, Study on Communication for Automation in Vertical Domains (Release 16), Dec 2018
94. 3GPP TR 22.186, Enhancement of 3GPP Support for V2X Scenarios; Stage 1 (Release 16), Dec 2018
95. 3GPP TR 38.824, Study on Physical Layer Enhancements for NR Ultra-Reliable and Low Latency Case (URLLC) (Release 16), Nov 2018
96. 3GPP TR 38.825, Study on NR Industrial Internet of Things (IoT) (Release 16), Nov 2018
97. GSMA White Paper, 3GPP Low Power Wide Area Technologies, available at https://www.gsma.com/iot/wp-content/uploads/2016/10/3GPP-Low-Power-Wide-Area-Technologies-GSMA-White-Paper.pdf
98. GSA Report, NB-IoT and LTE-M: Global Market Status, Aug 2018
99. VODAFONE Group Plc. GP-140421, New study item on cellular system support for ultra low complexity and low throughput internet of things, 3GPP TSG GERAN Meeting #62, Valencia Spain, 26–30 May 2014
100. Huawei, HiSilicon, GP-140322, Discussion on MTC evolution for cellular IoT, 3GPP TSG GERAN Meeting #62, Valencia Spain, 26–30 May 2014
101. Ericsson, GP-140297, GSM/EDGE optimization for internet of things, 3GPP TSG GERAN Meeting #62, Valencia Spain, 26–30 May 2014
102. Qualcomm Incorporated, GP-140839, Narrow band OFDMA based proposal for GERAN cellular IoT, 3GPP TSG GERAN Meeting #64, San Francisco, USA, 17–21 Nov 2014
103. 3GPP TR 45.820, Cellular System Support For Ultra-Low Complexity and Low Throughput Internet of Things (CIoT) (Release 13), Nov 2015
104. Ericsson, et.al., GP-150863, narrowband LTE-concept description, in 3GPP TSG GERAN Meeting #67, Yin Chuan, China, 10–13 Aug 2015
105. 3GPP PCG#34 Meeting Report, available at http://www.3gpp.org/ftp/PCG/PCG_34/report/
106. Qualcomm Incorporated, RP-151621, New work item: NarrowBand IOT (NB-IOT), 3GPP TSG RAN Meeting #69, Phoenix USA, 14–16 Sept 2015
107. Huawei, RWS-180023, 3GPP's Low-Power Wide-Area IoT Solutions: NB-IoT and eMTC, Workshop on 3GPP Submission Towards IMT-2020, Brussels, Belgium, 24–25 Oct 2018
108. Vodafone, Huawei, et al. RP-161324, New work item proposal; Enhancements of NB-IoT, 3GPP TSG RAN Meeting #72, Busan Korea, 13–16 June 2016
109. Huawei, HiSilicon, Neul, RP-170852, New WID on further NB-IoT enhancements, 3GPP TSG RAN Meeting #75, Dubrovnik, Croatia, 6–9 March 2017
110. Ericsson, RP-180581, Interim conclusions on IoT for Rel-16, 3GPP TSG RAN Meeting #79, Chennai, India, 19–22 March 2018
111. Vodafone, Provision of low-cost MTC UEs based on LTE, 3GPP TSG RAN Meeting #57, Chicago, USA, 4–7 Sept 2012
112. 3GPP TR 36.888, Study on Provision of Low-cost Machine-Type Communications (MTC) User Equipments (UEs) Based on LTE (Release 12)
113. Ericsson, Nokia Networks, RP-141660, New WI proposal: further LTE physical layer enhancements for MTC, 3GPP TSG RAN Meeting #65, Edinburgh Scotland, 9–12 Sept 2014
114. Deutsche Telekom, et al. RP-171467, Way forward on NB-IoT and eMTC evolution and UE capability, 3GPP TSG RAN Meeting #76, West Palm Beach, USA, 5–8 June 2017

115. Huawei, HiSilicon, RP-180888, Motivation for WI proposal on NR uplink enhancements for broadband MTC, 3GPP TSG RAN Meeting#80, La Jolla, USA, 11–14 June 2018
116. Huawei, HiSilicon, R1-1808080, Considerations and evaluation results for IMT-2020 for mMTC connection density, 3GPP TSG RAN1 Meeting#94, Gothenburg Sweden, 20–24 Aug 2018
117. RP-121772, Further Enhancements to LTE TDD for DL-UL Interference Management and Traffic Adaptation, 3GPP TSG RAN#28, 4–7 Dec 2012
118. Ericsson, RP-181869, Summary for WI ultra reliable low latency communication for LTE, 3GPP TSG RAN Meeting#81, Gold Coast, Australia, 10–13 Sept 2018

Chapter 3
5G Procedure, RAN Architecture, and Protocol

In this chapter, the newly introduced procedures (IAM/Beam management/Power control/HARQ) in 5G NR are first described by taking into account the characteristics of the NR design. The concept "beam" is used in NR and the beam-based operation is applied in the procedures such as initial access and beam management for both downlink and uplink. HARQ is enabled to be much more flexible to achieve the low latency requirement. Then RAN architecture is introduced considering different 5G deployment options including NR standalone and LTE-NR dual connectivity, and RAN functionality split. Finally, the protocols related to higher layer procedures including UE state transition, paging, mobility, etc. for standalone NR are described.

3.1 5G-NR New Procedures

3.1.1 Initial Access and Mobility (IAM)

Before being able to operate on a NR carrier, a UE needs to access to the network and establish an RRC connection with the network. There are two methods of accessing the system depending on whether NR is operating in non-standalone or standalone mode. In the non-standalone mode of operation, access to the 5G network can be performed on a LTE carrier that supports dual connectivity with NR. In this case, normal LTE initial access procedures apply (see Sect. 2.1.8), and once the network has learnt that the UE supports NR then it may configure the UE with a secondary cell group on a NR carrier frequency. The UE will not need to perform initial access on the NR carrier in this case, although the UE will still have to obtain time synchronization (system frame number (SFN) timing of the SCG) via the reception of the SS/PBCH block, and if beam sweeping is in operation on the NR carrier the UE will have to perform beam management procedures to obtain a transmit/receive beam pair for operating on the NR carrier. In non-standalone mode of operation for NR, the NR RRC information is delivered to the UE via its LTE PCell.

In standalone NR deployments, the UE will perform initial access directly on a NR carrier. The UE scans the NR frequencies for SS/PBCH blocks, and will try and access the network over the most favorable carrier using the SS/PBCH block that corresponds to the most favorable broadcast beam from the network. The SS/PBCH block that the UE decided to use for initial access provides resources that the UE will use for its first uplink transmission of the PRACH to the network. The PRACH resource corresponds to the SS/PBCH block index detected by the UE, which ensures that the network will receive the PRACH using the same transmit/receive beam pair that the UE used for detecting the most favorable downlink broadcast beam.

In order to find NR cells, the UE scans its known NR frequency bands. With the large number of wide bandwidth bands supported by NR below 6 GHz and between 24 and 52.6 GHz, it was important to ensure that the UE cell search can be performed in an energy-efficient manner. This led to the design where the synchronization raster for NR was decoupled from the channel raster of NR. The synchronization raster indicates the frequency positions of the SS block that can be used by the UE for system acquisition. For frequency bands above 3 GHz the gap between two frequency positions of the SS block in the synchronization raster is enlarged to 1.44 MHz compared to the LTE 100 kHz channel raster [1] in order to reduce the number of frequency positions that the UE has to search. For frequency bands below 3 GHz, the synchronization raster has a gap of 1.2 MHz between frequency positions. The UE will then test several hypothesis for the transmission of the SS block from the network, including the frequency raster position in the scanned band, the SS block subcarrier spacing and SS block pattern used for the SS block transmission (one or two hypothesis per band as defined by the global synchronization channel number (GSCN) parameters for the global frequency raster and by the applicable SS raster entries per operating band specified in [2]), and the cell identity where the number of NR physical cell identities is 1008 (twice that of LTE). The maximum number of SS/PBCH blocks within a SS burst set is $L = 4/8/64$ for frequency ranges 0–3 GHz/3–6 GHz/6–52.6 GHz, respectively.

Once the UE has acquired a SS/PBCH block, it will read the system information transmitted in the MIB (Master Information Block) as shown in Fig. 3.1. The MIB is transmitted with a periodicity of 80 ms and repetitions made within 80 ms. The MIB informs the UE of the basic physical layer parameters (frame timing reference, frequency domain offset between SSB and the overall resource block grid) that the UE needs to know in order to receive downlink control (PDCCH containing the scheduling RNTI, i.e., SI-RNTI, sent in the RMSI CORESET) that schedules the remaining system information (SIB1). The UE may be re-directed to another SS/PBCH block present in the same carrier at a different frequency position, if the detected SS/PBCH block is not a cell-defining SS/PBCH block. This is indicated in the PBCH by the GSCN offset/range that allows the UE to search for the cell-defining SS/PBCH block.

System information block 1 (SIB1) provides the UE with information that allows the UE to determine whether it is allowed to access this cell, the public

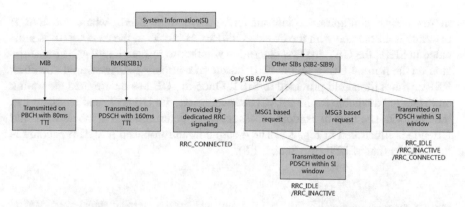

Fig. 3.1 System information acquisition

land mobile network (PLMN) identity, and cell-specific parameters configuration and other radio resource configuration (RRC) information common to all UEs on that cell. In particular, SIB1 tells the UE about downlink configuration, one or both of a normal uplink (NUL) and supplementary uplink (SUL) configurations, the initial downlink BWP common configuration, TDD DL/UL configuration, and physical resources reserved for coexistence. The downlink configuration in SIB1/ RMSI includes the absolute frequency position of the reference resource block,[1] a subcarrier spacing value, and at least one frequency band to which this carrier belongs. Note that for NR the UE cannot derive the carrier position in the frequency domain just from the detection of the SS/PBCH blocks, since contrary to LTE the NR SS/PBCH block is not necessarily at the center of a carrier. Similar information is provided in uplink configuration, with an optional configuration of frequency shift of 7.5 kHz for a SUL.

In addition, SIB1 includes the information on the availability and scheduling (e.g., mapping of SIBs to SI message, periodicity, SI-window size) of other SIBs with an indication whether one or more SIBs are only provided on-demand. In the case on-demand request for some SIBs are configured by SIB1, the SIBs can be requested on demand by the UE in the RRC_IDLE/RRC_INACTIVE state, either by message 1 (MSG 1) or message 3 (MSG 3) in RACH procedure, and OSI will be broadcast by PDSCH within a SI window. When on-demand request is not configured by SIB1 for some SIBs, these SIBs are broadcast in PDSCH within a SI window. For a UE in RRC_CONNECTED, the network can provide system information (e.g., SIB6/7/8) through dedicated RRC signaling.

At this point, the UE has acquired common RRC information, and is able to determine resources for downlink reception and uplink transmission, but the UE is still not connected to the cell. The UE must first perform the initial access procedure by transmitting message 1 (msg1) over the PRACH resource determined by the UE

---

[1] Common RB 0, whose lowest subcarrier is also known as "Point A" corresponding to the DL ARFCN on the downlink channel raster.

in the uplink configuration information[2] provided in SIB1, where the RACH resource is associated with the detected SS/PBCH block. In the case a SUL is indicated in SIB1, the UE must first determine whether to initiate its PRACH transmission on the normal UL or on the SUL, based on comparing its measured downlink RSRP with a threshold provided in SIB1. Once the UE has determined the uplink and resource available for its PRACH transmission, the UE will send message 1 to the network. The flow of random access resource determination at the network and UE side is illustrated in Fig. 3.2. The 4-step contention-based RACH procedure is the same as that of LTE (see Sect. 2.1.8).

Fig. 3.2 Flow of random access resource determination

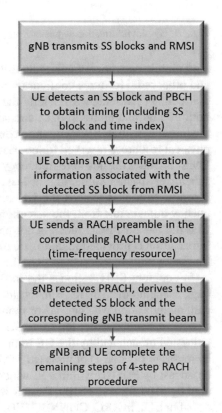

gNB transmits SS blocks and RMSI

UE detects an SS block and PBCH to obtain timing (including SS block and time index)

UE obtains RACH configuration information associated with the detected SS block from RMSI

UE sends a RACH preamble in the corresponding RACH occasion (time-frequency resource)

gNB receives PRACH, derives the detected SS block and the corresponding gNB transmit beam

gNB and UE complete the remaining steps of 4-step RACH procedure

---

[2]The uplink information is contained in RACH-ConfigCommon in BWP-UplinkCommon in UplinkConfigCommon in ServingCellConfigCommonSIB.

3.1.2   Beam Management

Supporting high-frequency bands in FR2 is a major feature for NR. The main enabling techniques for FR2 utilization include beam management to overcome much larger path loss and blockage probability and phase tracking reference signal (PT-RS) to compensate phase noise which is discussed in the section for PT-RS.

3.1.2.1   Downlink Beam Management

In Fig. 3.3, it shows the main steps for downlink beam management defined in NR. First, a set of beamformed downlink signals (CSI-RS and/or SS/PBCH blocks) are configured and transmitted to the UE so that proper beam(s) may be selected for transmissions of PDSCH and PDCCH. Configuration and transmission of beamformed CSI-RS and SS/PBCH blocks to facilitate downlink beam management are discussed in the sections for CSI-RS and SS/PBCH designs. UE then measures the set of downlink beamformed signals to select beam(s) and reports via CRI (CSI-RS resource indicator), SSBRI (SS/PBCH block indicator), and received power of the corresponding beam(s)/signal(s) as L1-RSRP as part of the CSI acquisition discussed in Sect. 2.5.3.2. Based on the UE reports, the network determines the transmission beam(s) for subsequent downlink transmissions (PDSCH, PDCCH, and CSI-RS for CSI acquisition, etc.) and indicates the transmission beam(s) to the UE so that proper receiver beam(s) is applied. Indication of transmission beam(s) is done through the spatial QCL (also known as Type-D QCL) information associated with each state of TCI (transmission configuration indicator) which is configured by RRC and dynamically indicated in DCI as discussed in Sect. 2.2.4.6. Due to mobility and blockage, the current beam(s) between the TRP and UE may be lost and a beam failure recovery mechanism is defined to quickly re-establish the connection via new beam(s). Details of beam failure recovery are discussed in the following part.

Note that during initial access stage, the best beam (of a SS/PBCH block) is indicated implicitly by the associated PRACH resource the UE uses to send PRACH signals for initial access.

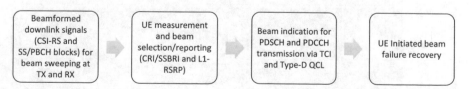

Fig. 3.3 Procedure for downlink beam management

3.1.2.2   Beam Failure Recovery

Beam failure recovery procedure, as shown in Fig. 3.4, includes four steps.

UE monitors periodic CSI-RS/SSB associated (as spatial QCL source reference signal) with the UE's CORESET(s) of PDCCH to detect if a beam failure trigger condition has been met, that is, whether its hypothetical PDCCH BLER is above a threshold for a number of consecutive instances. The UE also monitors configured SSB, CSI-RS, or a combination of SSB and CSI-RS to identify new candidate beam(s) with L1-RSRP above certain threshold. After the UE detects beam failure and identifies new candidate beam(s), it notifies gNB about the new candidate beam(s) by contention-free or contention-based RACH transmissions. Upon detection of beam failure recovery request, gNB responses on PDCCH. In the case that contention-free RACH is used for beam failure recovery request, the UE monitors PDCCH CORESET(s) for beam failure recovery and C-RNTI scrambled DCI as gNB response. In the case that contention-based RACH is used, the procedure is similar to that of initial access. If no response is detected within certain time window, the UE retransmits the beam failure recovery request. If beam failure recovery fails after a number (configured by the network) of attempts, higher layer entities are notified.

3.1.2.3   Uplink Beam Management

Uplink beam management can be done in two different approaches. The first approach as given in the figure below is similar to that of the downlink beam management procedure. Based on network configurations, UE transmits a set of SRS resources (with usage set as "Beam Management") with different transmitting

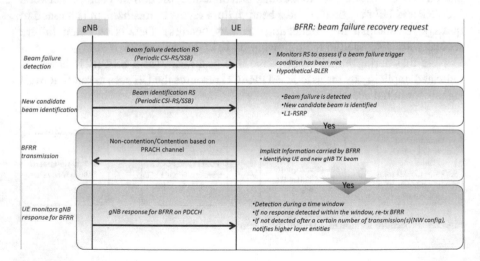

Fig. 3.4 Procedure for beam failure recovery

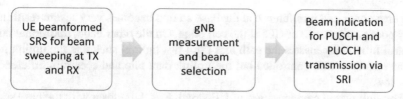

Fig. 3.5 Uplink beam management using SRS

beams. gNB then measures the set of SRS resources and selects the transmitting (and receiving) beam(s) accordingly. gNB indicates to the UE the selected transmitting beam(s) via SRI (SRS resource indicator) for PUSCH and PUCCH (Fig. 3.5).

For the second approach, UEs that support beam correspondence between its receiving and transmitting beams have the ability to select a corresponding beam for UL transmission based on DL measurements without relying on network-assisted UL beam management. Basically, the beam used for receiving SSB/CSI-RS can be used for transmitting corresponding signal/channel, and SSB/CSI-RS is contained in spatial relation indication for PUSCH and PUCCH transmissions. Requirements on beam correspondence are verified for FR2 in terms of the minimum peak EIRP and spherical coverage requirements to ensure small enough difference between the UE's receiving and transmitting beams.

3.1.3   Power Control

Power control is a key technique for cellular communication to improve channel quality, reduce interference, and save energy. On one hand, increasing transmission power may facilitate higher transmission data rate and lower error rate. On the other hand, high transmission power could contribute to severe interference in the system on channels on the same carrier as well as on the neighboring carriers. Especially for uplink, transmission without proper power control can dramatically reduce the handset battery life. Therefore, transmission power control needs to be designed carefully.

In NR, power control applies to both uplink and downlink and is very similar to that of LTE. Uplink power control is still based on fractional power control approach, and downlink part only has very limited specification impact and mainly relies on network implementation.

The principle of fractional power control (FPC) is extended in NR uplink power control by taking into account of the overall design and newly introduced features.

3.1.3.1   Fractional Power Control Design

Fractional power control was proposed to manage uplink interference for LTE system [3, 4]. To optimize spectrum efficiency, all neighboring cells occupy same time and frequency resource for uplink transmission. If UEs transmit at full power, the perceived inter-cell interference at the base station becomes very severe resulting in very poor performance of UEs at the cell edge. On the other hand, traditional power control fully compensates the path loss (and fast fading) such that the uplink inter-ference level is well controlled but the uplink data rate and spectrum efficiency is very low.

Through partial compensation of UE's path loss, UE closer with the base station will smaller path loss transmits with relatively smaller power to reduce its contribu-tion to interference to the neighbors but is still received with relatively higher power to support higher data rate. UE at the cell edge will transmit with higher (or even full) power to ensure good cell edge performance without unnecessarily increasing the overall interference level. A good trade off can be achieved between spectrum efficiency and cell edge performance.

Figure 3.6 [3] shows user throughput under different power control schemes. When path loss compensation factor (α) is set to zero, UE transmit at full power resulting in very high interference level and very poor cell edge performance. When $\alpha = 1$, all UE received at the same power level (via full path loss compensation) and the system performance is low. With fractional compensation, for example, $\alpha = 1/2$, good cell edge user throughput and system performance can be achieved simultane-ously. By choosing values for path loss compensation factor α and the open loop target received power, network can adopt different power control policy to adjust the uplink interference level and trade off edge and average performance.

In addition to the slow open loop fractional path loss compensation, two fast compensation factors were introduced in LTE. One is the closed loop compensation factor where base station send transmit power control (TPC) command in DCI to

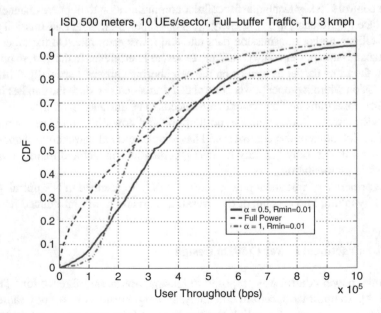

Fig. 3.6 User throughput under different power control schemes

quickly adjust uplink transmission power on top of path loss compensation. This give the system another mechanism to control the UE transmission power for interference management or data rate boosting based on the uplink traffic load, interference level, etc. Another fast compensation factor is an offset based on the uplink transmission MCS level and allocated bandwidth. This allows the system to control UE transmission power jointly with resource allocation and link adaptation.

3.1.3.2   NR Uplink Power Control Design Requirements and Framework

In NR, LTE fractional power control is extended to support new design aspects and features of NR. Comparing to LTE, one major change is the removing of cell common reference signal (CRS) for much higher flexibility and for eliminating the constant source of interference in the network and hence improving spectrum efficiency. Without CRS, path loss measurement for power control need to be based on other reference signal and the natural choices are CSI-RS and SS/PBCH.

Another unique feature of NR is to support high-frequency bands (FR2) by introducing beam management procedure to overcome its coverage issue. The introduction of beamforming complicates the estimation of path loss as the effective path loss depends on the transmitting and receiving beamforming gains at gNB and UE and the UE may communicate with the gNB using different pairs of transmitting and receiving beams on different slots and channels. Therefore, for different beams, UE needs to measure path loss via different downlink reference signals.

Furthermore, NR supports configurable numerology, frame structure, both CP-OFDM and DFT-s-OFDM uplink waveforms, various services (for example, eMBB and URLLC), etc. This flexibility requires that uplink power control to be very flexible for various scenarios and requirements. However, the potential number of combinations is very high and independent configuration for each of the scenarios can cause very high complexity for the UE and the network.

Uplink transmission power is calculated according to the following formula:

$$P = \min\left[P_{CMAX}, \{\text{Open loop PC}\} + \{\text{Closed Loop State}\} + \{\text{Other Offsets}\} \right]$$

$$= \min\left[P_{CMAX}, \{P_0(j) + \alpha(k) \cdot PL(q)\} + \{f(l)\} + \{10\log_{10} M + \Delta\} \right]$$

Among the key components of uplink power control, it is apparently path loss measurement and closed loop state have larger impacts on UE complexity while open loop parameters are relatively simple. The configuration framework of NR uplink power control is defined as the following to maintain relatively low complexity of the UE with yet sufficient flexibility for the network implementation. UE can be configured with a small number (up to 4) of downlink signal (CSI-RS or SS/PBCH block) for path loss measurement for possibly different TRP and/or transmitting and receiving beam pairs, a small number of (up to 2) closed loop power control states with corresponding TPC monitoring, and a relative large number (up to 32) of sets of open loop parameters (Po and α). The network configures (semi-statically via

RRC) and indicates (dynamically via DCI) the UE to apply combination of proper path loss, open loop parameters, and closed loop state to calculate uplink transmission power.

3.1.4 HARQ

HARQ in NR is adaptive and asynchronous for both the downlink and the uplink. On the downlink, a UE supports at most 16 HARQ processes but can be configured with fewer processes. Contrary to LTE where the time gap between the reception of PDSCH and the reporting of HARQ-ACK feedback is fixed to 4 ms, NR was designed to ensure a large amount of flexibility for the network and the UE in reporting HARQ-ACK feedback. A field in the DCI tells UE the number of slots and symbols between the end of the PDSCH reception and the transmission of HARQ-ACK feedback over PUCCH along with the corresponding PUCCH resource, or over PUSCH if data is transmitted on the uplink at the same time. This flexibility allows the network to request very fast HARQ-ACK feedback, e.g., for low latency services, or to delay the HARQ-ACK feedback as needed, e.g., to accommodate some other downlink or uplink transmission in the meantime.

The timing configurable for HARQ-ACK feedback by the network depends on the UE processing capability. In Release-15, NR supports a fast UE PDSCH processing capability. This is specified by the minimum number of uplink OFDM symbols (N1) between the last symbols of the PDSCH and the first symbol of the PUCCH which carries the HARQ-ACK feedback. The minimum time required by the UE to process the PDSCH, to compute the HARQ result and to prepare the PUCCH or PUSCH transmission is expressed in number of OFDM symbols depending on the subcarrier spacing configured on downlink and uplink, the PDSCH mapping type and the number of DM-RS symbols, and other conditions such as the number of component carriers in a cell group and timing difference between the component carriers, the number of symbols of the CORESET carrying the scheduling PDCCH, and depending on multiplexing HARQ-ACK feedback with other UCI in the PUCCH or PUSCH.

Since a slot consists of 14 OFDM symbols, it can be inferred from Table 3.1 that some UEs can support transmitting HARQ-ACK feedback in the same slot used for receiving the PDSCH. This allows for very low latency services.

The UE can be configured for TB-based HARQ feedback of CBG-based HARQ feedback. Codeblock-group (CBG) HARQ allows one transport block (TB) to be divided into codeblock groups where each codeblock group is transmitted with its own CRC check. This allows the UE to report that only some CBGs were not successfully received, and allows the network to retransmit only the CBGs that the UE failed to receive. This not only saves downlink resources, but also allows the gNB to puncture an on-going transmission for one UE with another shorter transmission for another UE, in case that other UE receives a service with lower latency

Table 3.1 Minimum UE PDSCH processing time N_1 (OFDM symbols)

	Minimum processing(capability #1)		Fast processing(capability #2)
	UE configured to handle only front-loaded DM-RS	UE configured to handle additional DM-RS	UE configured to handle only front-loaded DM-RS
15 kHz SCS	8 symbols [571 us]	13 symbols [929 us]	3 symbols [214 us]
30 kHz SCS	10 symbols [357 us]	13 symbols [464 us]	4.5 symbols [161 us]
60 kHz SCS	17 symbols [304 us]	20 symbols [357 us]	9 symbols [161 us] (for frequency range 1)
120 kHz SCS	20 symbols [179 us]	24 symbols [214 us]	

QoS, without corrupting the entire TB for the first UE. HARQ-ACK feedback for CBGs comes at the expense of higher feedback overhead. The UE is configured with the maximum number of CBGs per transport block and determines the exact number of CBGs depending on the number of codeblocks in the scheduled TB. CBG transmission information (CBGTI) is signaled to the UE in the downlink DCI that schedules the PDSCH. A DCI sent for a retransmission of a PDSCH initially transmitted with CBGs indicates which of the CBGs are being retransmitted, and can also indicate whether the previously transmitted CBGs where actually transmitted or not. This allows the UE to flush its buffer for the corresponding CBG, instead of assuming that some valid transmission was previously received for this CBG when the UE had in fact only received noise and interference. A CBG can only be retransmitted with the same codeblocks as in the initial transmission of the TB.

As in LTE, NR supports that the UE reports HARQ-ACK feedback for multiple HARQ processes in the same uplink transmission, with HARQ multiplexing and HARQ bundling. HARQ multiplexing supports semi-static HARQ codebook and dynamic HARQ codebook aided by DAI (downlink association index) signaling in DCI. HARQ-ACK bundling can be adapted in dynamic manner according to UCI payload.

On the uplink, HARQ was already very flexibly designed in LTE. This flexibility is enhanced in NR with similar features as on the downlink. While LTE supported an indication of HARQ ACK or NACK for a UL TB from the network in the downlink physical indication channel, all HARQ-ACK indication from the network can only be carried in a DCI in NR. This means the UE knows whether its uplink transmission was unsuccessful only if the UE receives another UL DCI granting the UE a retransmission of a TB on the uplink. Since the UL grant timing for PUSCH is very flexible in NR, so is the timing between two transmissions of the same TB by a UE. Uplink in NR also supports TB-based and CBG-based HARQ.

3.2   RAN Architecture Evolution and Protocol

3.2.1   Overall Architecture

The 5G system provides an enhanced end-to-end architecture to support multiple deployment options. The following options have been included in Release-15 to support the flexibility of network deployment for requirements for different markets:

- Option2: standalone NR
- Option3: E-UTRAN-NR Dual Connectivity
- Option4: NR-E-UTRA Dual Connectivity
- Option5: E-UTRA connecting to 5GC
- Option7: NG-RAN E-UTRAN-NR Dual Connectivity
- NR-NR Dual Connectivity

In addition, a CU-DU architecture with RAN split into different functional entities has been provided and the major design of the above RAN architecture is introduced in the following sections.

3.2.1.1   RAN Architecture Overview

In 5G, RAN network nodes are called NG-RAN, which includes gNB and ng-eNB. gNB provides NR control plane and user plane functions and ng-eNB provides EUTRA control plane and user plane functions.

The NG interface is to connect NG-RAN towards 5GC (5G Core network). For the control plane NG-RAN is connected to the access and mobility management function (AMF) via NG-C interface and for the user plane NG-RAN is connected to the user plane function (UPF) via NG-U interface. The gNBs and ng-eNBs are interconnected with each other via Xn interface. The overall architecture is shown in Fig. 3.7.

In terms of functionality, the NG-RAN provides functions for Radio Resource Management, IP header compression, encryption and integrity protection of data, selection of/routing of Control Plane data to AMF and routing of User Plane data to UPF, RRC connection management, interworking between NR and E-UTRA, etc.

The AMF/UPF/session management function (SMF) provides the functions from the core network. AMF provides mainly the control plane functions including Access Authentication and Access Authorization, mobility management, SMF selection, etc. UPF provides functions including packet routing and forwarding, QoS handling, etc. SMF mainly provides session management related functions.

The above functional split is summarized in Fig. 3.8 where grey boxes depict the logical nodes and blue boxes depict the main functions, and the details can also be seen in [Sect. 4.2 of [5]]

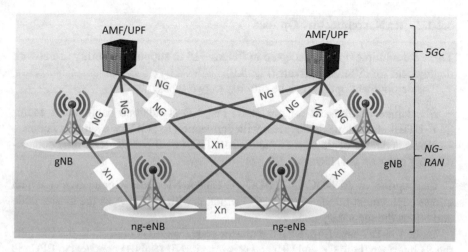

Fig. 3.7 The overall architecture for 5G

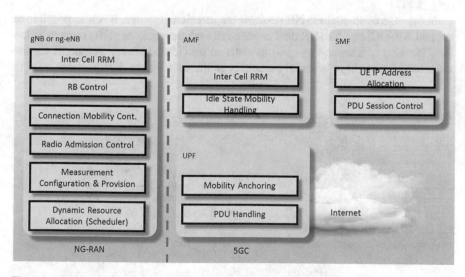

Fig. 3.8 Functional split between NG-RAN and 5GC

3.2.1.2 RAN Architecture Options

There are multiple options accepted in Release-15 to support flexibility of network deployment for different markets (Fig. 3.9).

The deployment options can be differed to two types:

– Standalone deployment for one RAT, i.e., NR standalone deployment and/or E-UTRA standalone deployment with connection to 5GC for each RAN node.
– Non-standalone deployment which is MR-DC, i.e., the multi-RAT dual connectivity between E-UTRA and NR.

One option of MR-DC is EN-DC (E-UTRA-NR Dual Connectivity) which allows deployment of NR without 5GC which is with E-UTRA as the master node and NR as the secondary node.

In the EN-DC case (Fig. 3.9), the master eNB (MeNB) connects to the EPC via the S1 interface (both CP and UP) and secondary gNB (SgNB) connects to EPC via S1-U interface (only UP). MeNB connects to en-gNB via X2 interface (both CP and UP) and en-gNB connects to en-gNB via X2-U interface (only UP).

In addition to EN-DC, there are also other multi-RAT deployment options which have connectivity to the 5GC: NG-RAN E-UTRA-NR Dual Connectivity (NGEN-DC) (Fig. 3.10) and NR-E-UTRA Dual Connectivity (NE-DC) (Fig. 3.11).

In the case of standalone NR deployment mentioned above, it is also possible to support NR-DC, i.e., NR-NR dual connectivity where a master gNB connects to the 5GC but also to a secondary gNB in order to provide additional resources for the same user. This option is useful especially in the case where the operators want to

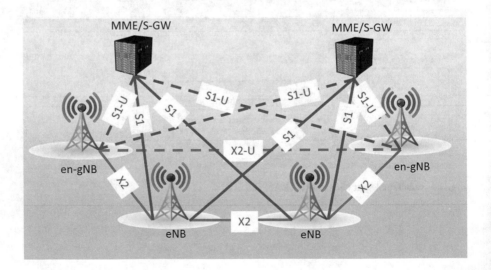

Fig. 3.9 The 5G EN-DC architecture

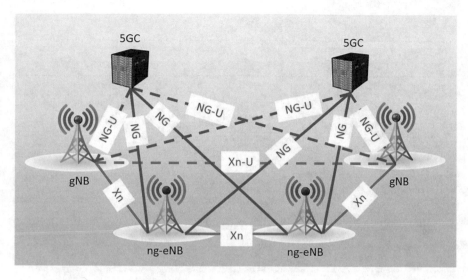

Fig. 3.10 The 5G NGEN-DC architecture

utilize FR1 and FR2 spectrum resources together. In 3GPP, NGEN-DC, NE-DC and NR-DC are grouped together and are known as multi-Radio dual connectivity.

3.2.1.3 CU-DU Split

Within an NG-RAN node, there are two functional entities: gNB-CU (gNB Central unit) and gNB-DU (gNB Distributed unit) shown in Fig. 3.12. A gNB consists of a gNB-CU and one or more gNB-DU(s), and the interface between a gNB-CU and a gNB-DU is called F1 interface. A more detailed discussion can be found in [14].

The CU provides the functions for RRC/PDCP/SDAP layers, and DU provides functions for PHY, MAC, and RLC layers. One CU is allowed to connect to multiple DUs while one DU can only connect to one CU. The F1 interface is only visible within one gNB, outside the gNB the interface between different CUs uses the Xn interface. The NG interface is also reused for the connection between CU and 5GC.

For a single CU, there is a further function split between its control plane (CP) and user plane (UP) (Fig. 3.13). The E1 interface is to connect CU-CP and CU-UP. The CU-CP is connected to the DU through the F1-C interface, and the CU-UP is connected to the DU through the F1-U interface.

3.2.1.4 RAN Protocol and Stack

3.2.1.4.1 NR Standalone Network Protocol Architecture

In standalone NR, the radio interface protocol is also split into control plane and user plane as in E-UTRA and the detailed design is introduced in the following sections.

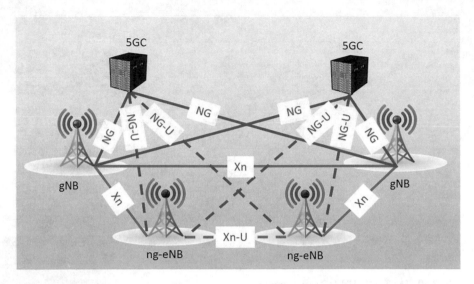

Fig. 3.11 The 5G NE-DC architecture

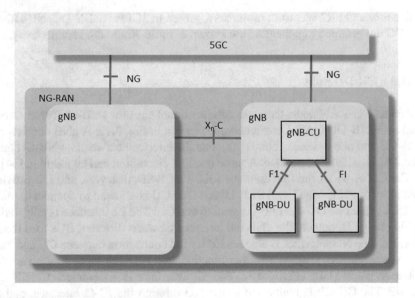

Fig. 3.12 Block diagram of the NR-RAN architecture

3.2.1.4.1.1   Control Plane

For the control plane, all layers shown in Fig. 3.14 provide similar functions than those in E-UTRA [5].

The NR RRC layer provides functions including system information, paging, RRC connection management, security, Signaling Radio Bearers (SRB) and Data Radio Bearers (DRB) establishment/modification, mobility management for UEs in

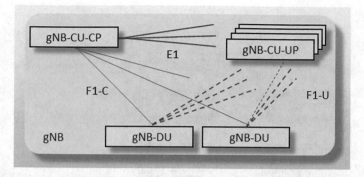

Fig. 3.13 CP and UP separation for CU

different RRC states, QoS management functions, UE measurement management, radio link failure detection and recovery, etc. If carrier aggregation or dual connectivity is supported, the RRC layer also provides support of addition, modification, and release of these features.

Compared with E-UTRA, the RRM measurements have been enhanced to support beam-based measurement and reporting. In addition, a new RRC state, RRC_INACTIVE, was introduced to save power and reduce the signaling overhead and latency for subsequent data transmission. The RRC layer also supports configuration of new concepts introduced in the physical layer like Bandwidth Part, Supplementary UL, Beam management, etc. The system information mechanism is further enhanced to reduce the broadcast capacity consumption, e.g., on-demand SI acquisition is introduced for NR [6].

3.2.1.4.1.2 User Plane

For the user plane (Fig. 3.15), medium access control (MAC) / Radio Link Control (RLC) / packet convergence protocol (PDCP) layer provides similar functions than those in E-UTRA. Additionally, NR introduced a new protocol: the Service Data Adaptation Protocol (SDAP).

MAC Layer

The MAC layer provides functions including mapping between logical channels and transport channels, (de)multiplexing of MAC SDUs, HARQ process, power headroom reporting, scheduling and priority handling for multiple UEs, etc. Compared with E-UTRA, the MAC layer is enhanced to support multiple numerologies, bandwidth parts, beam relevant functions (e.g., beam failure recovery), and supplementary UL handling. In the initial version of NR, only fundamental functions have been defined and some other functions might be supported in the future, e.g., broadcast/multicast services.

RRC is in control of the MAC configuration. The functions, timers, and parameters for different MAC entities in the UE are operated or configured independently in general. For dual connectivity case, each cell group has one MAC entity, i.e.,

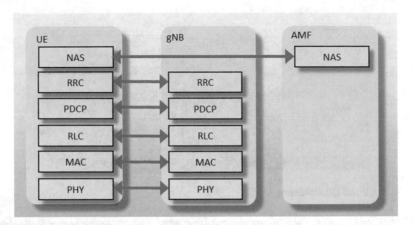

Fig. 3.14 Block diagram of the control plane

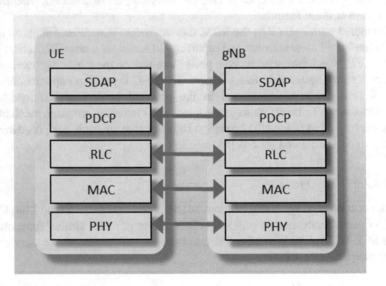

Fig. 3.15 Block diagram of the user plane

when the UE is configured with SCG, two MAC entities are configured to the UE: one for the MCG and one for the SCG. Figure 3.16 shows the structure of the MAC entity.

Figure 3.17 shows the structure for the MAC entities when MCG and SCG are configured.

Similar as E-UTRA, NR defines the transport channels and logical channels as in Tables 3.2 and 3.3. The MAC entity is responsible for mapping logical channels onto transport channels, which applies to both uplink and downlink, as also shown in Figs. 3.16 and 3.17.

The MAC PDU formats for DL and UL are shown in Figs. 3.18 and 3.19.The MAC PDU consists of multiple MAC subPDUs. For the DL MAC PDU, the MAC

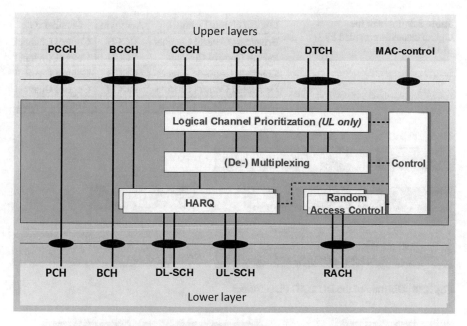

Fig. 3.16 Block diagram of the MAC structure overview

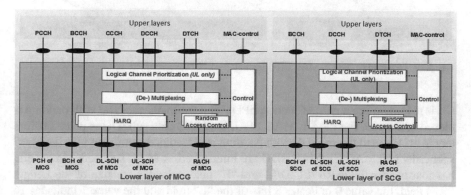

Fig. 3.17 Block diagram of the MAC structure overview with two MAC entities

Table 3.2 Table of the transport channels used by MAC

Transport channel name	Acronym	Link
Broadcast Channel	BCH	Downlink
Downlink Shared Channel	DL-SCH	Downlink
Paging Channel	PCH	Downlink
Uplink Shared Channel	UL-SCH	Uplink
Random Access Channel	RACH	Uplink

Table 3.3 Table of the logical channels provided by MAC

Logical channel name	Acronym	Channel type
Broadcast Control Channel	BCCH	Control Channel
Paging Control Channel	PCCH	Control Channel
Common Control Channel	CCCH	Control Channel
Dedicated Control Channel	DCCH	Control Channel
Dedicated Traffic Channel	DTCH	Traffic Channel

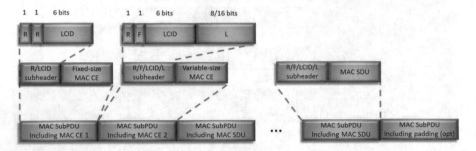

Fig. 3.18 Diagram of the DL MAC PDU format

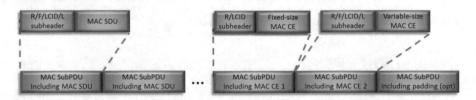

Fig. 3.19 Diagram of the UL MAC PDU format

CE is placed before any MAC data subPDU and the MAC subPDU with padding is placed in the end. For the UL MAC PDU, the MAC data subPDU is placed before any MAC CE, and the subPDU with padding is placed in the end. For NR, the MAC subheader and MAC CE/MAC SDU are interleaved. A MAC subheader in general has four header fields R/F/LCID/L, and "F" field indicates the size of the Length ("L") field, the value 0 indicates 8 bits of the "L" field and the value 1 indicates 16 bits of the "L" field. The LCID is the logical channel identifier to identify the logical channel instance of the corresponding MAC SDU or the type of the corresponding MAC CE which has 6 bits.

The MAC layer for NR is designed to support new features including bandwidth parts, supplementary uplink, beam recovery, and multiple numerologies. In addition, the NR MAC is different from that of E-UTRA as follows: the number of logical channels is expanded, only asynchronous HARQ is supported, and the scheduling mechanism is further enhanced. Each of these will now be discussed *in odinem*.

In E-UTRA, 10 LCIDs are reserved for Logical Channels: LCID 1 and LCID 2 are reserved for SRB1 and SRB2, and totally 8 DRBs can be supported. While in NR, as SRB3 is introduced, LCID 1–3 are reserved for the three SRBs, respectively, and up to 29 DRBs can be supported in a MAC entity.

In NR, only asynchronous HARQ is supported for both UL and DL. To enhance the reliability of data transmission, slot aggregation is introduced and the number of repetition is configured by RRC. To also improve the efficiency of retransmission, CBG-based retransmission is supported at physical layer, which is transparent to the MAC layer.

Both dynamic scheduling and semi-static scheduling are supported in NR. Furthermore, the Scheduling Request (SR) mechanism and Buffer Status Report (BSR) have been further improved compared with E-UTRA.

NR significantly enhanced the SR mechanism. Each SR configuration can now be associated with multiple PUCCH configurations across multiple bandwidth parts or serving cells. Furthermore, for a single MAC entity, multiple SR configurations are supported.

The BSR can be triggered with uplink data arrivals when UL data with higher priority arrives for a certain LCG (logical channel group) or when previously none of the logical channels within the LCG contain any uplink data. It can also be triggered when periodic BSR timer expires, or when there are padding bits left. Correspondingly, regular BSR, periodic BSR, or padding BSR are reported. The long BSR has been improved to use a variable size to support more efficient reporting. The long truncated BSR format has been introduced so that it can report, as many as possible, the buffered data for multiple LCGs.

This discussion in the differences between the NR and E-UTRA MAC has been cursory in nature. The interested reader can find further details in [7].

RLC Layer

Same as E-UTRA RLC, NR RLC also has three modes, AM (Acknowledge Mode), UM (Un-Acknowledged Mode), and TM (Transparent Mode), respectively, as shown in Fig. 3.20.

The RLC AM entity includes a transmitting side and a receiving side supporting retransmission. The RLC UM entity is configured as either a transmitting entity or a receiving entity. The RLC TM entity is either a transmitting entity or a receiving entity. System information and paging use RLC TM mode.

The most significant change in the NR RLC compared with the E-UTRA RLC is that RLC concatenation is removed and reordering is not performed in NR RLC. The former feature is to facilitate the pre-processing in NR such that it can meet the high demand in NR for latency. The latter feature is now supported at the PDCP layer. In this case, there is no need to duplicate the reordering functionality in the RLC layer, and when a complete packet is received, it is forwarded to PDCP layer immediately. More details can be seen in [8].

PDCP Layer

The PDCP layer provides header (de)compression, (de)ciphering and integrity protection and verification, as well as transfer of data for both user plane and control plane for various bearer types. The maximum supported size of a PDCP SDU and PDCP Control PDU is 9000 bytes.

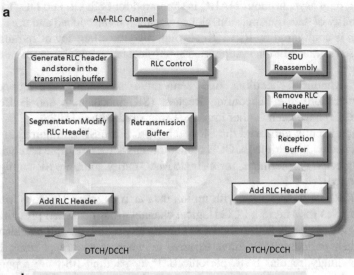

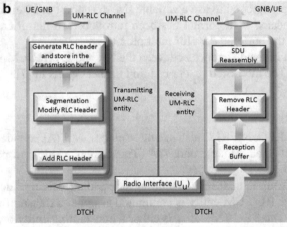

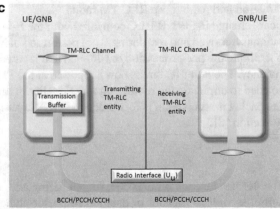

Fig. 3.20 Diagrams of the RLC AM/UM/TM modes

Unlike E-UTRA, out-of-order-delivery is supported in NR PDCP to allow ordering made only in upper layer. In addition, packet duplication is also supported to improve the reliability by transmission on different links, for both dual connectivity and carrier aggregation. Integrity protection is also newly introduced for data bearers in NR PDCP.

Figure 3.21 shows the functional overview of the PDCP layer. In transmitting operation for each PDCP SDU, the PDCP entity first starts the discard timer and assigns COUNT associated with the PDCP SDU, performs header compression and then integrity protection and ciphering which is configurable at a DRB level. If split bearer is configured, the PDCP PDU can be delivered to either the RLC primary path or the secondary path based on condition whether the total amount of data volume is equal or above the configured threshold.

In receiving operation, the PDCP entity determines the COUNT value of the received PDCP PDU and performs deciphering and integrity verification for the packet. Duplication detection and removal is then performed. If for a DRB, out-of-order-delivery is not configured, reordering is also performed. Then after header decompression, the data packet is ready to be delivered to upper layers. More details can be seen in [9].

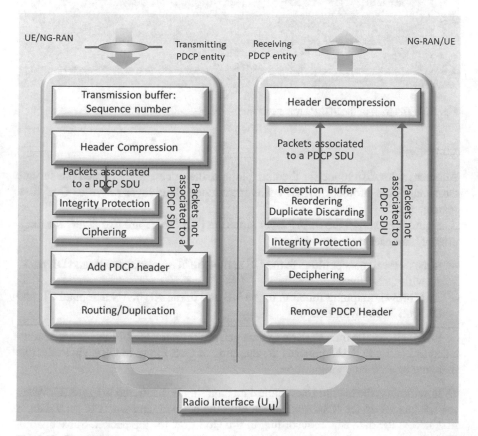

Fig. 3.21 The PDCP handling diagram

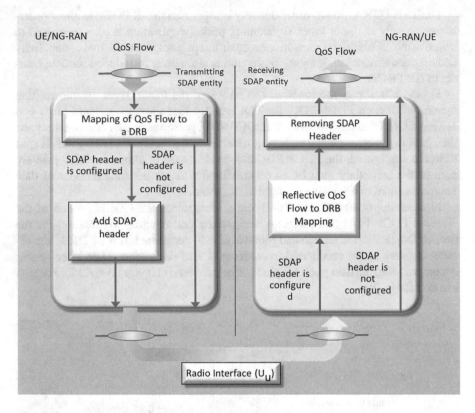

Fig. 3.22 Diagram of the SDAP layer structure

SDAP Layer

Another major difference for the user plane is that a new layer named the service data adaptation protocol (SDAP) was introduced. This layer provides QoS flow management.

Figure 3.22 shows the SDAP layer structure [10]. There are two layers for mapping of data packets to data bearer:

- At NAS level, packet filters in the UE and in the 5GC associate UL and DL packets with QoS Flows
- At AS-level, mapping function in the UE and in the NG-RAN associate UL and DL QoS Flows with DRBs

SDAP allows the transmission of data packets with the QoS Flow Identifier indicated by NAS and RAN manages the mapping of QoS Flows to DRB via Reflective mapping or explicit configuration.

In summary, the overall User Plane data handling is illustrated in Fig. 3.23. When the UE has set up the RRC connection with the network and wants to send data, it first selects the data bearer according to the QoS flow for the packet. Subsequently it applies header compression and security based on the network configuration.

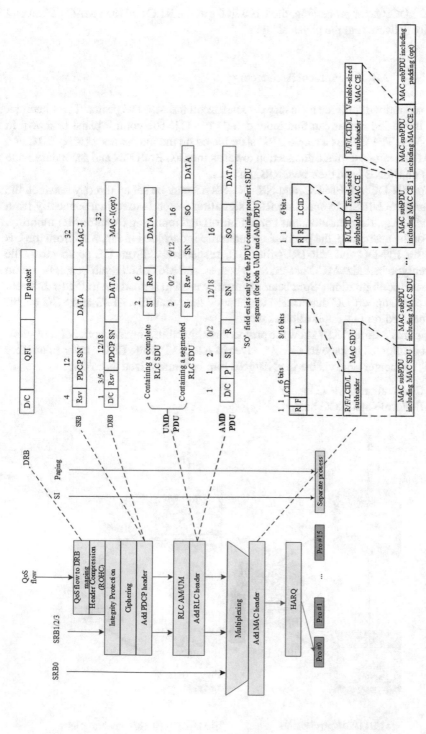

Fig. 3.23 Overview of user plane data handling

After PDCP layer processing, the UE will form the RLC and then MAC packet and finally delivers it to the physical layer.

3.2.1.4.2   MR DC Protocol Architecture

The section will provide a cursory discussion on the MR-DC protocol architecture. The interested reader can find more details in [11]. The control plane is shown in Fig. 3.24. The UE has a single RRC state based on the master node (MN) RRC and only has a single C-plane connection towards the CN. Both MN and secondary node (SN) gNB/eNB have their own RRC entities.

For EN-DC, NGEN-DC, and NR-DC, SRB3 may be set up directly between the UE and the NR SN to speed up RRC signaling which is used independently from the MN, e.g., for measurements configuration and reporting for intra-MN mobility.

For the user plane, the protocol architecture is designed in Fig. 3.25 from the UE side for EN-DC and MR-DC with 5GC, respectively. From UE perspective, the bearer type includes MCG bearer, SCG bearer, and MCG/SCG split bearer based on different configurations. Split bearer is mainly used to offload the traffic to different legs or to support DC duplication to enhance the reliability. MCG and SCG bearer can be used for CA duplication as well.

Figures 3.26 and 3.27 show the protocol architecture from network point of view. As it is allowed to have RLC/MAC layer in one RAT while PDCP layer in another RAT, the bearer type visible in eNB/gNB can be summarized as below:

– MN-terminated MCG bearer
– MN-terminated SCG bearer

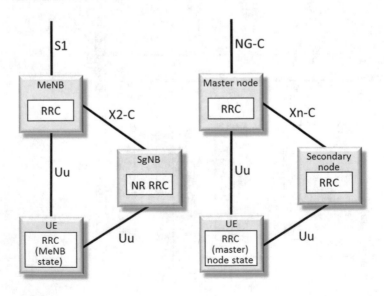

(a) EN-DC control plane   (b) MR-DC with 5GC control plane

Fig. 3.24 MR-DC control plane. (**a**) EN-DC control plane. (**b**) MR-DC with 5GC control plane

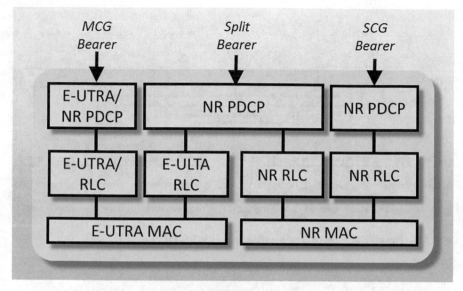

(a) EN-DC

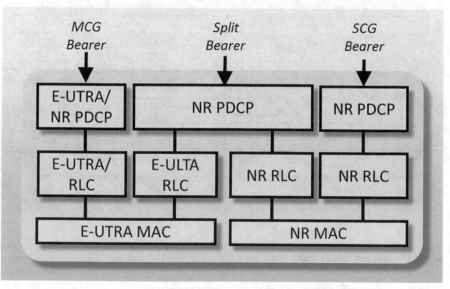

(b) MR-DC with 5GC

Fig. 3.25 NR user plane protocol architecture from the UE's point of view. (**a**) EN-DC. (**b**) MR-DC with 5GC

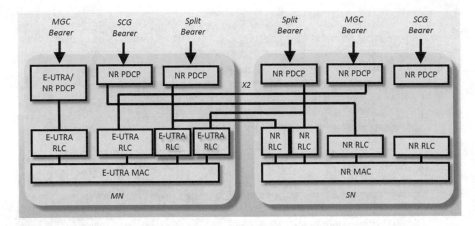

Fig. 3.26 Network side protocol termination options in EN-DC

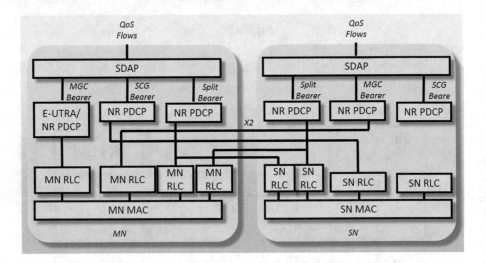

Fig. 3.27 Network side protocol termination options in MR-DC with 5GC

- MN-terminated Split bearer
- SN-terminated MCG bearer
- SN-terminated SCG bearer
- SN-terminated Split bearer

Like for E-UTRA, NR uses a RAN-level key specific to a UE-RAN node connection, from which ciphering and integrity protection keys are derived for SRBs and for each DRB. For MR-DC, this key is only used for MN-terminated bearers and SRB1/2 and the UE is allocated with a secondary key, which is used to derive ciphering and integrity protection keys for SN-terminated bearers and SRB3.

Apart from this aspect, the termination point is transparent to the UE as this only impacts the coordination between the eNB/gNBs.

3.2.2   Fundamental Procedures for NR Standalone

3.2.2.1   UE State Transition

In NR the state transition between RRC_CONNECTED and RRC_IDLE is the same as E-UTRA. In general the UE camps on a cell, reading the system information and listens to the paging periodically in idle mode. When the UE wants to initiate data transmission or responding to the paging, the UE initiates the random access procedure to enter RRC_CONNECTED state. The subsequent sections are organized in such an order to introduce basic procedures for each step. Further details for the RRC procedure can be found in [6].

Different from E-UTRA, a new UE state is introduced as NR RRC_INACTIVE. If the UE is in RRC_CONNECTED and the network sends RRCRelease message to the UE with suspend indication, the UE keeps the original context and enters RRC_INACTIVE state (Fig. 3.28). The next time when the UE has data or signaling (e.g., to perform RNA) to transmit, it sends the RRCResumeRequest message to the network and the RRC connection can be resumed. In this way the UE in RRC_INACTIVE can transmit and receive data with reduced signaling and latency compared to a UE moving from RRC_IDLE, because a UE in RRC_IDLE has to perform RRC connection establishment procedure every time when there is data to transmit. On the other hand, the UE in RRC_INACTIVE state also achieves the power consumption benefits from UE-based mobility and IDLE mode DRX compared with the case in RRC_CONNECTED state. The state transition is illustrated in Fig. 3.28.

Fig. 3.28 RRC state transition diagram

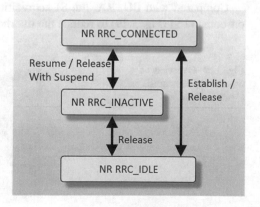

3.2.2.2   System Information Acquisition

Similar to E-UTRA, the system information is divided into MIB (Master Information Block) and SIBs (System Information Block). MIB is transmitted on BCH with a periodicity of 80 ms and repetitions made within 80 ms.

SIBs are further divided into SIB1 and other SIBs which are transmitted on DL-SCH. The minimum system information (that is the system information essential for the UE to perform an initial access to the cell, such as access control information and common channel configurations) is provided in MIB and SIB1, and is always broadcast. SIB1 is transmitted with a periodicity of 160 ms. The default transmission repetition periodicity of SIB1 is 20 ms and variable transmission repetition periodicity is up to network configuration. SIB1 also includes the scheduling information for other SIBs.

Other system information consists of system information not essential for initial access, and this may be optionally always provided by broadcast, or otherwise provided on-demand. In the initial version of NR, SIB2-SIB9 have been introduced:

– SIB2 contains cell reselection information, mainly related to the serving cell.
– SIB3 contains information about the serving frequency and intra-frequency neighboring cells relevant for cell reselection.
– SIB4 contains information about other NR frequencies and inter-frequency neighboring cells relevant for cell reselection.
– SIB5 contains information about E-UTRA frequencies and E-UTRA neighboring cells relevant for cell reselection.
– SIB6 contains an ETWS primary notification.
– SIB7 contains an ETWS secondary notification.
– SIB8 contains a CMAS warning notification.
– SIB9 contains GPS time and Coordinated Universal Time (UTC), by which the UE can obtain the UTC, the GPS, and the local time.

Other than SIB1, SIBs having the same periodicity can be included to the same SI message with its own time domain windows. Such SI windows do not overlap with different SI messages.

Compared with EUTRA, the SI acquisition procedure additionally introduced on-demand SI (Fig. 3.29) to reduce the overhead for the broadcast control channel.

Fig. 3.29 Overview of on-demand SI acquisition

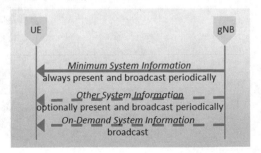

Fig. 3.30 Pectoral
representation of area-
based system information

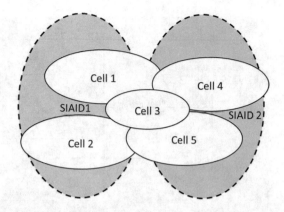

The network can at the UE (or UEs) request mute some SI messages for some
broadcast periods. On-Demand system information request can be sent based on
MSG 1 or MSG 3 in RACH procedure. In case of MSG1-based solution, a particular
RACH resource for SI request is mapped to a set of SI messages, while for MSG3-
based solution, the request for system information is indicated with a bitmap in
MSG3 of the RACH procedure.

Another difference than E-UTRA is that NR introduces the concept of area-based
SI. The system information (except for minimum system information in MIB and
SIB1) that has been acquired by the UE in a cell may be valid also in other cells. If, after
obtaining SIB1, the UE verifies the SIB provided on a new cell is within the same
System Information Area and the value tag is also the same, the UE would consider that
the stored SIB is valid and may use the stored SIB without reading the SI again. In this
way the UE can reduce the unnecessary reading of the SI which is helpful to reduce UE
power consumption, as well as broadcast overhead[3]. This is illustrated with Fig. 3.30.
When the UE moves from cell1 to cell2 or from cell4 to cell5, acquisition of all of the
SI is not needed since they are in the same SI area and have the same SI.

3.2.2.3   Paging and DRX

Paging is used for the network to transmit paging information to the UEs who are in
RRC_IDLE or RRC_INACTIVE, and to notify UEs in RRC_IDLE, RRC_
INACTIVE and RRC_CONNECTED state of system information change and
earthquake or emergency warning messages.

---

[3]The network may need to provide the on-demand SI less frequently if the UE can utilize system
information provided on one cell in multiple cells.

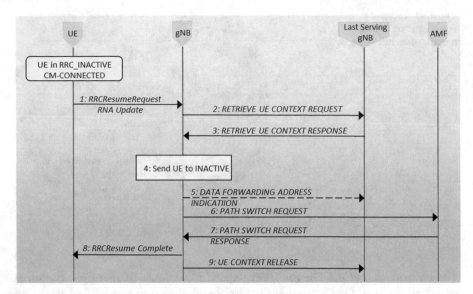

Fig. 3.31 Message flow diagram using the RNA update principle

- For UEs in RRC_IDLE, the paging is sent from the 5GC to the gNBs which are within that Tracking Area, and these gNBs then send these paging messages to the UE.
- For UEs in RRC_INACTIVE, a RAN-based notification area is configured. The gNB already knows which RAN notification area the UE is camping on and only sends the paging in that area to reduce the radio resources consumed on paging. The paging message is same for both RAN initiated paging and CN initiated paging.
- For indicating the notification of system information modification and ETWS/CMAS indication, NR has introduced short message which is transported by the DCI by PDCCH in the physical layer.

In case of paging for UEs in RRC_INACTIVE, a RAN area update procedure is introduced to allow the gNB to know whether the UE changes the RAN-based notification area (RNA). Such RNA update (RNAU) is periodically sent by the UE and is also sent when the UE selects a cell that does not belong to the configured RNA Fig. 3.31.

Discontinuous Reception (DRX) for idle mode and connected mode are supported in NR. The former one applies to the UE in RRC_IDLE and RRC_INACTIVE states in order to reduce power consumption. The definition of PF and PO is reused in NR. One Paging Frame (PF) is one Radio Frame and may contain one or multiple paging occasions (PO) or starting point of a PO. A PO is a set of PDCCH monitoring occasions and can consist of multiple time slots (e.g., subframe or OFDM symbol) where paging DCI can be sent [12]. The UE monitors one paging occasion (PO) per DRX cycle.

In multi-beam operations, the length of one PO is one period of beam sweeping and the UE can assume that the same paging message is repeated in all beams of the sweeping pattern and thus the selection of the beam(s) for the reception of the paging message is up to UE implementation.

3.2.2.4 Access Control

RAN supports various overload and access control functionalities such as RACH back off, RRC Connection Reject, RRC Connection Release, and UE-based access barring.

In NR, the unified access control (UAC) feature was introduced, so that operator-defined access categories can be used to enable differentiated handling for different slices. NG-RAN may broadcast barring control information (i.e., a list of barring parameters associated with operator-defined access categories) to minimize the impact of congested slices.

One unified access control framework applies to all UE states (RRC_IDLE, RRC_INACTIVE and RRC_CONNECTED) for NR. NG-RAN broadcasts barring control information associated with Access Categories and Access Identities (in case of network sharing, the barring control information can be set individually for each PLMN). The UE determines whether an access attempt is authorized based on the barring information broadcast for the selected PLMN, and the selected Access Category and Access Identity (ies) for the access attempt:

- For NAS-triggered requests, NAS determines the Access Category and Access Identity(ies).
- For AS-triggered requests, RRC determines the Access Category while NAS determines the Access Identity(ies).

Table 3.4 shows the mapping between Access Category and the type of access attempt [13]. The gNB handles access attempts with establishment causes "emergency," "mps-PriorityAccess," and "mcs-PriorityAccess" (i.e., Emergency calls, MPS, MCS subscribers) with high priority and responds with RRC Reject to these access attempts only in extreme network load conditions that may threaten the gNB stability.

3.2.2.5 Random Access Procedure

Normally when the UE has data to transmit and passes the access control, the UE can initiate random access procedure. To be more specific, there are various conditions to trigger random access procedure as below:

- Initial access from RRC_IDLE
- RRC Connection Re-establishment procedure
- Handover (HO)

Table 3.4 Access category

Access category number	Condition related to UE	Type of access attempt
0	All	MO signaling resulting from paging
1	UE is configured for delay tolerant service and subjected to access control for Access Category 1, which is a function of UE's HPLMN and the selected PLMN	All except for Emergency
2	All	Emergency
3	All except for the condition in Access Category 1	MO signaling on NAS level resulting from other than paging
4	All except for the condition in Access Category 1	MMTEL voice
5	All except for the condition in Access Category 1	MMTEL video
6	All except for the condition in Access Category 1	SMS
7	All except for the condition in Access Category 1	MO data that do not belong to any other Access Categories
8	All except for the condition in Access Category 1	MO signaling on RRC level resulting from other than paging
9–31		Reserved standardized Access Categories
32–63	All	Based on operator classification

- Arrival of DL/UL data during RRC_CONNECTED but UL is non-synchronized
- Arrival of UL data during RRC_CONNECTED when there are no PUCCH resources for SR available
- SR failure
- Request by RRC upon synchronous reconfiguration
- To establish time alignment at SCell addition
- Transition from RRC_INACTIVE
- Request for Other SI
- Beam failure recovery

The random access can be done in two ways: contention-based access and contention-free access as shown in Fig. 3.32. For contention-based random access, the UE randomly selects a preamble for access and consequently there are risks to have conflicts with other UEs. In this case the contention resolution procedure has to be performed. For contention-free access, as the preamble is assigned by the network to a specific UE, such risks can be avoided and the contention resolution is not needed.

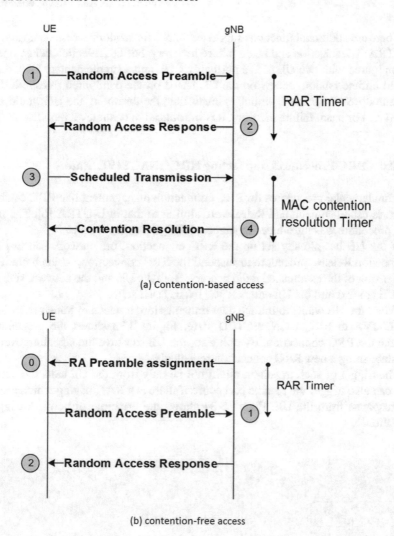

Fig. 3.32 Random access procedure. (**a**) Contention-based access, (**b**) contention-free access

In general the network configures the preambles mapped to each SSB, as well as the PRACH occasions associated with an SSB/CSI-RS in the system information. When the UE wants to initiate the random access, it selects an SSB/CSI-RS prior to RA resource selection. If there is an SSB/CSI-RS signal above with the configured threshold in the candidate list, the UE selects this SSB/CSI-RS; otherwise the UE selects any SSB which is up to UE implementation. The power ramping mechanism and back off timer also apply to NR random access.

There are additional functions supported in NR for random access compared with E-UTRA: UL selection and beam failure recovery. For UL selection, when one cell is configured with two UL (i.e., a normal NR UL and a supplementary UL), the UE would initiate random access on the UL based on the configured threshold. If the measured RSRP for the downlink is lower than the threshold, the UE would select the SUL. For beam failure recovery, it is introduced in beam level mobility section.

3.2.2.6 RRC Procedures Supporting RRC_INACTIVE State

The fundamental procedures for RRC connection management like RRC establishment, re-establishment, and Release are similar to that in E-UTRA [6]. The major new procedure was introduced to support RRC_INACTIVE state.

If the UE has already set up the RRC connection, the network can use RRC Connection Release procedure to suspend the RRC connection, which includes the suspension of the established radio bearers. Both the UE and the network side store the UE context and the UE enters RRC_INACTIVE state.

When the UE wants to initiate data transmission, it needs to transit from RRC_INACTIVE to RRC_CONNECTED state. Figure 3.33 shows the procedure to resume the RRC connection. By doing so, the UE can save the signaling overhead to establishing a new RRC connection repeatedly.

The trigger of such transition might not be only from the UE side, the network side can also trigger the resume procedure if there is a RAN paging which requires the response from the UE. Figure 3.34 shows the principle of network-triggered transition.

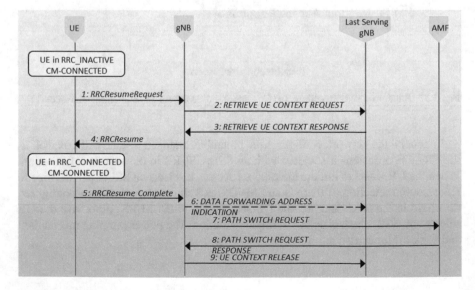

Fig. 3.33 UE-triggered RRC Resume procedure

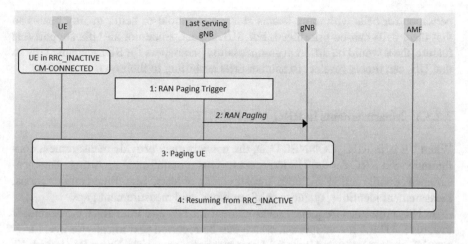

Fig. 3.34 Network-triggered RRC Resume procedure

3.2.3   Mobility Control

3.2.3.1   Cell selection

For cell selection the principle is the same as the generic cell selection procedures for E-UTRA [15]. The UE NAS layer selects and indicates the selected PLMN and equivalent PLMNs to the RRC layer, and the cell selection is based on existing principle. For cell selection in multi-beam operations, measurement quantity of a cell is up to UE implementation. If no suitable cell is found, the UE would then identify an acceptable cell to camp on.

3.2.3.2   Cell Reselection

Cell reselection within NR is also based on existing principle for E-UTRA with additional beam level measurements [15]. For cell reselection in multi-beam operations, the measurement quantity of this cell is derived among the beams corresponding to the same cell based on SS/PBCH block, either by deriving a cell measurement quantity using the highest beam measurement quantity value, or by deriving a cell measurement quantity as the linear average of the power values of up to a maximum configured highest beam measurement quantity value above a configured threshold. Intra-frequency cell reselection is based on the ranking of cells and inter-frequency cell reselection is based on priority-based cell reselection.

For inter-RAT cell reselection, currently only cell reselection between E-UTRA and NR is supported and priority-based inter-RAT cell reselection mechanism is reused.

Compared with E-UTRA, the main difference is that the ranking of the cells may be configured to take multi-beam aspect and SUL into account. There is some compensation for cells with more beams above a threshold or better quality beams so that these cells can be prioritized. For SUL aspect, since not all UEs support this feature, there would be different compensation parameters for SUL-capable UEs so that UEs can timely reselect to another cells according to their own capabilities.

3.2.3.3 Measurements in RRC_CONNECTED

When UE is in RRC_CONNECTED, the network may provide measurement configuration via RRC for the UE to report various measurement reporting. The measurement configuration includes measurement objects, reporting configurations, measurement identities, quantity configurations, and measurement gaps.

In NR, intra/inter-frequency measurement for NR and inter-RAT measurement towards E-UTRA are supported, which can be reported periodically or event triggered. For event-triggered inter-RAT measurements reporting, Event B1 (Inter-RAT neighbor becomes better than threshold) and B2 (PCell becomes worse than threshold1 and inter-RAT neighbor becomes better than threshold2) are supported. For intra-RAT measurement reporting in NR, the following event-triggered measurements are supported:

– Event A1 (Serving becomes better than threshold)
– Event A2 (Serving becomes worse than threshold)
– Event A3 (Neighbor becomes offset better than SpCell)
– Event A4 (Neighbor becomes better than threshold)
– Event A5 (SpCell becomes worse than threshold1 and neighbor becomes better than threshold2)
– Event A6 (Neighbor becomes offset better than SCell)

In NR the measurements can be performed based on SSB or CSI-RS. SMTC (SS/PBCH block measurement timing configuration) is introduced to configure measurement timing configurations, i.e., timing occasions at which the UE measures SSBs. The measurement gap needs also to be configured for inter-frequency or inter-RAT cases based on UE capabilities: if the UE only supports per-UE gap, the UE is not able to receive or transmit data in all serving cells; if the UE supports per-FR gap, the UE can continue receiving or transmitting data in the serving cells in FR2 when measuring FR1 cells.

The above measurements are mainly used for the purpose of connected mode mobility and SCG management in dual connectivity cases. In addition, CGI reporting for the purpose of ANR (Automatic Neighbor Relation) is also supported. Once configured by the network, the UE can report the cell identity to the network for a certain cell to help the network automatically maintain the neighbor cell list.

Another important measurement is called SFTD (SFN and Frame Timing Difference). This measurement is used to indicate the timing difference between two cells, which helps the network to make more accurate measurement configuration to the UEs especially considering the case where the neighbor cells are not synchronized with the serving cell.

More specifically, in MR-DC case, measurements can be configured independently by the MN and by the SN. The SN can only configure intra-RAT measurements and coordination via inter-node signaling between the MN and the SN is used to ensure consistent measurement configuration between these two nodes without exceeding the UE measurement capabilities. When SRB3 is configured, reports for measurements configured by the SN are sent on SRB3.

3.2.3.4   Inter-system/Inter-RAT Mobility in Connected Mode

In Release-15 NR mobility procedure reuses LTE handover procedures in a large extent with additional beam management. The mobility includes cell level mobility (Handover) and beam level mobility (intra-cell beam switch). The cell level mobility is based on the L3 cell level measurements while the beam level mobility is based on the L1 measurement on individual beams.

3.2.3.4.1   Beam Level Mobility

For the beam mobility management, Fig. 3.35 shows the corresponding procedure. The network first configures up to 64 beams which includes a list of SSB/CSR-RS configurations and the conditions for beam failure detection as well as the configuration for beam failure recovery. Upon receiving the configuration from the network, the UE keeps measuring beam quality. If the beam failure is detected at physical layer, PHY layer would inform the MAC layer. Once the criteria is fulfilled by the timer for beam failure instance and the threshold for the max number of beam failure instances, UE can use RACH procedure to request the recovery. The network in this case decides to reconfigure the UE with the updated beam configurations.

3.2.3.4.2   Handover

The process for intra/inter-frequency handover is shown in Fig. 3.36. The Handover Request message includes beam measurement information associated to SSB(s) and CSI-RS(s) for the reported cell. Flow-based QoS is taken into account for HO procedure.

The Handover command configures dedicated RACH resources associated either to the SSB(s) or to the CSI-RS(s), with which the UE can perform contention-free RACH procedure to the target cell. Since, common RACH resources can only be associated to the SSB(s), CSI-RS resource are configured with additional RACH occasions.

For EUTRA-NR interworking, the following scenarios are supported in Release-15:

- Intra-system intra-RAT handover

  - Both intra-NR handover and intra-EUTRA handover without CN change are supported.

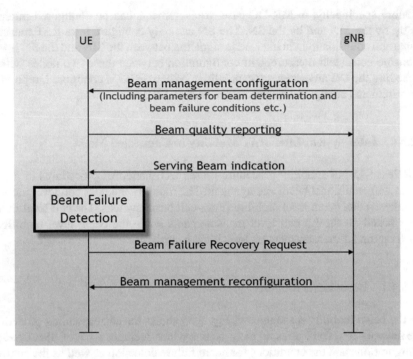

Fig. 3.35 Beam management procedure

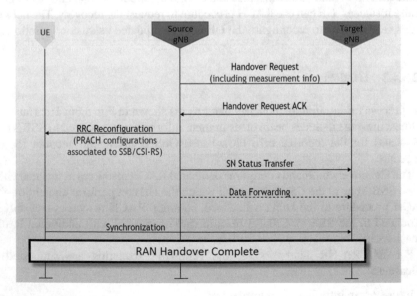

Fig. 3.36 Generic handover procedure

- Intra-system inter-RAT handover between EUTRA and NR

  • This only applies to the case where both RATs are connected to 5GC. Both Xn-based and NG-based handover are supported in this case.
- Inter-system intra-RAT handover

  • This case refers to the case that the EUTRA cells are connected to EPC and 5GC, respectively. When the UE moves between these two cells, inter-system handover is supported.
- Inter-system inter-RAT handover

  • This refers to the case that the EUTRA cell is connected to EPC and the UE moves between such a EUTRA cell and a NR cell. In this case, only CN-based handover is supported and in-sequence and lossless handover are not supported.

3.2.4   Vertical Support

3.2.4.1   Slicing

5G introduced slicing, which is an end-to-end function, to allow more flexible resource utilization and differentiated treatment depending on different customer requirements, Service Level Agreement (SLA), and subscriptions. This is a key feature that allows for the hyper-customization of the system to support the vertical services.[4]

A network slice always consists of a RAN part and a CN part. From the RAN point of view, traffic for different slices is handled by different PDU sessions. The network can realize the different network slices by scheduling and also by providing different L1/L2 configurations.NG-RAN supports awareness of the slicing and a differentiated handling for different network slices which have been pre-configured. The detailed resource management among different slices is up to the network implementation.

NG-RAN also supports the selection of the RAN part of the network slice, by assistance information provided by the UE or the 5GC.

The RAN awareness of Slice is introduced at PDU session level by indicating the slicing identifier corresponding to the PDU Session. NG-RAN is configured with a set of different configurations for different network slices by OAM so as to enable differentiated handling of traffic for network slices with different SLA. In addition, the slice-aware mobility is supported by RAN, which is to enable slice-aware admission and congestion control by transferring the slice identifier during mobility procedures.

To also help to find a suitable CN slice, the UE can provide an identifier which is assigned by the 5GC in MSG5 and the NG-RAN would use this identifier to route the messages to the suitable AMF.

---

[4]Q.v. Sect. 6.1.

3.2.4.2 URLLC

URLLC is another important feature introduced to 5G to support services which require ultra-high reliability and low latency. For higher layer, packet duplication and configured grant mechanisms are introduced: the former one is to provide duplication of the data transmission to enhance the reliability, and the latter one is to reduce latency for data transmission by pre-configuring the resources without dynamic scheduling.

3.2.4.2.1 Packet Duplication

In Release-15 two types of duplication are supported. One is CA duplication (Fig. 3.37), the other is DC duplication (Fig. 3.38). The duplication can apply to both SRB and DRBs.

More specifically, the PDCP layer generates two copies for each packet and then delivers them to two RLC entities. In case of CA duplication, the primary path and the secondary path must be on two different carriers to ensure transmission diversity, and up to eight DRBs and two SRBs can be configured with duplication at the

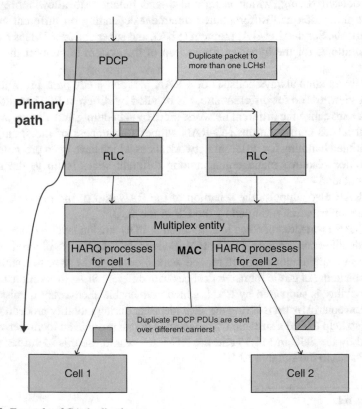

Fig. 3.37 Example of CA duplication

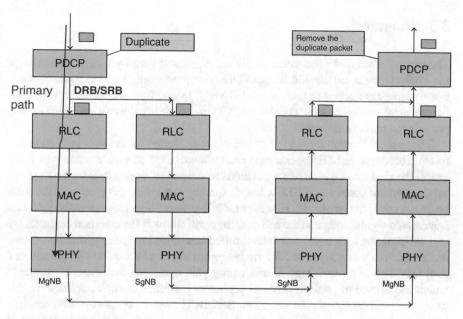

Fig. 3.38 Example of DC duplication

same time; in case of DC duplication, the primary path and secondary path are on MCG and SCG, respectively, and the network indicates the primary path and secondary path for the duplication in the RRC configuration.

3.2.4.2.2   Transmission/Reception Without Dynamic Grant

In NR, transmission/reception without dynamic grant has been enhanced to further reduce the latency brought by dynamic resource scheduling. For DL reception, SPS (semi-persistent scheduling) is supported which has the similar principle as in E-UTRA.

For UL transmission, two types have been introduced called Configured Grant Type 1 and Configured Grant type 2. For Type 1, the uplink grant is provided and activated by RRC and stored as configured grant; For Type 2, the configuration is provided by RRC and (de)activated via L1 signaling which is quite similar as E-UTRA UL SPS. For the same serving cell, either Type 1 or Type 2 can be configured.

Compared with Configured Grant Type 2, Configured Grant Type 1 saves more signaling on (de)activation as it is activated once configured by RRC which can further reduce the latency. Both types of configured Grant can also support repetition to improve the reliability by configuring the period of a transmission bundle for a transmission block, the number of repetition times and RV sequence for each repetition.

3.3   Summary

The newly introduced procedures in 5G are discussed here by taking into account the characteristics of the NR design. The new procedures include initial access, beam management, power control, and HARQ. In addition, the RAN architecture, fundamental procedures for standalone NR, and mobility control are discussed to provide the basic framework.

The concept "beam" is used in NR and the beam-based operation is applied in different procedures. In LTE, the common channels are based on a wide beam, which can enable the UEs located at different positions of the cell to receive it. Because the typical deployment spectrum of LTE is low-frequency band with less path loss, the wide beam can cover the whole cell. However for NR going to high frequency, the path loss is increased significantly and the cell coverage will shrink if the common channels are still based on the wide beam due to the limited beamforming gain. As multiple antennas are normally employed in NR, the beamforming can be used to compensate the path loss. The beam with high beamforming gain is narrow, and therefore multiple beams are needed to perform beam sweeping to guarantee the cell coverage. In this case, the common signals/channels, i.e., SS/PBCH (SSB) are conveyed on multiple beams, and the UE acquires the best SSB among the multiple SSBs to obtain the cell knowledge. Then the UE can obtain the RACH resources, the CORESET resources for RMSI, etc. which are associated with the detected SSB. Once the UE has these information, it will go to the normal communication procedure.

Beam management procedure is to overcome much larger path loss and blockage probability. Based on a set of configured beamformed downlink signals (CSI-RS and/or SSBs), the UE reports the detection quality of the downlink signals to network and the network selects the proper beam(s) for transmissions of PDSCH and PDCCH. Once the link for the associated beam is failed according to the detection result, the beam failure recovery procedure is triggered to look for a new proper beam for PDSCH/PDCCH transmission. For uplink, the UE transmits multiple beamformed SRS resource and the network can select the proper uplink beam for PUCCH/PUSCH transmission. Another method is that there is a beam correspondence between its receiving and transmitting beams. The UE selects a corresponding beam for UL transmission based on DL measurements without relying on network-assisted UL beam management.

In NR, power control applies to both uplink and downlink and is very similar to that of LTE. Uplink power control is still based on fractional power control approach, and downlink part only has very limited specification impact and mainly relies on network implementation.

In NR, adaptive and asynchronous HARQ is applied for both the downlink and the uplink. The HARQ timing is much flexible compared to LTE with fixed to 4 ms between data reception and HARQ-ACK feedback, and the HARQ timing is signaled in the DCI. A field in the DCI tells UE the number of slots and symbols between the end of the PDSCH reception and the transmission of HARQ-ACK feedback. The configuration of HARQ timing depends on the UE processing capability.

5G system supports multiple deployment options, including NR standalone and LTE-NR dual connectivity, etc. RAN is further split into different functional entities such as CU and DU. The design of RAN architecture is introduced in this part. In addition, the protocol for standalone NR including UE state transition, paging, random access procedure, etc. is discussed, which gives the detail of standalone NR operation.

The discussion in our 5G journey has so far focused on the physical layer and the associated RAN procedures and architecture. The multiple deployment options of the 5G system are dependent on the relationship between the RAN and the connected core network, which will be the topic of our next dissertation.

References

1. 3GPP TS 36.104, Base station (BS) radio transmission and reception (Release 15), June 2018
2. 3GPP TS 38.104, Base Station (BS) radio transmission and reception (Release 15), Sept 2018
3. Motorola, R1-060401, Interference Mitigation via Power Control and FDM Resource Allocation and UE Alignment for E-UTRA Uplink and TP, 3GPP TSG RAN1#44, Denver, USA, Feb 2006
4. W. Xiao, R. Ratasuk, R. L. Ghosh, Y. Sun, R. Nory, Uplink power control, interference coordination and resource allocation for 3GPP E-UTRA, in *Proceedings of IEEE Vehicle Technology Conference*, Sept 2006, pp. 1–5
5. 3GPP TS 38.300, NR; NR and NG-RAN Overall Description; Stage 2 (Release 15), Dec 2018
6. 3GPP TS 38.331, NR; Radio Resource Control (RRC) protocol specification (Release 15), Dec 2018
7. 3GPP TS 38.321, NR; Medium Access Control (MAC) protocol speciation (Release 15), Dec 2018
8. 3GPP TS 38.322, NR; Radio Link Control (RLC) protocol specification (Release 15), Mar 2019
9. 3GPP TS 38.323, NR; Packet Data Convergence Protocol (PDCP) specification (Release 15), Mar 2019
10. 3GPP TS 37.324, E-UTRA and NR; Service Data Adaptation Protocol (SDAP) specification (Release 15), Dec 2018
11. 3GPP TS 37.340, Evolved Universal Terrestrial Radio Access (E-UTRA) and NR; Multi-connectivity; Stage 2(Release 15), Dec 2018
12. 3GPP TS 38.213, NR; Physical Layer Procedures for Control (Release 15), Sept 2018
13. 3GPP, TS 22.261 Service requirements for the 5G system; Stage 1 (Release 16) V16.7.0, Mar 2019, https://portal.3gpp.org/desktopmodules/Specifications/SpecificationDetails.aspx?specificationId=3107
14. 3GPP TS 38.401, NG-RAN, Architecture description (Release 15), Dec 2018
15. 3GPP, TS 38.304 User Equipment (UE) procedures in Idle mode and RRC Inactive state (Release 15), v15.4.0, June 2019, [Online], https://portal.3gpp.org/desktopmodules/Specifications/SpecificationDetails.aspx?specificationId=3192

Chapter 4
5G System Architecture

Coauthored with: John Kaippallimalil and Amanda Xiang

This chapter provides a concise description of the 3GPP standards for the 5G System (5GS) and highlights the major enhancements over 4G. Network slicing, virtualization, and edge computing with enhanced connectivity, session/mobility management are some key enhancements. They cater to the requirements of a diverse set of services requiring low latency, high reliability, or a massive number of connections for short, intermittent periods over the same network.

It starts an overview of the end-to-end 5GS. It then describes the service-based architecture (SBA) that organizes core network functions into a set of service-oriented functions and inherent support for virtualization. Network slicing for supporting the broad range of services over the same network is described. An overview of connection and session management including new service and session continuity modes for mobility is given. It then outlines how the 5GC interworks with a 4G Evolved Packet Core (EPC). A detailed description of the control plane (CP) and user plane (UP) protocols in 5GC is then given. Support for virtualized deployments and edge computing is then elucidated. The chapter ends with a discussion on policy and charging for roaming and non-roaming cases.

4.1 5G System Architecture

As in the 4G (LTE/EPC) and previous generations, the 3GPP 5G system defines the architecture for communication between a User Equipment (UE) and an end point, such as an Application Server (AS) in the Data Network (DN), or another UE. The interaction between the UE and the Data Network is via the Access Network and Core Network as defined by 3GPP Standards. Figure 4.1 depicts a simple

Huawei Technologies, USA
Plano, TX

© Springer Nature Switzerland AG 2020
Wan Lei et al., *5G System Design*, https://doi.org/10.1007/978-3-030-22236-9_4

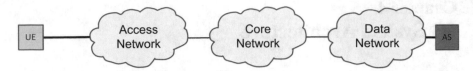

Fig. 4.1 End-to-end architecture

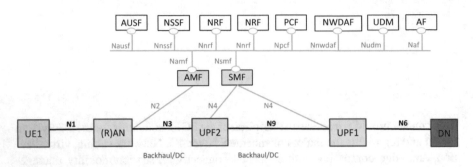

Fig. 4.2 5G System architecture

representation of an end-to-end architecture. In this chapter we focus on describing the 5G Core as defined by 3GPP 5G standards for PLMN [1–3]. The Access Network in 3GPP is referred to as Radio Access Network (RAN).

At a very high level the Core and RAN consist of several Network Functions which are associated with Control Plane and User Plane functionalities. The actual data (also refer it as user data) is normally transported via a path in the User Plane, while the Control Plane is used to establish the path in the User Plane. The Short Message Service (SMS) is an exception in which the data (short message) is communicated via the Control Plane.

The 5G System architecture (5GS) is represented in two ways in the 3GPP standards, one is a service-based representation in which the control plane network functions access each other's services, and the other is a reference point representation in which the interaction between the network functions is shown with point-to-point reference points. In this chapter we use the service-based representation since the 5G architecture is defined as service-based architecture. The 3GPP 5GS service-based non-roaming reference architecture in shown in Fig. 4.2. In Release-15 specifications the Service-based interfaces are defined within the Control Plane only. In 3GPP terminology, "a network function can be implemented either as a network element on a dedicated hardware, as a software instance running on a dedicated hardware, or as a virtualized function instantiated on an appropriate platform, e.g. on a cloud infrastructure."

The EPC in Release-14 was enhanced with an optional feature that allowed separation of control plane and user plane. In this feature, the Serving Gateway (SGW) and Packet Gateway (PGW) are divided into distinct control plane and user plane functions (e.g., SGW-C and SGW-U). This optional feature provided more flexibility and efficiency in network deployment—See [4] for details. In 5G architecture, the separation of control plane and user plane is an inherent capability. The Session

Management Function (SMF) handles the control plane functionality for setup and management of sessions while the actual user data is routed through the User Plane Function (UPF). The UPF selection (or re-selection) is handled by SMF. The deployment options allow for centrally located UPF and/or distributed UPF located close to or at the Access Network.

In EPC, the mobility management functionality and session management functionality are handled by Mobility Management Entity (MME). In 5GC, these functionalities are handled by separate entities. The Access and Mobility Management function (AMF) handles the mobility management and procedures. AMF is termination point for control plane connection from (Radio) Access Network ((R)AN) and UE. The connection between UE and AMF (which traversed through RAN) is referred to as Non-Access Stratum (NAS). The Session Management Function (SMF) handles the session management procedures. The separation of the mobility and session management functionalities allows for one AMF to support different Access Networks (3GPP and non-3GPP), while SMF can be tailored for specific Accesses.

Figure 4.3 shows a Roaming architecture with local breakout at the Visited PLMN (VPLMN). In this scenario the Unified Data Management (UDM), which includes the subscription information, and Authentication Server Function (AUSF), which includes authentication/authorization data, are located in the Home PLMN (HPLMN). There are Security Edge Protection Proxies (SEPP) that protect the communication between the Home and Visited PLMNS. UE communicates to Data Network (DN) via the User Plane Functions (UPF) in the VPLMN. The AMF and the Session Management Function (SMF) which handle the mobility and the session management for the UE are located in the VPLMN as well.

4.2   5G Core (5GC) Service-Based Architecture

A major change in the 5G Core (5GC) architecture compared to EPC and the previous generations is the introduction of the service-based architecture. In EPC architecture the control plane functions communicate with each other via the direct interfaces (or reference points) with a standardized set of messages. In the service-based architecture, the Network Functions (NF), using a common framework,

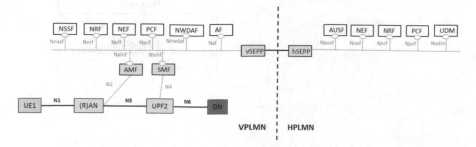

Fig. 4.3 Roaming 5G System architecture—Local breakout scenario

expose their services for use by other network functions. In the 5GC architecture model the interfaces between the networks functions are referred to as Service-Based Interfaces (SBI). The Service Framework defines the interaction between the NFs over SBI using a Producer-Consumer model. As such a service offered by a NF (Producer) could be used by another NF (Consumer) that is authorized to use the service. The services are generally referred to as "NF Service" in 3GPP specifications.

The interaction between the NFs may be a "Request-response" or a "Subscribe-Notify" mechanism. In the "Request-response" model a NF (consumer) request another NF (producer) to provide a service and/or perform a certain action. See Fig. 4.4. In "Subscribe-Notify" model a NF (consumer) subscribes to the services offered by another NF (producer) which notifies the subscriber of the result (Fig. 4.5).

As can be seen in Fig. 4.3, in the 5G System Architecture, each network function has an associated service-based interface designation. For example, "Namf" designates the services exhibited by the Access and Mobility Management function (AMF). 3GPP specifications define a set of Services that are offered/supported by each Network Function. For example, the NF services specified for AMF are shown in Table 4.1. The details for Service descriptions are described in [2].

There are three main procedures associated with the Service Framework as defined in 3GPP—see [1, 5] for details:

NF service registration and de-registration: to make the Network Repository Function (NRF) aware of the available NF instances and supported services.

NF service discovery: enables a NF (Consumer) to discover NF instance(s) (Producer) that provide the expected NF service(s). A NF typically performs a Services Discovery procedure with NRF for NF and NF service discovery.

NF service authorization: to ensure the NF Service Consumer is authorized to access the NF service provided by the NF Service Provider (Producer).

Fig. 4.4 "Request-response" NF Service illustration

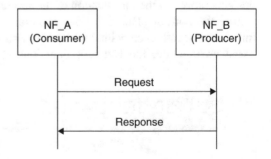

Fig. 4.5 "Subscribe-
Notify" NF Service
illustration 1

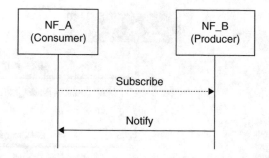

Table 4.1 Namf services

Service name	Description
Namf_communication	Enables an NF consumer to communicate with the UE and/or the AN through the AMF
	This service enables SMF to request EBI allocation to support interworking with EPS
Namf_EventExposure	Enables other NF consumers to subscribe or get notified of the mobility-related events and statistics
Namf_MT	Enables an NF consumer to make sure UE is reachable
Namf_Location	Enables an NF consumer to request location information for a target UE

4.2.1   Example of NF Service Registration

In this example (Fig. 4.6) AMF as the NF service consumer sends a HTTP PUT request to NRF with the resource URI representing the NF Instance. The request contains Nnrf_NFManagement_NFRegister request message (the NF profile of NF service consumer) to NRF to inform the NRF of its NF profile. The NF profile of NF service consumer includes information such as NF type, FQDN or IP address of NF, and Names of supported services. The NRF authorizes the request and upon success stores the NF profile of NF service consumer and marks the NF service consumer available. The NRF acknowledge the success of AMF Registration by returning a HTTP 201Created response containing the Nnrf_NFManagement_NF Register response (including the NF profile). See 3GPP TS 23.501 [1] and 3GPP TS 29.510 [5] for details.

4.2.2   Example of NF Service Discovery

In this example (Fig. 4.7) the AMF as NF service consumer intends to discover NF instances or services available in the network for a targeted NF type. The AMF sends HTTP GET request to NRF in the same PLMN by invoking Nnrf_NFDiscovery_Request. This request contains Expected NF service Name, NF Type of the expected NF instance, and NF type of the NF consumer and may also include

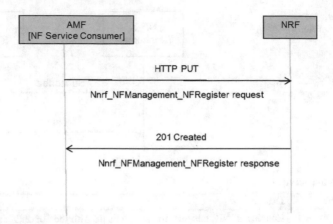

Fig. 4.6 Nnrf_NF Registration procedure

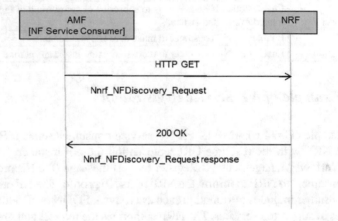

Fig. 4.7 Nnrf_NF service Discovery

other information/parameters such as Subscription Permanent Identifier (SUPI) and AMF Region ID. The NRF authorizes the request, and if allowed the NRF determines the discovered NF instance(s) or NF service instance(s) and provides the search results to the NF service consumer (e.g., AMF) in a HTTP 200 OK. See 3GPP TS 23.501 [1] and 3GPP TS 29.510 [5] for details.

4.3   Network Slicing

From the 3GPP point of view, a 5G network slice is viewed as a logical network with specific functions/elements dedicated for a particular use case, service type, traffic type, or other business arrangements with agreed-upon Service-level

Agreement (SLA). It is important to note that 3GPP only defines network slicing for 3GPP defined system architecture and does not address transport network slicing or resource slicing of components.

The most commonly discussed slice types in industry are enhanced Mobile Broadband (eMBB), Ultra-Reliable Low Latency Communications (URLLC), and massive IoT (mIoT). However, there could be many more network slices. In 4G systems (EPS) there is an optional feature called eDecor to support Dedicated Core Networks (DCNs) to allow selection of the core networks based on UE's subscription and usage type. The network slicing in 5GS is a more complete solution that provides capabilities for composing multiple dedicated end-to-end networks as slices.

An end-to-end Network Slice includes the Core Network Control Plane and User Plane Network Functions as well as the Access Network (AN). The Access Network could be the Next Generation (NG) Radio Access Network described in 3GPP TS 38.300 [6], or the non-3GPP Access Network with the Non-3GPP InterWorking Function (N3IWF). To emphasize that there could be multiple instances of a network slice, the 3GPP 5GS specifications define the term "Network Slice instance" as set of Network Function instances and resources (e.g., compute, storage, and networking resources) which form a Network Slice.

In 5GS, the Network Slice Selection Assistance Information (NSSAI) is a collection of identifications for network slices. A network slice is identified by a term referred to as Single-NSSAI (S-NSSAI). The S-NSSAI signaled by the UE to the network assists the network in selecting a particular Network Slice instance. An S-NSSAI comprises a Slice/Service type (SST) and an optional Slice Differentiator (SD) which may be used to differentiate among multiple Network Slices of the same Slice/Service type.

An S-NSSAI can have standard values or nonstandard values. The S-NSSAI with standard value means that it comprises an SST with a standardized SST value. An S-NSSAI with a nonstandard value identifies a single Network Slice within the PLMN with which it is associated.

3GPP has defined some standardized SST values in TS 23.501 [1]. These SST values are to reflect the most commonly used Slice/Service Types and will assist with global interoperability for slicing. The support of all standardized SST values is not required in a PLMN (Table 4.2).

Figure 4.8 shows an example of three Network Slices in 5GS. For Slice 1 and Slice 2, the Access and Mobility Management Function (AMF) instance that is

Table 4.2 Standardized SST values

Slice/service type	SST value	Characteristics
eMBB	1	Slice suitable for the handling of 5G enhanced mobile broadband
URLLC	2	Slice suitable for the handling of ultra-reliable low latency communications
MIoT	3	Slice suitable for the handling of massive IoT

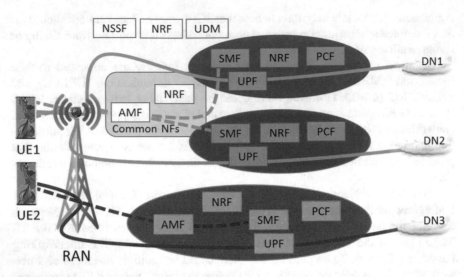

Fig. 4.8 Example of Network Slices in 5GS

serving the UE1 and UE2 is common (or logically belongs) to all the Network Slice instances that are serving them. The UE in Slice 3 is served by another AMF. Other network functions, such as the Session Management Function (SMF) or the User Plan Function (UPF) may be specific to each Network Slice.

The Network Slice instance selection for a UE is normally triggered as part of the registration procedure by the first AMF that receives the registration request from the UE. The AMF retrieves the slices that are allowed by the user subscription and may interact with the Network Slice Selection Function (NSSF) to select the appropriate Network Slice instance (e.g., based on Allowed S-NSSAIs, PLMN ID). The NSSF contains the Operators' policies for slice selection. Alternatively, the slice selection policies may be configured in the AMF.

The data connection between the UE and Data Network (DN) is referred to as PDU session in 5GS. In 3GPP Release-15 a PDU Session is associated to one S-NSSAI and one DNN (Data Network Name). The establishment of a PDU session is triggered when the AMF receives a Session Management message from UE. The AMF discovers candidate Session Management Functions (SMF) using multiple parameters (including the S-NSSAI provided in the UE request) and selects the appropriate SMF. The selection of the User Plane Function (UPF) is performed by the SMF. The Network Repository Function (NRF) is used for the discovery of the required Network Functions using the selected Network Slice instance—the detailed procedures are specified in 3GPP TS 23.502 [2]. The data transmission can take place after a PDU session to a Data Network is established in a Network Slice. The S-NSSAI associated with a PDU Session is provided to the (R)AN, and to the policy and charging entities, to apply slice specific policies.

For roaming scenarios, S-NSSAI values applicable in the Visited PLMN (VPLMN) are used to discover a SMF instance in the VPLMN and in Home–Routed deployments S-NSSAI values applicable in the Home PLMN (HPLMN) are also used to discover a SMF instance in the HPLMN.

4.4   Registration, Connection, and Session Management

This section provides an overview of the high-level features for registration, connection, and session management in 5GS.

4.4.1   Registration Management

A user (UE) registers periodically with the network to remain reachable, in case of mobility or to update its capabilities. During initial registration, the UE is authenticated and access authorization information based on subscription profile in UDM (Unified Data Management) is configured on AMF, and the identifier of the serving AMF is stored in UDM. When this registration process is complete, the state for the UE is 5GMM-REGISTERED (at the UE and AMF). In the 5GMM-REGISTERED state, the UE can perform periodic registration updates to notify that it is active, and mobility registration update if the serving cell is not in the list of TAI (Tracking Area Identifier) that was provisioned during registration. The UE/AMF state machine will transition to 5GMM-DEREGISTERED when the timers expire (and the UE has not performed periodic registration) or if the UE or network explicitly deregisters. When a UE is served by 3GPP and non-3GPP accesses of the same PLMN, the AMF associates multiple access specific registration contexts with the same 5G-GUTI (Globally Unique Temporary Identifier).

4.4.2   Connection Management

For signaling between the UE and AMF, NAS (Non-Access Stratum) connection management procedures are used. When the UE is in 5GMM-REGISTERED state and has no NAS connection established with the UE (i.e., in 5GMM-IDLE state), the UE will respond to paging (unless in MICO (Mobile Initiated Connections Only) mode) by performing a Service Request procedure and enter into 5GMM-CONNECTED mode. The UE will perform a Service Request procedure also and enter into 5GMM-CONNECTED mode if it has signaling or user data to send. The AMF enters into 5GMM-CONNECTED mode for that UE when the N2 connection (between Access Network and AMF) is established. Figure 4.9 shows the transitions between 5GMM-IDLE and 5GMM-CONNECTED for both the UE and the AMF.

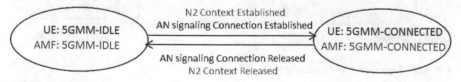

Fig. 4.9 Connection Management State Transitions

The UE can go from 5GMM-CONNECTED to 5GMM-IDLE when the Access Network (AN) signaling connection is released on inactivity (RRC Idle state). The AMF enters 5GMM-IDLE for the UE when the NGAP signaling connection (N2 context) and N3 user plane connection are released. When the UE is in RRC Inactive state and 5GMM-CONNECTED, the UE reachability and paging is managed by the RAN. The AMF provides assistance by configuring UE specific DRX (Discontinuous Reception) values, registration area, periodic registration update timer value and MICO mode indication. The UE monitors for paging with the 5G S-TMSI (Temporary Mobile Subscriber Identity) and RAN identifier.

Note: TS 23.501 [1] uses states RM-REGISTERED and RM-DEREGISTERED, while TS 24.501 [7] uses 5GMM-REGISTERED and 5GMM-DEREGISTERED for the same set of states. Similarly, 23.501 uses CM-IDLE and CM-CONNECTED, while TS 24.501 [7] uses 5GMM-IDLE and 5GMM-CONNECTED to refer to the same set of states.

4.4.3   Registration Call Flow

Figure 4.10 provides an overview[1] of the registration management procedure call flow. Initially, the UE and network states are RRC-Idle, 5GMM-IDLE, and 5GMM-DEREGISTERED. An RRC (Radio Resource Connection) layer is needed between

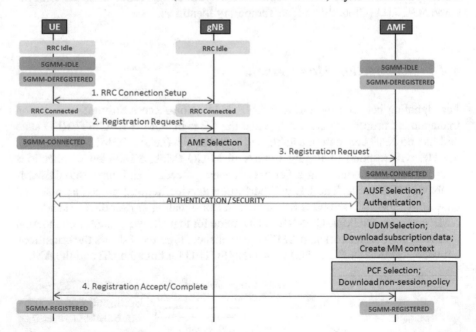

Fig. 4.10 Registration Management

---

[1] Complete call flow procedure and parameter details are found in 3GPP TS 23.502, 4.2.2 [2].

the UE and gNB to exchange messages. Following the radio link establishment, an RRC connection is established between the UE and gNB (1) (gNB is short term for next generation NodeB).

Once the RRC connection is set up, the UE is ready to start registering itself with the network. The Registration Request (2) includes registration type (initial, mobility registration or periodic registration, emergency registration), UE subscriber and network identifiers (SUCI/SUPI/5G-GUTI), security parameters, requested NSSAI (see details in Sect. 4.3), UE capability and PDU session information (status, sessions to be re-activated, follow-on request and MICO mode preference). The subsequent steps here are for an initial registration. The gNB uses SUPI (Subscription Permanent Identifier) and NSSAI to select an AMF and forward the Registration Request to the selected AMF (2).

The AMF selects an AUSF (Authentication Server Function) and initiates authentication based on SUPI or Subscription Concealed Identifier (SUCI). If it is an emergency registration, authentication is skipped. The AMF initiates NAS security functions and upon completion, the AMF initiates the NGAP procedure (for a logical, per UE association between a 5G access node and AMF). Following the security procedures, the UE is authenticated and gNB/access node stores the security context and uses it to protect messages exchanged with the UE (details are described in TS 33.501).

The AMF selects a UDM (Unified Data Management) and in turn a UDR (Unified Data Repository) instance based on the SUPI and retrieve access and mobility subscription data, SMF selection subscription data. The AMF then creates a MM Context based on the subscription data obtained. The AMF then selects a PCF (Policy Control Function) and requests non-session policy (access and mobility related, further described in the Sect. 4.10).

The AMF sends a Registration Accept to the UE with 5G-GUTI, Registration Area, Mobility restrictions, allowed NSSAI, Periodic Registration Update Timer, Local Area Data Network (LADN), MICO mode information and other session information. If the 5G-GUTI is new, the UE replies with a Registration Complete. The state in the UE and network for this UE is 5GMM-REGISTERED. Before the Periodic Registration Update Timer Expires, the UE can send a Registration Request (type: periodic update) to remain in 5GMM-REGISTERED state (the timer is also reset when the UE/network enter the 5GMM-CONNECTED state).

The UE can request the use of MICO (Mobile Initiated Connection Only) mode during regular (non-emergency) registration in a 3GPP access (gNB). In MICO mode, all NAS timers are stopped (except for periodic registration update timer and a few others) and the UE cannot be paged.

4.4.4 PDU Session Establishment Call Flow

The flow description in Fig. 4.11 below provides an overview of how a PDU session can be established following registration of the UE.

The UE initiates session establishment by sending to the AMF a PDU Session Establishment Request (1) with S-NSSAI request type (initial request, existing

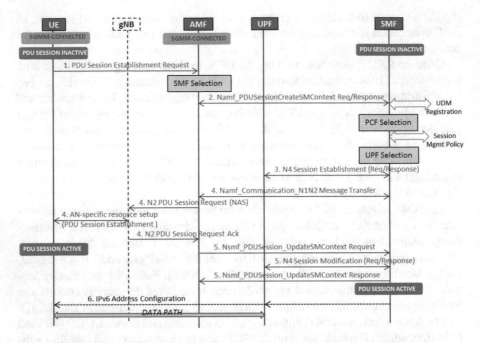

Fig. 4.11 PDU Session Establishment

PDU, emergency request), old PDU session id if one exists, and the N1 SM container (PDU Session Establishment Request).

When the AMF receives this request in (1), it determines the request type—in this case, an initial request—and selects an SMF based on S-NSSAI, subscription permissions on Local Breakout (LBO) roaming, selected Data Network Name (DNN) and per DNN local breakout /S-NSSAI permissions and local operator policies, load and access technology used by the UE. Since this is an initial request, the AMF initiates a create session context messaging request and response sequence (2) with the newly selected SMF. The AMF forwards the SM Container (PDU Session Establishment Request) and parameters along with the GUAMI (Globally Unique AMF Identifier). For an initial request, subscription parameters from UDM include authorized PDU type, authorized SSC modes (described further in Sect. 4.5), default 5QI (5G QoS Identifier), ARP (Allocation and Retention Priority), and subscribed session AMBR (Aggregate Maximum Bit Rate). If the UE request is compliant with subscription policies, the SMF responds back to the AMF (2, response).

The SMF then selects a PCF for dynamic policy, or may apply local policy if dynamic PCC (Policy Control and Charging) is not deployed. Session management related policy control includes gating, charging control, QoS, and per application policy (described further in Sect. 4.10).

Following dynamic policy installation, the SMF selects one or more UPFs (User Plane Functions). UPF selection includes deployment considerations (example central anchor and intermediate UPF close to access node), roaming consideration, configuration information from OA&M (such as capacity, location, supported capa-bilities), and dynamic conditions (such as UPF load). The SMF initiates the N4 Session Establishment request/response sequence (3) with the UPF (or UPFs) with packet detection, enforcement and reporting rules, CN tunnel information and user plane inactivity timer for deactivation. The SMF also allocates IPv4 /IPv6 addresses and installs forwarding rules. For an unstructured PDU session, neither MAC nor IP address is assigned by SMF.

Following UPF session establishment, the SMF configures session information at the gNB and UE (sequence of messages 4 in Figure SM). The SMF sends N2 SM Information (PDU Session ID, QFI, QoS Profile, CN Tunnel info, S-NSSAI, AMBR, PDU session type) and N1 SM Container (PDU Session Establishment Accept with QoS, slice and session parameters) in an N1N2Transfer Message to the AMF. The AMF sends the NAS message PDU Session Establishment Accept tar-geted to the UE, and N2 SM message to the gNB. The responses from the UE and gNB to the AMF contains the list of rejected QFIs and established AN Tunnel information.

The AMF sends Nsmf_PDUSession_UpdateSMContext Request (5) to the SMF with rejected QFI and AN tunnel information. The SMF performs an N4 session modification sequence (5) to update the AN tunnel and QoS information. At this point, the PDU session setup is complete.

IPv6 address Router advertisement (6) is initiated by the SMF and forwarded by the UPF for dynamic IP addresses (static IP addresses may be sent in earlier signal-ing). The interface identifier sent earlier is used to derive the complete IPv6 address. IP data traffic can be sent at this time.

4.4.5 Service Request

The Service Request procedure allows a UE to transition from a 5GMM-IDLE to 5GMM-CONNECTED state. For example, when the UE is in 5GMM-IDLE state (and not in MICO mode), the network may page the UE to indicate that it has down-stream data (which is temporarily buffered at the UPF). Once the service request procedure is executed, the UE and network transition to 5GMM-CONNECTED state for that UE and control and data plane paths are established. In the case above, downstream data that has been buffered at the UPF can then be delivered to the UE.

Service requests can be UE triggered or network triggered. In a UE-triggered service request, the UE in 5GMM-IDLE state requests the establishment of a secure connection to the AMF. The UE triggering service request maybe as a result of receiving a paging request from the network, or when the UE wants to send uplink signaling messages or data. After receiving the service request, the network initiates procedures to set up the control plane and the user plane. The service request proce-dure may support independent activation of the UP connections for existing PDU sessions.

Network-triggered service request procedure is used when the network needs to signal (N1 signaling) or deliver mobile terminating user data to a UE (e.g., SMS). The network-triggered service request may be invoked in 5GMM-IDLE or 5GMM-CONNECTED state.

4.4.6   Other Procedures

There are a number of procedures that 5GS provides to support various session management capabilities. For session management, in addition to PDU Session Establishment (described above), they include PDU Session Modification and Session Release.

UE connection, registration, and mobility procedures include all the registration, service request procedures, UE configuration, AN release, and N2 signaling procedures. SMF and UPF procedures are used to set up and manage PDU session state in UPF (setup, modify, delete, reporting, charging). User profile management procedures are used to notify subscriber data updates, session management subscription notifications and purge of subscriber data in AMF.

Details of these and other procedures can be found in 3GPP TS 23.502, section 4 (System Procedures) [2].

4.5   Session and Service Continuity in 5GC

5G support for a range of services from IoT to critical communications introduces various requirements on packet data and user plane including degrees of mobility and session continuity for connections with varying levels of latency, bandwidth, and reliability. Thus, the connectivity service (PDU sessions) modes support classical session continuity with a central anchor as in 4G systems, or newer forms where a PDU session can be retained until after establishing another PDU session to the same data network (DN), or where a PDU session is released prior to a new PDU session establishment to that DN. Handling of IP addresses for these PDU sessions, subnets and the layout of IP networks and gateways are important to consider when discussing how session and service continuity is handled. An outline of how the 5G control plane and user plane entities behave to support the modes of session and service continuity is discussed below.

In 5GC, PDU sessions and the IP addresses that identify it are anchored at a UPF PSA (PDU Session Anchor). There are three session continuity modes supported by the 5G system to satisfy the continuity requirements of different applications: SSC Mode 1, Mode 2, and Mode 3. PDU sessions established with SSC Mode 1 maintain the same session anchor for the length of the session. Thus, with SSC mode 1 the network selects a centrally located UPF PSA so that it is able to serve the PDU session even when the UE moves across radio access locations. This is shown in Fig. 4.12.

Figure 4.12 also shows an outline of a handover in SSC Mode 1. The UE initially has a PDU session to DN (external data network) with UPF2 as ULCL (Uplink Classifier) and UPF 1 as the PSA. As the UE moves to new radio networks and based on location information from the AMF, the SMF decides to relocate the ULCL from UPF2 to UPF3. The PDU session continues to be anchored at UPF1 since it is SSC Mode 1.

With SSC Mode 2 and Mode 3, there is the option to relocate the PDU session anchor as the PDU session can be released and a new one established to continue the connection. In the case of SSC Mode 2, the UPF anchor (PSA) is released and a new anchor is selected and programmed by the SMF. In the case of SSC Mode 3, the UPF anchor (PSA) is released only after a new anchor is programmed. Applications that need continuous availability of a connection path may select SSC Mode 3, while applications that can tolerate a break in connectivity before a new path can select SSC Mode 2.

In addition to the PDU session management, the 5GC also needs to manage IP addresses. When the SMF selects a new anchor UPF for the PDU session, the IP address assigned to that PDU session also needed to be re-assigned to a topologically

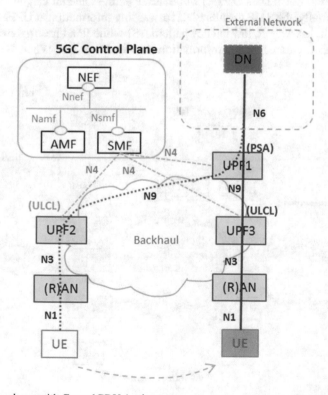

Fig. 4.12 Handover with Central PDU Anchor

correct one. For example, an application that uses TCP transport over SSC Mode 2 PDU connection may stream data over a TCP connection and when moving establish a new TCP connection as a result of a new IP address/PDU connection establishment. The application would have means to associate the two TCP connections by means of a cookie or other state information. SSC Mode 3 and applications using MPTCP (Multi-Path TCP) and multiple connection paths can on the other hand use different IP addresses and connection paths in parallel. During the handover process, both the connection path segment from the source UPF-PSA and the connection path segment from the target UPF-PSA can be handled by MPTCP. IP address management may also have a bearing on privacy (inadvertent leakage of location) related to the location of the UPF-PSA / IP address anchor. IP addresses assigned to PDU sessions can be tracked externally because they reveal topological information as to where the UPF PSA is located. Many applications are sensitive to privacy and inadvertent leakage of fine-grained location information via the IP address can be avoided by assigning an IP address centrally. While this takes care of privacy concerns, sessions that require low latency are not served well as the path to the central anchor can be long (relative to an anchor that is closer). In Fig. 4.13, an example of how IP address/location privacy can be maintained along with low latency paths is shown.

Figure 4.13 shows an example of handover with PDU session anchors near local data networks (Local Data Center) while the IP address anchor can be more central (UPFa). Initially, the SMF installs PDU forwarding information at UPFa and UPFb. Thus, the UE has a short, low latency path to AS1 while IP address gateway at UPFa does not leak accurate location information to external entities. When the UE moves

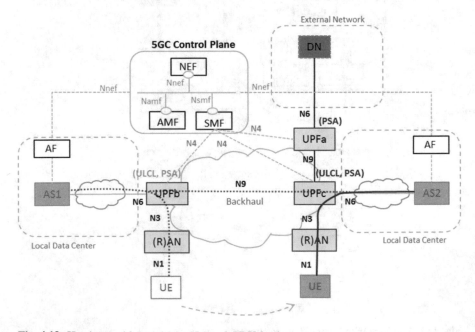

Fig. 4.13 Handover with Local Data Network PDU Anchors

to another radio access, the SMF reprograms the forwarding information at UPFc (add forwarding info) and UPFb (remove forwarding info). The SMF also provides events to AF (via NEF (Network Exposure Function), if AF cannot directly subscribe to UE information from SMF). The AF can use events received on UE session relocation to provision or migrate the application session from AS1 to AS2.

For session continuity in 5GC as the UE moves, the SMF reprograms UPF forwarding state of the PDU sessions as described in this section. In addition to the change in forwarding path, transport of packets with minimal loss and re-ordering is needed. The UPFs buffer packets and use end-markers to avoid re-ordering during the transition from one UPF to another. End-to-end congestion and flow control mechanisms in TCP and QUIC are also managed to prevent congestion collapse during the handover—especially with short round trip times and low latency flows.

4.6   Interworking with EPC

The network deployments may comprise 5GC and EPC as well as coexistence of UEs supporting 5G and 4G within the same network (i.e., within one PLMN). The UEs that supports 5G may also be supporting EPC NAS (Non-Access Stratum) procedures to provide service in legacy 4G networks when roaming to those networks. 3GPP standards have defined architecture options to support interworking and migration from EPC to 5GC. Figure 4.14 shows an example of a non-roaming architecture for interworking between EPC and 5GC. For migration scenarios, it is generally assumed that the subscriber database is common between 5GC and EPC. That is, the UDM in 5GC and the HSS in EPC is a common database. Optionally it could be further assumed that PCRF (Policy and Charging Rules Function) in EPC and PCF in 5G, PGW Control plane function (PGW-C) in 4G and SMF in 5G, PGW User plane function (PGW-U) in 4G and UPF in 5G are collocated, respectively, and dedicated for interworking between EPC and 5GC. These are referred to as HSS + UDM, PCF + PCRF, SMF + PGW-C, and UPF + PGW-U. For User Plane management and connectivity when interworking with EPC, the SMF + PGW-C provides information over N4 to the UPF + PGW-U related to the handling of traffic over S5-U.

An optional interface, N26, is defined to provide an inter Core Networks interface between EPC and 5GC by interconnecting MME and AMF. Since N26 is optional, the networks may provide interworking without N26 interface as well. Interworking procedures with N26 provide IP address continuity on inter-system mobility to UEs that support single registration mode as well as both 5GC NAS and EPC NAS (When the N26 interface is used for interworking, the UE operates in single-registration mode). Networks that support interworking procedures without N26 provide IP address continuity on inter-system mobility to UEs operating in both single-registration mode and dual-registration mode. The N26 interface enables the exchange of mobility and session management states (MM and SM) between the source and target network. In these interworking scenarios the MM state for the UE

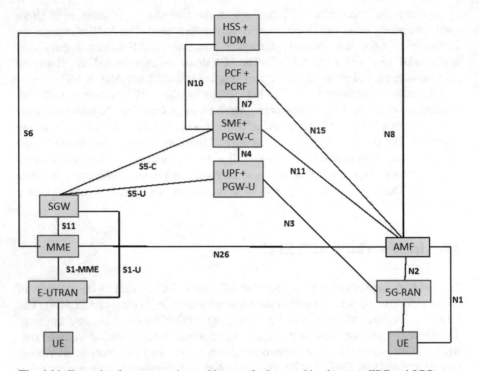

Fig. 4.14 Example of a non-roaming architecture for interworking between EPC and 5GC

is either kept in the AMF or MME, and the AMF or the MME is registered in the HSS + UDM. The details on interworking between EPC and 5GC are defined in TS 23.501clause 4.3 and 5.17.

4.7   CP and UP Protocols in 5G Core

5G core (5GC) protocols are used for registering the UE, managing its access and connections, transporting user data packets, and signaling among 5G network functions to manage and control various aspects for the user. The most significant change is the introduction of SBA (Service-Based Architecture) with virtualization and disaggregation—and the resulting changes in signaling protocols.

4.7.1   CP Protocol Stack

The Control plane (CP) protocol stack comprises the CP stack between 5GAN and the 5G core (N2 interface), CP stack between the UE and 5GC, CP stack between the network functions in 5GC, and the CP stack for untrusted non-3GPP access. Figure 4.15 below shows the CP stack from UE to network.

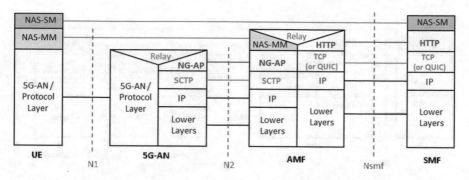

Fig. 4.15 Control Plane Protocol Stack

The set of protocols on N1 interface across UE–5GAN depends on the access network. In the case of NG-RAN, the radio protocol between the UE and NG-RAN (eNB, gNB) is specified in TS 36.300 and TS 38.300 [6]. For non-3GPP accesses, EAP-5G/IKEv2 and IP are used over the non-3GPP radio (WLAN) for establishing the IPSec SA (Security Association) and the NAS is sent over the established IPSec connection.

The N2 interface between 5G-AN and AMF has an SCTP/IP transport connection over which the NG-AP protocol runs. The control plane interface between 5G-AN and 5GC supports connection of different kinds of AN (3GPP RAN, N3IWF) with the same protocol. The AN–SMF protocols where N2 SM (Session Management) messages are relayed via AMF uses NG-AP between 5G-AN and AMF. In addition, the NAS protocol from the UE side is also relayed over NG-AP.

The NAS MM layer between UE and AMF is used for Registration Management (RM) and Connection Management (CM) as well as for relaying Session Management (SM) messages. The NAS-SM layer carries session management messages between UE–SMF. 5G NAS protocols are defined in TS 24.501 [7].

The Nsmf interface and protocol between AMF and SMF is based on Service-Based Architecture (SBA) using HTTPS protocol over a transport layer of TCP. It is expected that as QUIC transport is standardized and stable in IETF, 3GPP will adopt it as the SBA signaling transport layer. Functions in the core network interact with each other over in the SBA architecture using 1:N (many) bus over HTTPS. Service discovery in SBA is via the NRF (Nnsf interface) that resolves the destination service function based on a number of request criteria. Control plane signaling between SMF–UPF however uses the N4 interface and extensions to PFCP (Packet Flow Control Protocol) defined in TS 29.244 [8].

4.7.2 User Plane Protocol Stack

The User Plane (UP) protocol stack consists of the protocol stack for a PDU session and the user plane for an untrusted, non-3GPP access (Fig. 4.16).

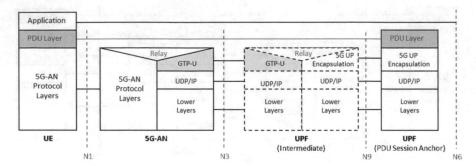

Fig. 4.16 User Plane Protocol Stack

User plane protocol stack is primarily to transport PDU sessions across from the UE to the UPF (PDU session anchor). The PDU layer can be IPv4, IPv6, or IPv4v6. For IPv4 and IPv6, the SMF is responsible for allocating and managing the IP address. When the PDU session type is unstructured, Ethernet is used. In the case of Ethernet, the SMF does not allocate MAC or IP addresses. Since the 3GPP access is an NBMA (Non-broadcast Multiple Access) type network, tunneling is used to carry packets to the PDU session anchor. GTP-U (GPRS Tunneling Protocol—User Plane) multiplexes user data over N3 between 5G-AN—intermediate UPF (e.g., UPF that performs uplink classification) and over N9 towards a PDU session anchor UPF. Unlike 4G where the GTP-U tunnels correspond to a bearer (and a UE may have multiple bearers), all UE packets of a UE across an N3 or N9 interface are transported over a single GTP-U connection. QFI/ flow marking associated with a QoS flow is signaled explicitly in this connection to indicate the level of QoS in the IP transport layer.

4.8   Support for Virtualized Deployments

The 5GS embraces Network Function virtualization (NFV) and cloudification for its architecture design. The 5GS supports different virtualization deployment scenarios, such as

- A Network Function instance can be deployed as fully distributed, fully redundant, stateless, and fully scalable NF instance that provides the services from several locations and several execution instances in each location.
- A Network Function instance can also be deployed such that several network function instances that are present within a NF set are fully distributed, fully redundant, stateless and scalable as a set of NF instances.

The network slicing feature supported by 5GS is also enabled by virtualization. Network function instances can be created, instantiated and isolated with each other in the virtualized environment, into different network slices in order to serve different services.

In order to manage the life cycle of the virtualized 5GS functions and its instances as well as the virtual resource of network slice, 5G OAM provides means to integrate with virtualized network function management and orchestration capability, as well as providing standardized life cycle management interfaces with other virtualized function management and orchestration system which are defined by other standards, such as ETSI ISG NFV, and other open source project, such as ONAP.

Figure 4.17 illustrates the integration of 5G OAM system with ETSI NFV Management and orchestration (MANO) system [9].

In this illustration, 5G management system provides the 5G service and functionality management, such as NM plays one of the roles of OSS/BSS and provides the functions for the management of mobile network which includes virtualized network functions; and EM/DM is responsible for FCAPS management functionality for a VNF on an application level and physical NE on a domain and element level. ETSI ISG NFV MANO provides the virtualized resource and life cycle management for those virtualized 5G functions and the network service and network slice composed by those functions.

4.9 Support for Edge Computing

The 5GS architecture was designed taking the Edge Computing requirements in mind from the start. As such, edge computing is considered a key technology for efficient routing to the application servers as well as addressing the low latency requirements.

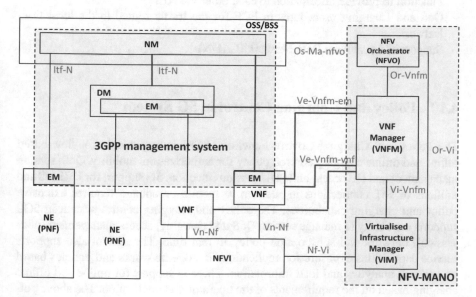

Fig. 4.17 Integration of 5G OAM system with ETSI NFV MANO system

In the context of 3GPP, the Edge Computing refers to the scenarios that the services need to be hosted close to the access network of the UE (e.g., at or close to the RAN). As described earlier, in 5GS the routing of the data (or user) traffic is done via UPF interface to a Data Network. The 5G core network supports the capability to select a UPF that allows routing of traffic to a local Data Network that is close to the UE's access network. This includes the local breakout scenarios for roaming UEs and non-roaming scenarios.

The decision for selection (or reselection) of UPF for local routing may be based on the information from an Edge Computing Application Function (AF) and/or to other criteria such as the subscription, location, and policies. Depending on the operator's policy and arrangements with the third parties, an AF may access the 5G core directly or indirectly via the Network Exposure Function (NEF). For example, an external AF at the edge data center could influence the routing of the traffic by altering the SMF routing decisions via its interaction with the Policy Control Function (PCF).

The 3GPP TS 23.501 clause 5.13 [1] defines several enablers that can support Edge Computing:

- User plane selection and re-selection for UPF to route the user traffic to the local Data Network
- Local Routing and Traffic Steering to the applications in the local Data Network
- Session and service continuity to enable UE and application mobility
- Application Function influencing UPF (re)selection and traffic routing via PCF or NEF
- Network capability exposure between 5G Core Network and Application Function to provide information to each other via NEF
- QoS and Charging procedures in PCF for the traffic routed to the local Data Network
- Support of Local Area Data Network (LADN)

4.10   Policy and Charging Control in 5G System

The Policy and Charging Control Function (PCF) is responsible for flow-based offline and online charging control, policy for authorization, mobility, QoS, session management and UE access and PDU session selection. 5G support for URLLC and millions of IoT connections in addition to mobile broadband requires extensive policy and charging capabilities. The policy and charging control system in 5GS supports the ability to manage session QoS and charging, access management (non-session), network and subscription policy in real time. The system also supports service capabilities exposure for applications in edge-networks and policies based on network analytics and load information. There is support for online and offline charging based on the requirements of the operator and application. The above policy control can be broadly classified as session management policy and non-session management policy. This is described in further detail followed by the functions and architecture for non-roaming and roaming with local breakout or home routed sessions. Non-session management related policy control includes access and mobility management policy, access and PDU selection policy. Access and mobility policies are installed by the PCF when the UE initially registers with AMF. AMF provides the PCF with the SUPI (Subscription Permanent Identifier), user location information, time zone, serving radio access of the UE, and parameters received from the UDM including service area restrictions, RSFP (RAT Frequency Selection Priority) index and GPSI (Generic Public Subscriber Identifier). The PCF makes policy decisions and provides UE access network discovery and selection policy, URSP (UE Route Selection Policy) and revised access and mobility policy (RFSP and service area restrictions). These policies can be modified as a result of changes and notified by the PCF, or re-evaluated if triggered by the AMF.

Session management related policy control includes gating control (to discard packets that do not match any policy), charging control and QoS for PDU session as well as SDF (Service Data Flow) and per application policy (service data flow and application policy is only set up as needed). QoS control policies may be service based, subscription based, or predefined policies that can be applied at the PDU session level or per SDF. SDF filters include flow, precedence, provider identifier, charging key, charging method (online, offline), and measurement method (volume, time, event, or combinations of these). Gating policies allow gate open/closed for SDF, 5QI, reflective QoS, GBR (Guaranteed Bit Rate), and AMBR (Aggregate Maximum Bit Rate) parameters.

The session-based policies also include usage monitoring control, application detection, service capability exposure, and traffic steering. Usage monitoring policies may be applied to a PDU session, service data flow or have rules for PDU session monitoring that excludes specific data flows.

The PCF may subscribe to Network Data Analytic Function (NWDAF), or initiate a request-response sequence for load level information of a network slice instance to make the policy decisions.

Figure 4.18 shows a service-based representation of the policy and charging architecture for the non-roaming scenario. The PCF and AMF interact over the Npcf/Namf interfaces to create an AM Policy association and provision non-session policy for access and mobility management. The PCF and SMF interact over the Npcf/Nsmf interfaces to create an SM Policy association and provision policy for sessions and charging control. The interactions between SMF and CHF enable offline charging, while online charging is handled at the OCS (N28 interface).

The PCF may access policy control related subscription information at the UDR (Npcf/Nudr interfaces). The UDR may notify the PCF on subscription changes. The PCF may also subscribe to the UDR for AF requests targeting a DNN and S-NSSAI or a group of UEs identified by an internal group identifier.

The PCF and AF interact to transport application level session information to the AF including IP filter information (or Ethernet packet filter information) for policy control or differentiated charging, media bandwidth for QoS control and for sponsored data connectivity. The Npcf and Naf enable the AF to subscribe to notifica-

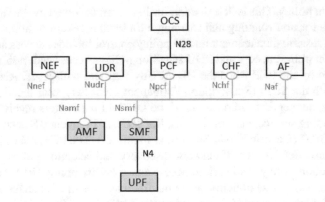

Fig. 4.18 Non-Roaming Policy and Charging Control Architecture

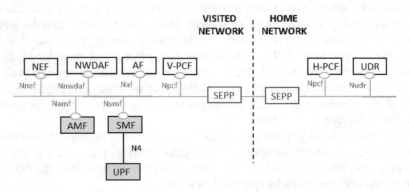

Fig. 4.19 LBO Roaming Policy and Charging Control Architecture

tions of PDU session events. The PCF and AF may also interact over Rx (Diameter) interface for IMS based and Mission Critical Push to Talk (MCPTT) services. For deployments where the AF belongs to a third party, AF requests to the PCF and SMF may not be allowed to interact directly. In this case, these requests and notifications are handled by the NEF.

Figure 4.19 shows a home routed LBO (Local Breakout) roaming architecture. In the LBO roaming architecture, the H-PCF and UDR are in the home PLMN while other functions are in the visited PLMN. In this case the V-PCF (PCF in the visited PLMN) uses locally configured policies according to the roaming agreement. The V-PCF obtains UE access selection and PDU session selection information from the H-PCF using either the Npcf or N24 interface. There is no offline charging support in this scenario. It maybe noted that an SEPP (Security Edge Protection Proxy) may be used for filtering and policing messages on the interPLMN control plane interface.

In a home routed roaming architecture, the access management functions are in the visited PLMN while session and policy management are in the home PLMN (Fig. 4.20).

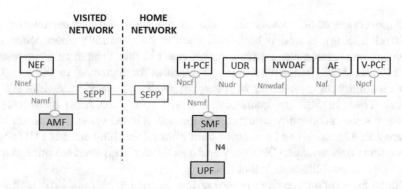

Fig. 4.20 Home Routed Roaming Policy and Charging Control Architecture

4.11 Summary

The 3GPP standards that define the 5G System (5GS) architecture provide enhanced connection, session, and mobility management services with major enhancements from 4G to support network slicing, virtualization, and edge computing. These capabilities are designed to provide support for a range of services requiring low latency, high reliability, high bandwidth, or a massive amount of connectivity over the same network.

As described in this section, the 3GPP standards define the 5GS architecture which consist of network functions and interfaces between them. In the 5GS service-based architecture the service-based interfaces (SBI) are defined within the control plane to allow the network functions to access each other's services using a common framework. The Service Framework defines the interaction between the NFs over SBI using a Producer-Consumer model. The 5GS embraces Network Function virtualization and cloudification for its architecture design. The 5GS supports different virtualization deployment scenarios where for example a NF instance can be deployed as fully distributed, fully redundant, stateless, and fully scalable NF instance.

A major feature of 5GS is network slicing. In 3GPP view network slice is a logical network with specific functions/elements dedicated for a particular use case, service type, traffic type, or other business arrangements. The most commonly discussed slice types in industry are enhanced Mobile Broadband (eMBB), Ultra-Reliable Low Latency Communications (URLLC), and massive IoT (mIoT). In this chapter we provided an overview of network slicing as defined in 3GPP 5G core standards.

The 5GS architecture has been designed taking the Edge Computing requirements in mind. In the context of 3GPP, the Edge Computing refers to the scenarios that the services need to be hosted close to the access network of the UE. The 5G core network supports the capability to select a User Plane Function (UPF) that allows routing of traffic to a local Data Network that is close to the UE's access network.

An overview of 5G architecture, network functions, and new capabilities (e.g., Network Slicing), as well as high-level features for registration, connection management, and session management are provided in this chapter. In comparison to 4G/EPS, the 5G System defines more capabilities for degrees of mobility and session continuity for connections with varying levels of latency, bandwidth and reliability. Thus, in 5GS the connectivity service (PDU sessions) modes support classical session continuity with a central anchor as in 4G systems, or newer forms where a PDU session can be retained until after establishing another PDU session to the same data network (DN), or where a PDU session is released prior to a new PDU session establishment to that DN.

Since 5G supports services requiring low latency, high reliability, high bandwidth, or massive connectivity (i.e., URLLC, eMBB and mMTC) over the same network, performance capabilities have to be carefully engineered and evaluated. The next section provides a comprehensive view on the performance evaluation methodologies, metrics and system level simulations.

References

1. 3GPP TS 23.501, Technical Specification Group Services and System Aspects; System Architecture for the 5G System; Stage 2
2. 3GPP TS 23.502, Technical Specification Group Services and System Aspects; Procedures for the 5G System; Stage 2
3. 3GPP TS 23.503, Technical Specification Group Services and System Aspects; Policy and Charging Control Framework for the 5G System; Stage 2
4. 3GPP TS 23.214, Technical Specification Group Services and System Aspects; Architecture enhancements for control and user plane separation of EPC nodes; Stage 2
5. 3GPP TS 29.510, Technical Specification Group Core Network and Terminals; 5G System; Network Function Repository Services; Stage 3
6. 3GPP TS 38.300, Technical Specification Group Radio Access Network; NR; NR and NG-RAN Overall Description; Stage 2
7. 3GPP TS 24.501, Technical Specification Group Core Network and Terminals; Non-Access-Stratum (NAS) protocol for 5G System (5GS); Stage 3
8. 3GPP TS 29.244, Technical Specification Group Core Network and Terminals; Interface between the Control Plane and the User Plan Nodes; Stage 3
9. 3GPP TS 28.500, Technical Specification Group Services and System Aspects; Telecommunication management; Management concept, architecture and requirements for mobile networks that include virtualized network functions

Chapter 5
5G Capability Outlook: ITU-R Submission and Performance Evaluation

In this chapter, the capability of 5G is evaluated against the eMBB, URLLC, and mMTC performance requirements using the evaluation method as defined in Report ITU-R M.2412 and the detailed evaluation method introduced in Sect. 5.3 as well as field trials. The performance evaluation is conducted to the key 5G technical components, including 5G wideband frame structure and physical channel structure, the physical layer key features including NR massive MIMO, multiple access and waveform, and the LTE/NR coexistence (UL/DL decoupling).

It is no doubt that 3GPP's 5G technology shows strong capability on fulfilling all the IMT-2020 requirements as defined by ITU-R. Such capability is the basis for further progressing of 3GPP 5G to more industry applications and bridging the gaps of future service requirements and the realistic grounds.

5.1 Overview of 5G Requirements

As discussed in Sect. 1.3, 5G is extended from enhanced mobile broadband (eMBB) to massive machine type communication (mMTC) and ultra-reliable and low latency communication (URLLC). In ITU-R, the candidate 5G technology will be evaluated against the technical performance requirements to test the support of the above three usage scenarios.

The ITU technical performance requirements are summarized in Tables 5.1, 5.2, and 5.3. The related test environments as defined in Report ITU-R M.2412 are also discussed. It is seen that the test environments cover the major range of application scenarios for each of the usage scenario (eMBB, URLLC, and mMTC). For eMBB, indoor, dense urban, and rural are defined as test environments, while for URLLC and mMTC, urban macro are defined as test environment.

To reach the 5G vision defined by ITU-R, 3GPP further studied the deployment scenarios and the related requirements associated with the three usage scenarios as documented in 3GPP TR 38.913 [1]. These requirements are usually higher than

© Springer Nature Switzerland AG 2020

Wan Lei et al., *5G System Design*, https://doi.org/10.1007/978-3-030-22236-9_5

Table 5.1 eMBB technical performance requirements

Technical performance requirement	Test environment	DL	UL	Comparison to IMT-advanced requirement
Peak data rate	All eMBB test environments	20 Gbit/s	10 Gbit/s	*~6x LTE-A (Release-10)*
Peak spectral efficiency	All eMBB test environments	30 bit/s/Hz	15 bit/s/Hz	*2x IMT-Advanced*
User experienced data rate (5th percentile user data rate)	Dense urban—eMBB	100 Mbit/s	50 Mbit/s	–
5th percentile user spectral efficiency	Indoor, dense urban, rural	~3x IMT-Advanced	~3x IMT-Advanced	*~3x IMT-Advanced*
Average spectral efficiency	Indoor, dense urban, rural	~3x IMT-Advanced	~3x IMT-Advanced	*~3x IMT-Advanced*
Area traffic capacity	Indoor	10 Mbit/s/m²	–	–
Energy efficiency	All eMBB test environments	High sleep ratio and long sleep duration under low load		–
Mobility class With traffic channel link data rates	Indoor, dense urban, rural	–	Up to 500 km/h, with 0.45 bit/s/Hz	*1.4x mobility class 1.8x mobility link data rate*
User plane latency	All eMBB test environments	4 ms	4 ms	*>2x reduction compared to IMT-Advanced*
Control plane latency	All eMBB test environments	20 ms	20 ms	*>5x reduction compared to IMT-Advanced*
Mobility interruption time	All eMBB test environments	0	0	*Much reduced*

Table 5.2 URLLC technical performance requirements

Technical performance requirement	Test environment	DL	UL	Comparison to IMT-advanced requirement
User plane latency	All URLLC test environment	1 ms	1 ms	*>10x reduction compared to IMT-Advanced*
Control plane latency	All URLLC test environment	20 ms	20 ms	*>5x reduction compared to IMT-Advanced*
Mobility interruption time	All URLLC test environment	0	0	*Much reduced*
Reliability	Urban Macro—URLLC	99.999% within 1 ms	99.999% within 1 ms	–

Table 5.3 mMTC technical performance requirements

Technical performance requirement	Test environment	DL	UL	Comparison to IMT-advanced requirement
Connection density	Urban Macro—mMTC	–	1,000,000 devices/km²	–

ITU's technical performance requirement, showing 3GPP's ambition of providing higher capability than what the ITU required.

In the following subsections, the detailed definitions of evaluation metrics for eMBB, URLLC, and mMTC technical performance requirements, and the related calculation method for 3GPP 5G technology are elucidated.

5.2 Overview of Evaluation Methodologies

Evaluation methodologies for the technical performance requirements as defined in Sect. 5.1 are well defined in Report ITU-R M.2412 [2].

Generally, simulation, analysis, and inspection are employed. The evaluation methodology employed in [2] for each of the technical performance requirement is listed in Table 5.4.

5.2.1 System-Level Simulation for eMBB Technical Performance Requirements

System-level simulation is well established during IMT-Advanced period. Report ITU-R M.2412 defines the following principles and procedures for system-level simulation for eMBB requirements:

Table 5.4 Evaluation method for technical performance requirements

Usage scenario	Technical performance requirement	Evaluation method
eMBB	Peak data rate	Analysis
	Peak spectral efficiency	Analysis
	User experienced data rate	Analysis, or *system-level simulation* (for multi-layer)
	5th percentile user spectral efficiency	*System-level simulation*
	Average spectral efficiency	*System-level simulation*
	Area traffic capacity	Analysis
	Energy efficiency	Inspection
	Mobility	*System-level simulation plus link-level simulation*
eMBB, URLLC	User plane latency	Analysis
	Control plane latency	Analysis
	Mobility interruption time	Analysis
URLLC	*Reliability*	*System-level simulation plus link-level simulation*
mMTC	*Connection density*	*System-level simulation plus link-level simulation, or Full system-level simulation plus link-level simulation*
General	Bandwidth and Scalability	Inspection

- Users are dropped independently with a certain distribution over the predefined area of the network layout throughout the system.
- UEs (User Equipment) are randomly assigned LOS and NLOS channel conditions according to the applicable channel model.
- Cell assignment to a UE is based on the proponent's cell selection scheme.
- Packets are scheduled with an appropriate packet scheduler(s), or with non-scheduled mechanism when applicable for full buffer and other traffic models separately. Channel quality feedback delay, feedback errors, PDU (protocol data unit) errors, and real channel estimation effects inclusive of channel estimation error are modeled and packets are retransmitted as necessary.
- The overhead channels (i.e., the overhead due to feedback and control channels) should be realistically modeled.
- For a given drop, the simulation is run and then the process is repeated with UEs dropped at new random locations. A sufficient number of drops is simulated to ensure convergence in the UE and system performance metrics.
- All cells in the system shall be simulated with dynamic channel properties and performance statistics are collected taking into account the wraparound configuration in the network layout, noting that wraparound is not considered in the indoor case.

System-level simulation is applied to 5th percentile user spectral efficiency and average spectral efficiency evaluation. The system-level simulation for the above eMBB technical performance requirements (evaluation metrics) assumes the user terminal is already in connected mode, and in Report ITU-R M.2412, full buffer traffic is employed for system-level simulation for the above eMBB metrics, i.e., the data packet for this user terminal always exist in the simulation. This case matches well for heavy load network, and are usually used in spectral efficiency evaluation, where the test aims to find out the maximum potential capability of the network to utilize the spectrum resource when providing high data rate.

However, in many 3GPP evaluations, burst buffer traffic also plays an important role in system-level simulation. The burst buffer traffic can demonstrate the performance assessment of a candidate technology in low load to median load network. For burst buffer traffic, the user-perceived throughput (see, e.g., [3]) is usually employed for a specific packet. In this case, the waiting time of packet schedule can be well taken into account in the throughput assessment, which more precisely reflects the perceived data rate of the user on downloading a file. However, it is difficult to test the network capability under "extreme" situations. Therefore, both full buffer and burst buffer traffic are useful in the evaluation for system-level simulation.

In addition, for multi-layer or multi-band evaluation of user experienced data rate, system-level simulation should also be used, instead of using analytical method, as is for single layer single band evaluation. For single layer single band case, once the spectral efficiency is derived, the user data rate could be analytically derived by multiplying the spectral efficiency with system bandwidth. However, for multi-layer or multi-band evaluation, the cell association of a user would impact the number of users on a specific band of a specific cell. It would in turn impact the user data rate, and the data rate cannot be directly derived from a single band single layer simulation. Therefore a new system-level simulation is needed to test the multi-layer or multi-band case.

5.2.2   Full System-Level Simulation and System plus Link-Level Simulation for Connection Density Evaluation

Simulation method is used for connection density evaluation for mMTC usage scenario. Full system-level simulation and system-level simulation plus link-level simulation are two alternatives for connection density evaluation.

5.2.2.1   Overview of Full System-Level Simulation

The full system-level simulation is similar to the system-level simulation defined for eMBB evaluation metrics. The principles and procedures are reused.

However, for connection density evaluation, the full system-level simulation assumes the user terminal is in idle mode before a packet is arrived. This is because the data packet arrival rate is small (e.g., up to one packet every 2 h). In this case, the user terminal will return to idle mode after its data transfer. This is different from system-level simulation for spectral efficiency evaluation for eMBB, where connected mode is assumed for user terminals. Thus, the user access procedure from idle mode to connected mode needs to be modeled in the system-level simulation.

In the evaluation, the idle mode synchronization and system information acquisition performance and delays are taken into account (see detailed access procedure provided in Sect. 5.3 and [4]). The chosen system access procedure, the uplink data transmission and the connection release procedures are modeled.

In addition, the burst buffer traffic is used in full system-level simulation. This is because the connection density is defined based on the certain system load with appropriate QoS requirement. In Report ITU-R M.2412, it is assumed that each user will have one packet every 2 h, and the packet should be correctly received within 10s. If full buffer traffic is assumed, the packet arrival rate is too dense, and it would be very difficult to guarantee the packet delay of 10s under such heavy system load. Therefore burst buffer traffic is assumed in the evaluation.

It is noted that if one models the full details of each step of the connection procedure, it would overload the evaluation effort to a significant extent. Therefore the link-level abstraction of "SINR-to-delay" model is proposed for the synchronization, PRACH transmission, etc., before the user terminal start the data transmission. However, the resource allocation and possible resource collision (e.g., PRACH collision, PDSCH collision) are still simulated through the resource allocation modeling like in conventional system-level simulator. Such modeling ensures an acceptable level of accuracy and complexity.

5.2.2.2   Overview of System-Level plus Link-Level Simulation

The other alternative is to use the system-level simulation followed by the link-level simulation. This is a simplified method that allows the evaluation to derive a first-order assessment on the capability of connection density for the candidate technology.

This evaluation method employs a full buffer system-level simulation in the first step to derive the uplink SINR distribution for a candidate technology. In a second step link-level simulation is performed to determine the uplink spectral efficiency and data rate as functions of SINR. When combined these three functions, the expected long-term time-frequency resources required for each SINR is calculated to support the specified traffic model.

Finally, the connection density is conceptually derived by the system bandwidth, declared for the candidate technology, divided by the average required frequency resource.

This evaluation method is targeted to evaluate the connection density in terms of the throughput of the uplink data transmission. The capacity calculation is based on an assumption of ideal resource allocation among the multiple packets and users (e.g., there is no collision on resource allocation). The packet delay calculation does not consider the delays introduced by the connection access procedure.

5.2.3   *System-Level plus Link-Level Simulation for Mobility and Reliability*

For mobility evaluation and reliability evaluation, system-level plus link-level simulation are employed. This is also a simplification for evaluation purpose.

In the first step, the full buffer system-level simulation is employed to derive the SINR (either uplink or downlink) distribution of the candidate technology. For mobility evaluation, the 50%-tile SINR is used for link-level simulation to derive the user normalized data rate. For reliability evaluation, the 5%-tile SINR is used for link-level simulation to derive the block error rate (BLER) within the given latency.

Such evaluation does not consider the resource allocation impact on data rate and reliability. However, it takes into account the user distribution and the SINR condition under the given distribution. Therefore it is a compromise of the evaluation complexity and accuracy.

5.2.4   *Analysis Method*

Analysis method is employed for peak data rate, peak spectral efficiency, area traffic capacity, user plane latency, control plane latency, and mobility interruption time.

Analysis method means that a calculation or mathematical analysis is applied without a simulation. The analytical method captures the most essential impacts for a technical performance requirement. For example, for latency evaluation, the impact of frame structure (including the time slot length, downlink/uplink ratio, etc.) is captured by the analysis method. However, it does not capture system-level impacts. Furthermore, taking user plane latency as example, the scheduling delay cannot be captured by this analytical way. To reflect the scheduling delay, system-level simulation will be required.

5.2.5   Inspection Method

Inspection is used for energy efficiency and bandwidth. Inspection is accomplished by reviewing the functionality and parameterization of the candidate technology.

5.3   Detailed Definition of Evaluation Metrics and Evaluation Method

To define the performance assessment for the abovementioned technical performance requirements (usually referred to as evaluation metrics), the detailed definition and evaluation method for each of the evaluation metrics are described in detailed. The basis is according to what is defined in ITU-R, which is applicable to any of the candidate technologies. However, complementary details are provided to evaluate 3GPP 5G technologies.

5.3.1   Evaluation Metrics for eMBB Requirements

The following evaluation metrics are used to assess the technical performance for eMBB usage scenario.

5.3.1.1   Peak Spectral Efficiency

As defined in Report ITU-R M.2410, peak spectral efficiency is the maximum data rate under ideal conditions normalized by channel bandwidth (in bit/s/Hz), where the maximum data rate is the received data bits assuming error-free conditions assignable to a single mobile station, when all assignable radio resources for the corresponding link direction are utilized (i.e., excluding radio resources that are used for physical layer synchronization, reference signals or pilots, guard bands and guard times).

It is noted that, in case of multiple discontinuous "carriers" (one carrier refers to a continuous block of spectrum), the peak spectral efficiency should be calculated per carrier. This is because for the carriers on different frequency range, the spectral efficiency could be significantly different.

The generic formula for peak spectral efficiency for FDD and TDD for a specific component carrier (say jth CC) is given by

$$SE_{p_j} = \frac{v_{Layers}^{(j)} \cdot Q_m^{(j)} \cdot R_{max} \cdot \dfrac{N_{PRB}^{BW(j),\mu} \cdot 12}{T_s^{\mu}} \cdot \left(1 - OH^{(j)}\right)}{BW^{(j)}} \tag{5.1}$$

wherein

- $R_{max} = 948/1024$ is the maximum coding rate supported by 5G data channel.
- For the jth CC,

  - $v_{Layers}^{(j)}$ is the maximum number of layers
  - $Q_m^{(j)}$ is the maximum modulation order
  - μ is the numerology which is related to the subcarrier spacing as defined in 3GPP TS38.211, that is, subcarrier spacing is given by $15 \times 2^\mu$ kHz; e.g., $\mu = 0$ if subcarrier spacing is 15 kHz, and $\mu = 1, 2, \ldots$ if subcarrier spacing is 30 kHz, 60 kHz, ...
  - T_s^∞ is the average OFDM symbol duration in a subframe for numerology μ, i.e., $T_s^\mu = \dfrac{10^{-3}}{14 \cdot 2^\mu}$. Note that normal cyclic prefix is assumed.
  - $N_{PRB}^{BW(j),\mu}$ is the maximum RB allocation in bandwidth $BW^{(j)}$ with numerology μ, as given in 3GPP TR 38.817–01 Sect. 4.5.1, where $BW^{(j)}$ is the UE supported maximum bandwidth in the given band or band combination.
  - $OH^{(j)}$ is the overhead calculated as the average ratio of the number of REs occupied by L1/L2 control, Synchronization Signal, PBCH, reference signals and guard period (for TDD), etc. with respect to the total number of REs in effective bandwidth time product as given by $\left(\alpha^{(j)} \cdot BW^{(j)} \cdot \left(14 \times T_s^\mu\right)\right)$.

    $\alpha^{(j)}$ is the normalized scalar considering the downlink/uplink ratio; for FDD $\alpha^{(j)} = 1$ for DL and UL; and for TDD and other duplexing $\alpha^{(j)}$ for DL and UL is calculated based on the DL/UL configuration.

    For guard period (GP), 50% of GP symbols are considered as downlink overhead, and 50% of GP symbols are considered as uplink overhead.

It is observed that the peak spectral efficiency can be increased by reducing the overhead on the whole bandwidth. For example, in 3GPP 5G NR design, the control signal overhead and reference signal overhead can be reduced on a larger bandwidth. This implies the NR large bandwidth should provide a higher peak spectral efficiency.

5.3.1.2   Peak Data Rate

As defined in [2], peak data rate is the maximum achievable data rate under ideal conditions (in bit/s), which is the received data bits assuming error-free conditions assignable to a single mobile station, when all assignable radio resources for the corresponding link direction are utilized (i.e., excluding radio resources that are used for physical layer synchronization, reference signals or pilots, guard bands and guard times).

The DL/UL peak data rate for FDD and TDD over Q component carriers can be calculated as below

$$R = \sum_{j=1}^{Q} W_j \times SE_{p_j} = \sum_{j=1}^{Q} \left(\alpha^{(j)} \cdot BW^{(j)} \right) \times SE_{p_j} \qquad (5.2)$$

where W_j and SEp_i $(j = 1,\ldots, Q)$ are the effective bandwidth and spectral efficiencies on component carrier j, respectively, $\alpha^{(j)}$ is the normalized scalar on component carrier j considering the downlink/uplink ratio on that component carrier; for FDD $\alpha^{(j)} = 1$ for DL and UL; and for TDD and other duplexing $\alpha^{(j)}$ for DL and UL is calculated based on the frame structure, and $BW^{(j)}$ is the carrier bandwidth of component j.

5.3.1.3   Fifth Percentile User Spectral Efficiency and Average Spectral Efficiency

As defined in [2], the 5th percentile user spectral efficiency is the 5% point of the CDF of the normalized user throughput. The normalized user throughput is defined as the number of correctly received bits, i.e., the number of bits contained in the SDUs delivered to Layer 3, over a certain period of time, divided by the channel bandwidth and is measured in bit/s/Hz.

The channel bandwidth is defined as the effective bandwidth times the frequency reuse factor, where the effective bandwidth is the operating bandwidth normalized appropriately considering the uplink/downlink ratio.

With $R_i(T_i)$ denoting the number of correctly received bits of user i, T_i the active session time for user i and W the channel bandwidth, the (normalized) user through-put of user i, r_i, is defined according to

$$r_i = \frac{R_i(T_i)}{T_i \cdot W} \qquad (5.3)$$

On the other hand, average spectral efficiency is the aggregate throughput of all users (the number of correctly received bits; i.e., the number of bits contained in the SDUs delivered to Layer 3, over a certain period of time) divided by the channel bandwidth of a specific band divided by the number of TRxPs and is measured in bit/s/Hz/TRxP.

Let R_i (T) denote the number of correctly received bits by user i (downlink) or from user i (uplink) in a system comprising a user population of N users and M TRxPs. Furthermore, let W denote the channel bandwidth and T the time over which the data bits are received. The average spectral efficiency SE_{avg} is then defined according to

$$SE_{avg} = \frac{\sum_{i=1}^{N} R_i(T)}{T \cdot W \cdot M} \tag{5.4}$$

As required by Report ITU-R M.2412, 5th percentile user spectral efficiency shall be assessed jointly with average spectral efficiency using the same simulation.

The spectral efficiency is related to quite a lot of design factors. Similar to peak spectral efficiency, the reduced overhead including control overhead and reference signal overhead will contribute to increase spectral efficiency. Therefore, 5G NR's capability on reducing overhead on large bandwidth can benefit spectral efficiency improvement.

On the other hand, massive MIMO will contribute a lot to achieve the three times spectral efficiency improvement target defined by ITU-R. 5G NR design on massive MIMO will benefit this aspect.

Finally, innovative channel state information (CSI) feedback mechanism together with massive MIMO will guarantee the performance of multiple antenna could reach its maximum. 5G NR design on CSI feedback would be helpful.

5.3.1.4 User Experienced Data Rate

As defined in [2], the user experienced data rate is the 5% point of the cumulative distribution function (CDF) of the user throughput. User throughput (during active time) is defined as the number of correctly received bits, i.e., the number of bits contained in the service data units (SDUs) delivered to Layer 3, over a certain period of time.

The user experience data rate is related to the edge user data throughput. For uplink user experienced data rate, the challenge is how to achieve high edge user data rate at the constraint of maximum transmit power. This is because high data rate usually requires large user bandwidth (the bandwidth occupied by a user). However, for uplink transmission, the transmit power is limited, and therefore for edge users the transmit power could not be increasing proportionally to the occupied bandwidth if the propagation loss of the edge user is large such that the fractional power control makes the desired transmit power exceed the maximum transmit power. In this case, the transmit power density would be decreasing with the increased bandwidth. And hence the large bandwidth under this situation could not help to increase the data rate as the spectral efficiency would be decreasing with decreased transmit power density.

For 5G NR, it would face such challenge since the early available 5G spectrum is most likely on the range of 3.5 GHz. In this case, DL/UL decoupling described in Sect. 5.2.4 would help.

5.3.1.5   Area Traffic Capacity

As defined in [2], the area traffic capacity is the total traffic throughput served per geographic area (in Mbit/s/m^2). The throughput is the number of correctly received bits, i.e., the number of bits contained in the SDUs delivered to Layer 3, over a certain period of time.

The area traffic capacity can be derived for a particular use case (or deployment scenario) of one frequency band and one TRxP layer, based on the achievable average spectral efficiency, network deployment (e.g., TRxP (site) density), and bandwidth [2].

Let W denote the channel bandwidth and ρ the TRxP density (TRxP/m^2). The area traffic capacity C_{area} is related to average spectral efficiency SE_{avg} as [2]:

$$C_{area} = \rho \times W \times SE_{av} \tag{5.5}$$

In case bandwidth is aggregated across multiple bands, the area traffic capacity will be summed over the bands.

It can be seen from the definition that larger bandwidth, higher spectral efficiency, lower overhead (either control or reference signal overhead), and higher network node density could provide larger area traffic capacity. 5G NR will demonstrate strong capability on the above aspects to achieve 10 times to 100 times, or even more area traffic capacity compared to Release-10 LTE-Advanced.

5.3.1.6   User Plane Latency

As defined in [2], user plane latency is the contribution of the radio network to the time from when the source sends a packet to when the destination receives it (in ms). It is defined as the one-way time it takes to successfully deliver an application layer packet/message from the radio protocol layer 2/3 SDU ingress point to the radio protocol layer 2/3 SDU egress point of the radio interface in either uplink or downlink in the network for a given service in unloaded conditions, assuming the mobile station is in the active state.

The evaluation of 3GPP 5G NR user plane latency is based on the procedure illustrated in Fig. 5.1. It takes into account the case of retransmission. If one assumes the latency for initial transmission is T_0, the latency for initial transmission and one-time retransmission is T_1, and the latency for initial transmission plus n-time retransmission is T_n, the expected user plane latency is given by

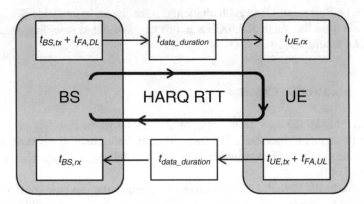

Fig. 5.1 User plane procedure for evaluation

$$T(1) = p_0 T_0 + p_1 T_1 + \ldots + p_N T_N \tag{5.6}$$

where p_n is the probability of n-times retransmission ($n = 0$ means initial transmission only), and $p_0 + p_1 + \ldots + p_N = 1$. The probability of retransmission is related to SINR, coding scheme, modulation order, etc. For simplicity, it is usually assumed that $p_2 = \ldots p_N = 0$ ($N > 2$).

It is noted herein that the exact value of $T(l)$ is dependent on l, which is the index of the OFDM symbol in one slot when the data packet arrives. This is obviously true for TDD band since, if the DL data packet arrives in an UL slot, it of course needs more waiting time for the next DL slot than the case of arriving in the DL slot. For NR, this dependency is also valid for FDD. This is because NR allows sub-slot processing, and if the packet arrives in the later part of the slot, it may need to wait until the beginning of the next slot to further proceed, while if the packet arrives in the early part of the slot, it may probably proceed within the slot. To remove the dependency, an averaged value of user plane latency is helpful. This is defined as follows:

$$T_{UP} = \frac{1}{14 \times N} \sum_{l=1}^{14 \times N} T(l) \tag{5.7}$$

where N is the number of slots that constitutes one period of DL/UL pattern, and 14 is the number of OFDM symbols in one slot. For example, for FDD, $N = 1$, and for TDD pattern "DDDSU," $N = 5$. T_{UP} is used in the evaluation.

For downlink evaluation, the downlink procedure is abstracted in Table 5.5, where the notations representing each of the step component are given. The method of value calculation for 5G NR in each step as well as the total latency is presented. It should be noted that it is assumed the data packet may arrive at any time of any OFDM symbol in one slot. Therefore the latency is an average of the cases. This will be employed in the performance evaluation as shown in Sect. 5.4.

In Table 5.5, the frame alignment time T_{FA} and the waiting time T_{wait} are mentioned in Step 1.2 and Step 2.2. The frame alignment time occurs due to the fact that

Table 5.5 DL user plane procedure for NR

Step	Component	Notations	Values
1	DL data transfer	$T_1 = (t_{BS,tx} + t_{FA,DL}) + t_{DL\_duration} + t_{UE,rx}$	
1.1	BS processing delay	$t_{BS,tx}$ The time interval between the data is arrived, and packet is generated	$T_{proc,2}/2$, with $d_{2,1} = d_{2,2} = d_{2,3} = 0$. ($T_{proc,2}$ is defined in Section 6.4 of 3GPP TS38.214)[a,b]
1.2	DL Frame alignment (transmission alignment)	$t_{FA,DL}$ It includes frame alignment time, and the waiting time for next available DL slot	$T_{FA} + T_{wait}$, T_{FA} is the frame alignment time within the current DL slot; T_{wait} is the waiting time for next available DL slot if the current slot is not DL slot
1.3	TTI for DL data packet transmission	$t_{DL\_duration}$	Length of one slot (14 OFDM symbol length) or non-slot (2/4/7 OFDM symbol length), depending on slot or non-slot selected in evaluation
1.4	UE processing delay	$t_{UE,rx}$ The time interval between the PDSCH is received and the data is decoded	$T_{proc,1}/2$ ($T_{proc,1}$ is defined in Section 5.3 of 3GPP TS38.214), $d_{1,1} = 0$; $d_{1,2}$ should be selected according to resource mapping type and UE capability. $N_1 =$ the value with "No additional PDSCH DM-RS configured"[c,d]
2	HARQ retransmission	$T_{HARQ} = T_1 + T_2$ $T_2 = (t_{UE,tx} + t_{FA,UL}) + t_{UL\_duration} + t_{BS,rx}$ (For Steps 2.1 to 2.4)	
2.1	UE processing delay	$t_{UE,tx}$ The time interval between the data is decoded, and ACK/NACK packet is generated	$T_{proc,1}/2$ ($T_{proc,1}$ is defined in Section 5.3 of 3GPP TS38.214), $d_{1,1} = 0$; $d_{1,2}$ should be selected according to resource mapping type and UE capability. $N_1 =$ the value with "No additional PDSCH DM-RS configured"[c,e]
2.2	UL frame alignment (transmission alignment)	$t_{FA,UL}$ It includes frame alignment time, and the waiting time for the next available UL slot	$T_{FA} + T_{wait}$, T_{FA} is the frame alignment time within the current UL slot; T_{wait} is the waiting time for next available UL slot if the current slot is not UL slot
2.3	TTI for ACK/NACK transmission	$t_{UL\_duration}$	1 OFDM symbol
2.4	BS processing delay	$t_{BS,rx}$ The time interval between the ACK is received and the ACK is decoded	$T_{proc,2}/2$ with $d_{2,1} = d_{2,2} = d_{2,3} = 0$[a,f]

(continued)

Table 5.5 (continued)

Step	Component	Notations	Values
2.5	Repeat DL data transfer from 1.1 to 1.4	T_1	
–	Total one way user plane latency for DL	$T_n = T_1 + n \times T_{HARQ} + 0.5T_s$ where n is the number of retransmissions ($n \geq 0$), T_s is the symbol length, and $0.5T_s$ is added as the "average symbol alignment time" at the beginning of the procedure due to the fact that the packet arrives at any time of any OFDM symbol	

[a]Here the value of $T_{proc,2}$ is employed for evaluation purpose. The value of $T_{proc,2}$ is defined in Section 6.4 in 3GPP TS 38.214 for UE PUSCH transmission preparation time. It is assumed that gNB PDSCH preparation time is identical to UE's PUSCH preparation time. However, gNB processing delay would vary depending on implementation. It is further noted that the gNB PDSCH preparation time $T_{proc,2}$ contains two parts for the case of retransmission: one is the processing time of ACK/NACK preparation, and the other is the processing time of preparing the data packets. Step 1.1 is for the second part. It is assumed that the second part would consume $0.5T_{proc,2}$, whereas the first part (represented by Step 2.4) consumes the other $0.5T_{proc,2}$
[b]For the case of a TDD band (30 kHz SCS) with an SUL band (15 kHz SCS), the value of this step is $T_{proc,2}(\mu = 30$ kHz$)/2$ for Initial transmission, and $T_{proc,2}(\mu = 15$ kHz$)/2$ for retransmission
[c]The value of $T_{proc,1}$ is defined in Section 5.3 in TS 38.214 for UE PDSCH reception processing time
[d]For the above case, the UE is processing PDSCH reception on TDD band with 30 kHz SCS, and it is assumed that the value of this step is $T_{proc,1}(\mu = 30$ kHz$)/2$
[e]For the above case, the value of this step is $T_{proc,1}(\mu = 15$ kHz$) - T_{proc,1}(\mu = 30$ kHz$)/2$
[f]For the above case, the value of this step is $T_{proc,2}(\mu = 15$ kHz$)/2$

the end of the previous step may end at the middle of a slot. In this case, the procedure should wait until the available starting OFDM symbol. The next available starting OFDM symbol is related to resource mapping type and the length of the OFDM symbols that are needed for the data transmission. If S is the index of the starting OFDM symbol, and L is the scheduled length of the OFDM symbols for data transmission, then the selection of S should guarantee that $S + L \leq 14$. That is to say, the selected starting OFDM symbol should guarantee that the data transmission is not exceeding the boundary of one slot.

The waiting time is accounted for the TDD band where the next slot might not be available for the desired transmission direction. For example, if the procedure is now in Step 1.2 in Table 5.5, it implies the DL transmission is desired. However, if the next slot is an UL slot, the waiting time would occur until the DL slot arrives.

The illustration of frame alignment time T_{FA} and waiting time T_{wait} is shown in Fig. 5.2. In this plot, it is assumed that the next available OFDM symbol is at the beginning of the next slot.

For uplink evaluation, the grant-free mechanism is employed. In this case, the UL data could be transmitted once it arrives at the UE side; thus, the latency is reduced due to the fact that the UE does not need to wait for the grant for UL data transmission. This procedure is abstracted in Table 5.6. The method of value calculation for 5G NR in each step as well as the total latency is presented. This will be employed in the performance evaluation as shown in Sect. 5.4.

Table 5.6 UL user plane procedure for NR

Step	Component	Notations	Value
1	UL data transfer	$T_1 = (t_{UE,tx} + t_{FA,UL}) + t_{UL\_duration} + t_{BS,rx}$	
1.1	UE processing delay	$t_{UE,tx}$ The time interval between the data is arrived, and packet is generated	$T_{proc,2}/2$ ($T_{proc,2}$ is defined in Section 6.4 of 3GPP TS38.214), with $d_{2,1} = d_{2,2} = d_{2,3} = 0$
1.2	UL Frame alignment (transmission alignment)	$t_{FA,UL}$ It includes frame alignment time, and the waiting time for next available UL slot	$T_{FA} + T_{wait}$ T_{FA} is the frame alignment time within the current UL slot T_{wait} is the waiting time for next available UL slot if the current slot is not UL slot
1.3	TTI for UL data packet transmission	$t_{UL\_duration}$	Length of one slot (14 OFDM symbol length) or non-slot (2/4/7 OFDM symbol length), depending on slot or non-slot selected in evaluation[a]
1.4	BS processing delay	$t_{BS,rx}$ The time interval between the PUSCH is received and the data is decoded	$T_{proc,1}/2$ ($T_{proc,1}$ is defined in Section 5.3 of 3GPP TS38.214), $d_{1,1} = 0$; $d_{1,2}$ should be selected according to resource mapping type and UE capability. $N_1 =$ the value with "No additional PDSCH DM-RS configured"; It is assumed that BS processing delay is equal to UE processing delay as for PDSCH[b]
2	HARQ retransmission	$T_{HARQ} = T_2 + T_1$ $T_2 = (t_{BS,tx} + t_{FA,DL}) + t_{DL\_duration} + t_{UE,rx}$ (For Steps 2.1 to 2.4)	
2.1	BS processing delay	$t_{BS,tx}$ The time interval between the data is decoded, and PDCCH preparation	$T_{proc,1}/2$ ($T_{proc,1}$ is defined in Section 5.3 of 3GPP TS38.214), $d_{1,1} = 0$; $d_{1,2}$ should be selected according to resource mapping type and UE capability. $N_1 =$ the value with "No additional PDSCH DM-RS configured"
2.2	DL Frame alignment (transmission alignment)	$t_{FA,DL}$ It includes frame alignment time, and the waiting time for next available DL slot	$T_{FA} + T_{wait}$ T_{FA} is the frame alignment time within the current DL slot T_{wait} is the waiting time for next available DL slot if the current slot is not DL slot
2.3	TTI for PDCCH transmission	$t_{DL\_duration}$	1 OFDM symbols for PDCCH scheduling the retransmission
2.4	UE processing delay	$t_{UE,rx}$ The time interval between the PDCCH is received and decoded	$T_{proc,2}/2$ ($T_{proc,2}$ is defined in Section 6.4 of 3GPP TS38.214), with $d_{2,1} = d_{2,2} = d_{2,3} = 0$

(continued)

Table 5.6 (continued)

Step	Component	Notations	Value
2.5	Repeat UL data transfer from 1.1 to 1.4	T_1	
	Total one way user plane latency for UL	$T_n = T_1 + n \times T_{HARQ} + 0.5T_s$ where n is the number of retransmissions ($n \geq 0$), T_s is the symbol length, and $0.5T_s$ is added as the "average symbol alignment time" at the beginning of the procedure due to the fact that the packet arrives at any time of any OFDM symbol	

Note:
[a]The grant-free transmission is assumed to use the following start symbols according to 3GPP specification:

1. For 2-symbol PUSCH, the start symbol can be symbols {0, 2, 4, 6, 8, 10, 12} for PUSCH resource mapping type B
2. For 4-symbol PUSCH, the start symbol can be:
   (i) For PUSCH resource mapping type B: symbols {0,7}
   (ii) For PUSCH resource mapping type A: symbol 0
3. For 7-symbol PUSCH, the start symbol can be:
   (i) For PUSCH resource mapping type B: symbols {0, 7}
   (ii) For PUSCH resource mapping type A: symbol 0
4. For 14-symbol PUSCH, the start symbol can be at symbol #0 for PUSCH resource mapping type A and B

[b]Here the value of $T_{proc,1}$ is employed for evaluation purpose. The value of $T_{proc,1}$ is defined in Section 5.3 in 3GPP TS 38.214 for UE PDSCH reception processing time. It is assumed that gNB PUSCH reception processing time is identical to UE's PDSCH reception time. However, gNB processing delay would vary depending on implementation. It is further noted that the gNB PUSCH reception time $T_{proc,1}$ contains two parts for the case of retransmission: one is the processing time of PUSCH reception, the other is the preparation for PDCCH. Step 1.4 is for the first part. It is assumed that the first part would consume $0.5T_{proc,1}$, whereas the second part (represented by Step 2.1) consumes the other $0.5T_{proc,1}$.

In Table 5.6, the frame alignment time T_{FA} and the waiting time T_{wait} are mentioned in Step 1.2 and Step 2.2. The definitions and calculations are the same as in DL user plane latency evaluation, and an illustration is found in Fig. 5.2.

It can be seen from the above procedure that, for TDD system, there is frame alignment time which is used to wait for the available time resource (time slot) for the desired link direction, e.g., Step 1.2 and 2.2 for DL or UL procedure. This would result in larger delay of the DL and UL user plane data transmission. Especially, if the downlink dominant frame structure is used, the uplink user plane latency will be poor. DL/UL decoupling mechanism will be helpful in this case.

5.3.1.7   Control Plane Latency

As defined in [2], control plane latency refers to the transition time from a most "battery efficient" state (e.g., Idle state) to the start of continuous data transfer (e.g., Active state).

For 3GPP 5G NR, control plane latency can be evaluated from RRC_INACTIVE state to RRC_CONNECTED state. Figure 5.3 provides the control plane flow employed in evaluation.

The detailed assumption of each step as shown in Fig. 5.3 is provided in Table 5.7. The evaluation is for UL data transfer.

It should be noted that the waiting time for the available downlink or uplink time resource for the desired link direction should be calculated in Step 2, 4, 6, and 8. It depends on the detailed DL/UL configuration. Again, if the downlink dominant frame structure is used on a TDD band, the control plane latency might be large. DL/UL decoupling mechanism will be helpful in this case.

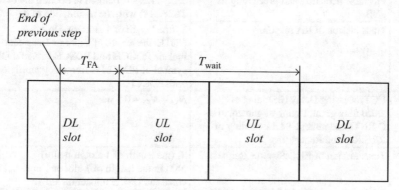

Fig. 5.2 User plane procedure for evaluation

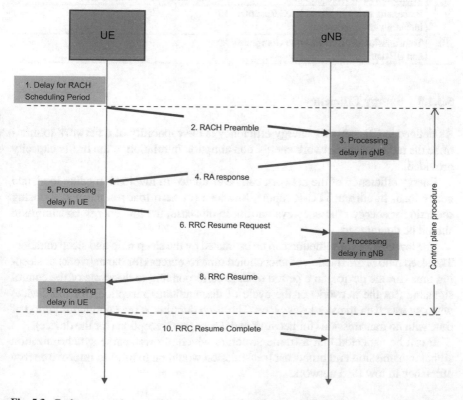

Fig. 5.3 C-plane procedure employed in control plane latency evaluation

Table 5.7 Assumption of C-plane procedure for NR

Step	Description	Value
1	Delay due to RACH scheduling period (1TTI)	0
2	Transmission of RACH Preamble	Length of the preamble according to the PRACH format as specified in Section 6 of 3GPP TS 38.211
3	Preamble detection and processing in gNB	$T_{proc,2}$ ($T_{proc,2}$ is defined in Section 6.4 of 3GPP TS38.214) with the assumption of $d_{2,1} = 0$
4	Transmission of RA response	T_s (the length of 1 slot/non-slot) NOTE: the length of 1 slot or 1 non-slot include PDCCH and PDSCH (the first OFDM symbol of PDSCH is frequency multiplexed with PDCCH)
5	UE Processing Delay (decoding of scheduling grant, timing alignment and C-RNTI assignment + L1 encoding of RRC Resume Request)	$N_{T,1} + N_{T,2} + 0.5$ ms
6	Transmission of RRC Resume Request	T_s (the length of 1 slot/non-slot) NOTE: the length of 1 slot or 1 non-slot is equal to PUSCH allocation length
7	Processing delay in gNB (L2 and RRC)	3
8	Transmission of RRC Resume	T_s (the length of 1 slot/non-slot)
9	Processing delay in UE of RRC Resume including grant reception	7
10	Transmission of RRC Resume Complete and UP data	0

5.3.1.8   Energy Efficiency

As defined in [2], network energy efficiency is the capability of a network to mini-mize the radio access network energy consumption in relation to the traffic capacity provided.

Energy efficiency of the network can be evaluated in low load, median load, and heavy load. In current ITU-R report, low load (or zero load) is the first focusing scenario for energy efficiency evaluation. In this case, the low energy consumption should be guaranteed.

The low energy consumption can be estimated by the sleep ratio and sleep duration. The sleep ratio is the fraction of unoccupied time resources (for the network) or sleep-ing time (for the device) in a period of time corresponding to the cycle of the control signaling (for the network) or the cycle of discontinuous reception (for the device) when no user data transfer takes place. The sleep duration is the continuous period of time with no transmission (for network and device) and reception (for the device).

It can be expected that a frame structure which allows sparse synchronization signal transmission and broadcast transmission would be helpful to improve energy efficiency in low load network.

5.3.1.9   Mobility

As defined in [2], mobility is the maximum mobile station speed at which a defined QoS can be achieved (in km/h). The QoS is defined as normalized traffic channel link data rate.

The following classes of mobility are defined:

- Stationary: 0 km/h
- Pedestrian: 0–10 km/h
- Vehicular: 10–120 km/h
- High speed vehicular: 120–500 km/h.

The following steps are defined to evaluate the mobility requirement [2]:

Step 1: Run uplink system-level simulations, where the user speeds are assumed to be predefined value (e.g., 10 km/h for indoor hotspot, 30 km/h for dense urban, 120 or 500 km/h for rural), using link-level simulations and a link-to-system interface appropriate for these speed values, and collect overall statistics for uplink *SINR* values, and construct CDF over these values for the test environment.

Step 2: Use the uplink SINR CDF to save the 50th-percentile *SINR* value.

Step 3: Run new uplink link-level simulations for either NLOS or LOS channel conditions using the associated speeds as input parameters, to obtain link data rate and residual packet error ratio as a function of *SINR*. The link-level simulation shall take into account retransmission, channel estimation, and phase noise impact.

Step 4: Compare the uplink spectral efficiency values (link data rate normalized by channel bandwidth) obtained from *Step 3* using the associated SINR value obtained from *Step 2*, with the corresponding threshold values (requirement).

Step 5: The proposal fulfills the mobility requirement if the spectral efficiency value is larger than or equal to the corresponding threshold value and if also the residual decoded packet error ratio is less than 1%.

For massive MIMO configuration, the number of physical element is usually large (e.g., up to 256 elements might be used). But usually the number of radio frequency (RF) chain is limited (e.g., 16 RF chains are used for 256 element antenna). In this case, it is proposed to use 256 element antenna in system-level simulation with 16 RF chains to derive the uplink SINR distribution. And in link-level simulation, 16 RF chains (or 16 "element" antenna) are assumed to simplify the link-level simulator computational load.

It is seen that the mobility evaluation is conducted by assuming high user speed. Therefore the frame structure with larger subcarrier spacing is helpful for high mobility evaluation since it can provide good performance to combat with Doppler spread. And MIMO can enhance the SINR under high mobility which is also beneficial to keep a good normalized data rate, given that MIMO can well handle the CSI feedback under high mobility scenarios.

5.3.1.10   Mobility Interruption Time

As defined in [2], mobility interruption time is the shortest time duration supported by the system during which a user terminal cannot exchange user plane packets with any base station during mobility transitions.

The mobility interruption time includes the time required to execute any radio access network procedure, radio resource control signaling protocol, or other message exchanges between the mobile station and the radio access network, as applicable to the candidate technology.

Inspection is used for the evaluation.

5.3.2   Evaluation Metrics for mMTC Requirements

5.3.2.1   Connection Density

As specified in [2], connection density is the system capacity metric defined as the total number of devices fulfilling a specific quality of service (QoS) per unit area (per km^2) with 99% grade of service (GoS).

There are two alternative evaluation methods to evaluate connection density:

- Full system-level simulation using burst buffer traffic.
- Full-buffer system-level simulation followed by link-level simulation.

The following steps are used to evaluate the connection density based on full system-level simulation [2]. Burst buffer traffic is employed, e.g., one packet every 2 h for each user terminal, and the packet arrival follows Poisson arrival process.

Step 1: Set system user number per TRxP as N.
Step 2: Generate the user packet according to the traffic model.
Step 3: Run non-full buffer system-level simulation to obtain the packet outage rate. The outage rate is defined as the ratio of the number of packets that failed to be delivered to the destination receiver within a transmission delay of less than or equal to 10s to the total number of packets generated in *Step 2*.
Step 4: Change the value of N and repeat *Step 2–3* to obtain the system user number per TRxP N' satisfying the packet outage rate of 1%.
Step 5: Calculate connection density by equation $C = N'/A$, where the TRxP area A is calculated as $A = \text{ISD}^2 \times \text{sqrt}(3)/6$, and ISD is the inter-site distance.

The following steps are used to evaluate the connection density based on full-buffer system-level simulation followed by link-level simulation. Full buffer traffic is used for system-level simulation part. And the traffic used to calculate the connection density is bursty traffic, e.g., one packet every 2 h for each user terminal.

Step 1: Perform full-buffer system-level simulation using the evaluation parameters for Urban Macro-mMTC test environment, determine the uplink *SINRi* for each percentile $i = 1...99$ of the distribution over users, and record the average allocated user bandwidth W_{user}.

Step 2: Perform link-level simulation and determine the achievable user data rate R_i for the recoded *SINRi* and W_{user} values.

Step 3: Calculate the packet transmission delay of a user as $D_i = S/R_i$, where S is the packet size.

Step 4: Calculate the traffic generated per user as $T = S/T_{inter-arrival}$, where $T_{inter-arrival}$ is the inter-packet arrival time.

Step 5: Calculate the long-term frequency resource requested under *SINRi* as $B_i = T/(R_i/W_{user})$.

Step 6: Calculate the number of supported connections per TRxP, $N = W/mean(B_i)$. W is the simulation bandwidth. The mean of B_i may be taken over the best 99% of the *SINRi* conditions.

Step 7: Calculate the connection density as $C = N/A$, where the TRxP area A is calculated as $A = ISD^2 \times sqrt(3)/6$, and ISD is the inter-site distance.

The technology should ensure that the 99th percentile of the delay per user D_i is less than or equal to 10s.

It can be observed that the key issue for connection density is the capability of transmitting the data packet (usually small) in a short delay from the user terminal to the base station, especially when the user terminals are at the cell edge. In this case, the narrow band transmission would be beneficial because the power can be concentrated for uplink and therefore guarantees the reasonable delay of the uplink packet transmission. Narrow-band (NB) IoT is capable of providing this capability.

5.3.3   Evaluation Metrics for URLLC Requirements

5.3.3.1   User Plane Latency

User plane latency is defined as in Sect. 5.3.1.6. Similar evaluation method can be applied for URLLC. However, one difference is that for URLLC, usually 99.999% of successful transmission ratio is required. If the air interface design guarantees the initial transmission achieves 99.999% successful ratio, then one can assume that there is no retransmission. In this case, the average delay can be approximated by the delay of initial transmission.

5.3.3.2   Control Plane Latency

Control plane latency is defined as in Sect. 5.3.1.7. The same evaluation method can be applied for URLLC.

5.3.3.3 Reliability

As defined in [2], reliability is the success probability of transmitting a layer 2/3 packet within a required maximum time, which is the time it takes to deliver a small data packet from the radio protocol layer 2/3 SDU ingress point to the radio protocol layer 2/3 SDU egress point of the radio interface at a certain channel quality.

The following steps are applied to evaluate the reliability requirement using system-level simulation followed by link-level simulations [2]:

Step 1: Run downlink or uplink full buffer system-level simulations of candidate technology, and collect overall statistics for downlink or uplink *SINR* values, and construct CDF over these values.

Step 2: Use the SINR CDF to save the 5th percentile downlink or uplink *SINR* value.

Step 3: Run corresponding link-level simulations for either NLOS or LOS channel conditions to obtain success probability, which equals to $(1 - P_e)$, where P_e is the residual packet error ratio within maximum delay time as a function of *SINR* taking into account retransmission.

Step 4: The technology should guarantee that at the 5th percentile downlink or uplink *SINR* value of *Step 2* and within the required delay, the success probability derived in *Step 3* is larger than or equal to the required success probability.

5.3.3.4 Mobility Interruption Time

Mobility interruption time is defined as in Sect. 5.3.1.10. The same evaluation method can be applied for URLLC.

5.4 5G Performance Evaluation

Using the abovementioned evaluation methodologies, the capability of 3GPP 5G NR is evaluated against the eMBB, mMTC, and URLLC performance requirements.

The key technical components are considered: 5G NR design of carrier and channels as introduced in Sect. 2.2, including wideband frame structure and physical channel structure, reference signal design, etc.; LTE/NR coexistence design as introduced in Sect. 2.4; new physical layer design as introduced in Sect. 2.5, including NR massive MIMO, multiple access, and waveform, etc. The contribution of the above technical components to the fulfillment of the requirements or the improvement of performance are evaluated.

5.4.1 5G Wideband Frame Structure and Physical Channel Structure

5G NR supports wideband frame structure and wideband physical channel structure (see Sect. 2.2). The wideband frame structure is characterized by the support of multiple subcarrier spacings, the flexible organization of OFDM symbols to non-slot, and the reduced guard band. The NR wideband physical channel structure is featured by PDCCH design, PDSCH and PUSCH resource mapping, and reference signal structure.

The wideband frame and physical channel structure is designated to efficiently utilize the wideband channel such that the overhead is reduced. With this capability, spectral efficiency is improved. The support of multiple subcarrier spacing enables the possibility to use wider subcarrier spacing when short latency and when high Doppler spread is observed. The use of less number of OFDM symbols to form a non-slot data scheduling can reduce the latency of small amount of data transmissions. The PDCCH and PDSCH sharing can reduce the PDCCH overhead for wideband cases. And the resource mapping type B defined in NR can make immediate transmission within one slot. These capabilities can be used in low latency and high mobility scenarios. Reliability can also benefit since the reduced one-time transmission latency can introduce more retransmissions within given latency budget. These benefits will be evaluated in this subsection.

5.4.1.1 Contribution to Overhead Reduction and Spectral Efficiency/Data Rate Improvement

5G NR frame structure and physical channel structure supports the reduced overhead and guard band ratio, which in turn enhances the spectral efficiency and cell data rate. Recall that both spectral efficiency and cell data rate are related to the user throughput $R_i(T)/T$ (see Sects. 5.3.1.3, and 5.3.1.5, etc.), and the transmitted user bits $R_i(T)$ are related to guard band ratio $\bar{\eta}$ and overhead OH through

$$R_i\left(T\right)=\left[N_{RE}^{BW} \times\left(1-\bar{\eta}\right)\times\left(1-OH\right)\right]\times M \times CR \times N_{layer} \qquad (5.8)$$

where $N_{RE}^{BW} = N_{RB}^{BW} \times 12 \times N_{OS}$ is the number of resource elements (REs) on the whole system bandwidth, N_{RB}^{BW} is the number of resource blocks (RBs) on the whole system bandwidth, 12 is the number of subcarriers in one RB, N_{OS} is the number of OFDM symbols during the time duration T, $\eta = \left(1-\bar{\eta}\right)Z$ is the spectrum utilization ratio, N_{RE} is the number of REs on available bandwidth (excluding the guard band),

$$N_{RE} = N_{RE}^{BW} \times\left(1-\bar{\eta}\right)= N_{RE}^{BW} \times\eta \qquad (5.9)$$

OH is the overhead which is given by OH= N_{RE}^{data}/N_{RE}, and N_{RE}^{data} is the number of REs that can be occupied by data transmission, M is the modulation order, CR is the

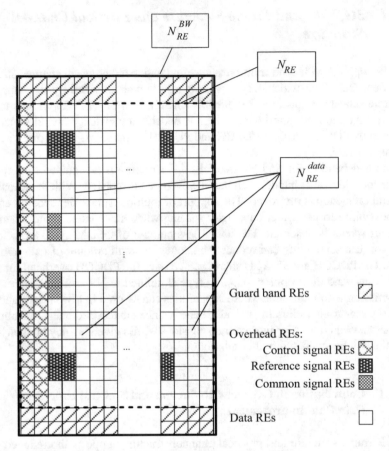

Fig. 5.4 Illustration of guard band REs, overhead REs, and data REs

coding rate, N_{layer} is the number of layers of this user. The value of M, CR, and N_{layer} are related to the post-processing SINR that could be enhanced by advanced multi-antenna processing schemes.

The overall overhead by taking into account both guard band and control/reference signal overhead is defined as

$$\Gamma = 1 - (1 - \bar{\eta}) \times (1 - OH) \tag{5.10}$$

Figure 5.4 illustrates the guard band REs, overhead REs, and data REs, and how the value of N_{RE}^{BW}, N_{RE}, and N_{RE}^{data} can be calculated.

It can be therefore observed that by reducing the guard band ratio $\bar{\eta}$ and the overhead OH the spectral efficiency and data rate can be increased. If two technologies provide the same post-processing SINR (that is, the same modulation order, the same coding rate, and the same number of layers per user is supported), the one with smaller guard band ratio and overhead would outperform the other one.

Table 5.8 NR spectrum utilization ratio

(A) Frequency range 1 (below 6 GHz)

SCS (kHz)		5 MHz	10 MHz	15 MHz	20 MHz	25 MHz	30 MHz	40 MHz	50 MHz	60 MHz	70 MHz	80 MHz	90 MHz	100 MHz
15	N_{RB}	25	52	79	106	133	160	216	270	N/A	N/A	N/A	N/A	N/A
	η	90.0%	93.6%	94.8%	95.4%	95.8%	96.0%	97.2%	97.2%	N/A	N/A	N/A	N/A	N/A
30	N_{RB}	11	24	38	51	65	78	106	133	162	189	217	245	273
	η	79.2%	86.4%	91.2%	91.8%	93.6%	93.6%	95.4%	95.8%	97.2%	97.2%	97.7%	98.0%	98.3%
60	N_{RB}	N/A	11	18	24	31	38	51	65	79	93	107	121	135
	η	N/A	79.2%	86.4%	86.4%	89.3%	91.2%	91.8%	93.6%	94.8%	95.7%	96.3%	96.8%	97.2%

(B) Frequency range 2 (above 24 GHz)

SCS [kHz]		50 MHz	100 MHz	200 MHz	400 MHz
60	N_{RB}	66	132	264	N/A
	η	95.0%	95.0%	95.0%	N/A
120	N_{RB}	32	66	132	264
	η	92.2%	95.0%	95.0%	95.0%

Table 5.9 LTE spectrum utilization ratio (SCS = 15 kHz)

Channel bandwidth $BW_{Channel}$ [MHz]	1.4	3	5	10	15	20
N_{RB}	6	15	25	50	75	100
η	77.1%	90.0%	90.0%	90.0%	90.0%	90.0%

In the following, the guard band ratio (or spectrum utilization ratio) and over-head of NR and LTE are analyzed.

The NR guard band ratio reduction is achieved by appropriate processing on frequency or time domain filtering. Guard band ratio can be simply calculated by the given subcarrier spacing (SCS), SCS, the number of subcarriers, N_{SC}, that are allowed within the system bandwidth BW

$$\bar{\eta} = 1 \quad \eta = 1 - \frac{SCS \times N_{SC}}{BW} \tag{5.11}$$

It is noted that the number of subcarriers is related to the number of resource blocks (RBs), N_{RB}, that are allowed within the system bandwidth BW, by

$$N_{SC} = N_{RB} \times 12 \tag{5.12}$$

where 12 is the number of subcarriers in one RB.

According to 3GPP TS38.104, the number of RBs supported for different SCS and system bandwidth is given in Table 5.8, by taking into account the appropriate processing of frequency or time domain filtering. Based on the given values, the spectrum utilization ratio is calculated and is shown in the same table. It can be observed that NR frame structure provides a better spectrum utilization ratio com-pared to LTE for large bandwidth because of the reduced guard band. The spectrum utilization ratio for LTE is 90% for 15 kHz SCS for 10 MHz or 20 MHz system bandwidth (see Table 5.9).

On the other hand, NR wideband frame structure and physical channel structure helps to reduce the control and reference signal overhead. For control signaling overhead saving, it can be intuitively understood because the control signal will not be increased linearly with the increasing bandwidth. If the size of the control signal-ing is kept on a similar level for larger bandwidth compared to smaller bandwidth, the overhead will be reduced. Specifically, for an NR system (say, NR system 1) with bandwidth of BW_1, and another NR system (say, NR system 2) with bandwidth of BW_2, if the number of served users in both systems are the same, and the SINR is not changed much (e.g., the interference level is similar in the two systems despite the bandwidth difference), then the number of OFDM symbols occupied by PDCCH in the two systems is approximately related by

$$N_{PDCCH,1} BW_1 = N_{PDCCH,2} BW_2 \tag{5.13}$$

or,

Table 5.10   Downlink overhead summary for NR and LTE

		NR	LTE
Common signals	SSB	For synchronization signal	–
	TRS	For tracking reference signal	–
	PSS/ SSS	–	For primary and secondary synchronization signal
	PBCH	–	For broadcast signal
Control signal	PDCCH	For downlink control signal. The signaling is applicable to up to 100 MHz bandwidth for frequency range 1 (below 6 GHz)	For downlink control signal. The signaling is applicable to up to 20 MHz bandwidth per component carrier (CC)
Reference signal	DM-RS	Reference signal for demodulation	Reference signal for demodulation
	CSI-RS	For channel state estimation	For channel state estimation
	CSI-IM	For interference measurement	For interference measurement
	CRS	–	For common reference signal
Guard period	GP	Guard period overhead for TDD frame structure; if GP occupies N OFDM symbols, downlink GP overhead slot is $N/2$ OFDM symbols	Guard period for TDD

$$N_{PDCCH,2} = N_{PDCCH,1} \frac{BW_1}{BW_2} \qquad (5.14)$$

where $N_{\text{PDCCH,1}}$ and $N_{\text{PDCCH,2}}$ are the number of OFDM symbols for PDCCH for NR system 1 and NR system 2, respectively.

This model will be used in NR overhead evaluation when the bandwidth is larger than 10 MHz for FDD, and 20 MHz for TDD. In this case, $BW_1 = 10$ MHz for FDD and $BW_1 = 20$ MHz for TDD, and is called reference bandwidth.

For reference signal overhead, NR removes the common reference signal (CRS) which always exists in non-MBSFN subframes in LTE. This is helpful to reduce the reference signal overhead. In addition, NR DM-RS design can help to reduce the overhead when larger number of layers are employed. NR can configure the transmission period of synchronization signal blocks (SSBs). Therefore, even using the beam-based transmission of such common signals, the overhead can be well controlled to a low level.

The control and reference signal overhead for downlink NR and LTE are summarized in Table 5.10.

The downlink overhead for NR and LTE can be calculated as

$$OH = \frac{N_{RE}^{data}}{N_{RE}} = \frac{N_{RE} - N_{RE}^{OH}}{N_{RE}} = 1 - \frac{N_{RE}^{OH}}{N_{RE}} \qquad (5.15)$$

where

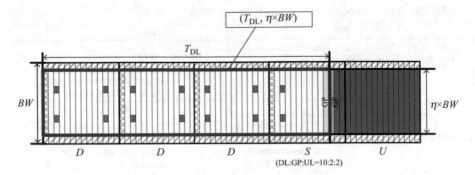

Fig. 5.5 Illustration of time-frequency resource block (T_{slot}, $\eta \times$ BW) in overhead calculation

- $N_{RE} = N_{RB} \times 12 \times N_{OS}$ is the number of REs within the time-frequency resource block (T_{DL}, $\eta \times$ BW),

  - T_{DL} represents the length of N_{OS} OFDM symbols for downlink transmission in one period of the given DL/UL pattern, e.g., every 10 slots (for NR) or 10 TTIs (for LTE),
  - η is the spectrum utilization ratio, and BW is the system bandwidth,
  - N_{RB} is the number of RBs within the bandwidth $\eta \times$ BW, and 12 is the number of subcarriers in one RB,
  - Note that ($\eta \times$ BW) is the bandwidth that excludes the guard band, and half of the GP symbols are regarded as for downlink resource so that they should be taken into account when calculating N_{OS} (see Fig. 5.5 for illustration);

- N_{RE}^{data} is the number of data transmission REs that can be used for downlink data transmission within the same time-frequency resource (T_{DL}, $\eta \times$ BW), and
- N_{RE}^{OH} is the number of overhead REs that are used for control signals, common signals, and reference signal within the same time-frequency resource (T_{DL}, $\eta \times$ BW).
- $N_{RE} = N_{RE}^{data} + N_{RE}^{OH}$

The illustration of time-frequency resource block (T_{DL}, $\eta \times$ BW) is shown in Fig. 5.5. In this example, the DL/UL pattern is assumed to be "DDDSU," that is, three full downlink slots, one special slot, and one uplink slot, with each slot consisting of 14 OFDM symbols. The one special slot is constituted of 10 downlink OFDM symbols, two GP symbols, and two uplink OFDM symbols. One GP symbol is regarded as downlink resource, and the other GP symbol is regarded as uplink resource. Therefore, T_{DL} consists of $3 \times 14 + 11 = 53$ OFDM symbols in one DDDSU period (5 slots). The number of REs, N_{RE}, N_{RE}^{data}, and N_{RE}^{OH} will be calculated within the 53 OFDM symbols. If one employs 10 slots as the DL/UL period, then T_{DL} will include 106 OFDM symbols.

According to Table 5.10, the NR overhead REs for downlink transmission is comprising of REs occupied by SSB, TRS, PDCCH, DM-RS, CSI-RS, CSI-IM, and GP, while the LTE overhead REs is consisting of REs occupied by PSS/SSS, PBCH,

PDCCH, DM-RS, CSI-RS, CSI-IM, CRS, and GP. The value of N_{RE}^{OH} is the sum of the number of REs for the above signals or physical channels.

Once the guard band ratio $\bar{\eta}$ and the overhead OH is derived, the overall overhead Γ can be calculated by $\Gamma = 1 - (1 - \bar{\eta}) \times (1 - OH)$. If one further assumes that the post-processing SINR is the same between two candidate technologies, the spectral efficiency gain or cell data rate gain brought by overhead reduction itself can be expressed as

$$G_{OH} = \frac{1 - \Gamma_1}{1 - \Gamma_0} - 1 \tag{5.16}$$

where Γ_0 is the overall overhead of the reference technology, and Γ_1 is the overall overhead of the comparing technology.

Using the above analytical method, the downlink overhead of NR and LTE for a typical FDD and TDD configuration are presented in Tables 5.11 and 5.12, respectively. In both analyses, full buffer transmission is assumed, that is, the data transmission is required in every of the slots or TTIs. It is observed that the overall overhead of NR FDD wideband configuration can be reduced to 18% for FDD 40 MHz as compared to 33.7% for LTE 10 MHz, and for NR TDD 100 MHz the overhead can be reduced to 28% as compared to 40% for LTE 20 MHz. With these overhead reduction, NR wideband frame structure itself can bring more than 20% gain for wideband configuration.

5.4.1.2 Contribution to Latency

The NR wideband frame and physical channel structure supports large subcarrier spacing with reduced OFDM symbol length, and less than 14 OFDM symbols (OSs) can be organized as one non-slot to transmit small amount of data. Both of the above features can be beneficial to reduce the user plane latency and the control plane latency. Table 5.13 lists the average length of one OFDM symbol, the length of 2/4/7 OS non-slot, and the length of one slot with 14 OSs. Here "average" means that the 14 OFDM symbols are assumed to be of equal length. Although this is not the case in 5G NR frame structure (in NR, the first OFDM symbol is a bit longer than the other OFDM symbols like in LTE), this approximated assumption would not result in much loss of accuracy in the evaluation.

According to the frame structure of 5G NR, 30 kHz SCS will reduce the symbol length by 50%, and 60 kHz SCS can reduce the length by 75%. This would help to reduce the latency of the steps related to air interface transmission for both user plane latency and control plane latency.

Besides, 5G NR physical channel design supports two resource mapping types for PDSCH and PUSCH transmission: Mapping type A and mapping type B. This is introduced in details in Sects. 2.2.3.2 and 2.2.3.4. For resource mapping type A, PDSCH transmission can be started in the first 3 OFDM symbols of a slot with duration of 3 symbols or more until the end of a slot. For resource mapping type B, PDSCH

Table 5.11 Downlink overhead analysis for FDD NR and LTE with 8Tx

Item		LTE 8 Tx, MU layer = 4–8, with 6 MBSFN subframes	NR 8 Tx, MU layer = 8, SCS = 15 kHz 10 MHz BW	NR 8 Tx, MU layer = 8, SCS = 15 kHz 40 MHz BW	NR 8 Tx, MU layer = 8, SCS = 30 kHz 40 MHz BW	Remarks on Assumptions in overhead evaluation
System bandwidth [MHz]		10	10	40	40	
SCS [kHz]		15	15	15	30	
Total number of RBs on the whole system bandwidth, N_{RB}^{BW}		55.6	55.6	222.2	111.1	
Total number of DL slots/TTIs in 10 slots/TTIs		10	10	10	10	
Number of DL OFDM symbols in the length of T_{DL}, N_{OS}		140	140	140	140	
Total number of DL REs on the whole system bandwidth, N_{RE}^{BW}		93333.3	93333.3	373333.3	186666.7	
Total number of RBs excluding guard band, N_{RB}		50	52	216	106	
Total number of DL REs excluding guard band, N_{RE}		84,000	87,360	362,880	178,080	
Common signals						
SSB	Number of SSB slot (in 10 slots)	–	0.5	0.5	0.25	1 SSB per 20 ms period is assumed
	SSB REs in 10 slots	–	480	480	240	
	SSB OH	–	0.51%	0.13%	0.13%	

TRS	Number of TRS slot (in 10 slots)	–	0.5	0.5	0.25	20 ms period, 2 burst slots, and 2× (3×2) REs/PRB are assumed
	TRS REs in 10 slots	–	312	312	156	
	TRS OH	–	0.33%	0.08%	0.08%	
PSS/SSS	PSS/SSS REs in one TTI	144	–	–	–	PSS/SSS period is assumed to be 5 ms
	Number of TTIs for PSS/SSS (in 10 TTIs)	2	–	–	–	
	PSS/SSS REs in 10 TTIs	288	–	–	–	
	PSS/SSS OH	0.31%	–	–	–	
PBCH	PBCH	240	–	–	–	PBCH period is 10 ms
	PBCH REs in 10 TTIs	240	–	–	–	
	PBCH OH	0.26%	–	–	–	

(continued)

Table 5.11 (continued)

Item		LTE 8 Tx, MU layer = 4–8, with 6 MBSFN subframes	NR 8 Tx, MU layer = 8, SCS = 15 kHz 10 MHz BW	NR 8 Tx, MU layer = 8, SCS = 15 kHz 40 MHz BW	NR 8 Tx, MU layer = 8, SCS = 30 kHz 40 MHz BW	Remarks on Assumptions in overhead evaluation
Control channel						
PDCCH	Number of OFDM symbols for PDDCH per slot	2	2	0.5	0.5	2 OFDM symbols per slot is assumed. For NR with larger bandwidth (BW2) than BW1 = 10 MHz, the number of OFDM symbol is assumed to be 2 x (BW1/BW2) = 2 x (10/BW2) (BW2 > 10, unit: MHz)
	PDCCH REs in 10 slots	12,000	12,480	12,960	6360	
	PDCCH OH	12.86%	13.37%	3.47%	3.41%	
Reference signals						
DM-RS	Number of slots that transmit DM-RS (in 10 slots)	10	10	10	10	Type II DM-RS pattern is assumed for NR
	Number of MU layers	8	8	8	8	
	Number of DM-RS REs per RB	12	16	16	16	
	DM-RS REs in 10 slots	6000	8320	34,560	16,960	
	DM-RS OH	6.43%	8.91%	9.26%	9.09%	

CSI-RS	Number of slots that transmit CSI-RS (in 10 slots)	2	2	2	2	CSI-RS period is assumed to be every 5 slots or TTIs
	Number of CSI-RS REs per RB	8	8	8	8	Cell-specific CSI-RS is assumed (e.g., non-beamformed CSI-RS)
	CSI-RS REs in 10 slots	800	832	3456	1696	
	CSI-RS OH	0.86%	0.89%	0.93%	0.91%	
CSI-IM	Number of CSI-IM slot (in 10 slots)	2	2	2	2	
	Number of CSI-IM REs per RB	4	4	4	4	
	CSI-IM REs in 10 slots	400	416	1728	848	
	CSI-IM OH	0.43%	0.45%	0.46%	0.45%	

(continued)

Table 5.11 (continued)

Item		LTE	NR	NR	NR	NR	Remarks on Assumptions in overhead evaluation
		8 Tx, MU layer = 4–8, with 6 MBSFN subframes	8 Tx, MU layer = 8, SCS = 15 kHz 10 MHz BW	8 Tx, MU layer = 8, SCS = 15 kHz 40 MHz BW	8 Tx, MU layer = 8, SCS = 15 kHz 40 MHz BW	8 Tx, MU layer = 8, SCS = 30 kHz 40 MHz BW	
CRS	Number of MBSFN subframe (TTI) in 10 TTIs	6	–	–	–	–	
	CRS port number	2					
	Number of CRS REs per PRB	12					
	Number of TTIs that transmits CRS (in 10 TTIs)	4					
	CRS REs in 10 slots	2400	–	–	–	–	
	CRS OH	2.57%	–	–	–	–	
Total number of OH REs in 10 slots and excluding guard band		22,128	22,840	53,496	26,260		–
Total number of OH REs in 10 slots (including guard band REs)		31461.3	28813.3	63949.3	34846.7		
Total overhead, OH		26.3%	26.1%	14.7%	14.7%		
Guard band ratio, $\bar{\eta}$		10.00%	6.40%	2.80%	4.60%		
Overall overhead, $\Gamma = 1 - (1 - OH)(1 - \bar{\eta})$		33.71%	30.87%	17.13%	18.67%		
Gain introduced by reduction of overall overhead		Baseline	4.3%	25.0%	22.7%		

Table 5.12 Downlink overhead analysis for TDD NR and LTE with 32Tx (DL/UL pattern = "DDDSU")

Item	LTE 32 Tx, MU layer = 4~8, with 4 MBSFN subframes	NR 32 Tx, MU layer = 8, SCS = 30 kHz 20 MHz BW	NR 32 Tx, MU layer = 8, SCS = 30 kHz 40 MHz BW	NR 32 Tx, MU layer = 12, SCS = 30 kHz 100 MHz BW	Remarks on Assumptions in overhead evaluation
System bandwidth [MHz]	20	20	40	100	
SCS [kHz]	15	30	30	30	
Total number of RBs on the whole system bandwidth, N_{RB}^{BW}	111.1	55.6	111.1	277.8	
Total number of DL slots/TTIs in 10 slots/TTIs	8	8	8	8	For LTE: DSUDD For NR: DDDSU S slot configuration is DL:GP:UL = 10:2:2
Number of DL OFDM symbols in the length of T_{slot}, N_{os}	106	106	106	106	
Total number of DL REs on the whole system bandwidth, N_{RE}^{BW}	141333.3	70666.7	141333.3	353333.3	
Total number of RBs excluding guard band, N_{RB}	100	51	106	273	
Total number of DL REs on the whole system bandwidth, N_{RE}	127,200	64,872	134,832	347,256	
Common signals					
SSB Number of SSB slot (in 10 slots)	–	0.25	0.25	0.25	1 SSB per 20 ms period is assumed
SSB REs in 10 slots	–	240	240	240	
SSB OH	–	0.34%	0.17%	0.07%	

(continued)

Table 5.12 (continued)

Item		LTE 32 Tx, MU layer = 4–8, with 4 MBSFN subframes	NR 32 Tx, MU layer = 8, SCS = 30 kHz 20 MHz BW	NR 32 Tx, MU layer = 8, SCS = 30 kHz 40 MHz BW	NR 32 Tx, MU layer = 12, SCS = 30 kHz 100 MHz BW	Remarks on Assumptions in overhead evaluation
TRS	Number of TRS slot (in 10 slots)	–	0.25	0.25	0.25	20 ms period, 2 burst slots, and 2 x (3x2) REs/PRB are assumed
	TRS REs in 10 slots	–	153	156	156	
	TRS OH	–	0.22%	0.11%	0.04%	
PSS/SSS	PSS/SSS REs in one TTI	144	–	–	–	PSS/SSS period is assumed to be 5 ms
	Number of TTIs for PSS/SSS (in 10 TTIs)	2	–	–	–	
	PSS/SSS REs in 10 TTIs	288	–	–	–	
	PSS/SSS OH	0.2%	–	–	–	
PBCH	PBCH	240	–	–	–	PBCH period is 10 ms
	PBCH REs in 10 TTIs	240	–	–	–	
	PBCH OH	0.17%	–	–	–	
Control channel						
PDCCH	Number of OFDM symbols for PDDCH per slot	2	2	1	0.4	2 OFDM symbols per slot is assumed. For NR with larger bandwidth (BW_2) than $BW_1 = 20$ MHz, the number of OFDM symbol is assumed to be $2 \times (BW_1/BW_2) = 2 \times (20/BW_2)$ ($BW_2 > 20$, unit: MHz)
	PDCCH REs in 10 slots	19,200	9792	10,176	10483.2	
	PDCCH OH	13.58%	13.86%	7.20%	2.97%	

Reference signals

DM-RS						
	Number of slots that transmit DM-RS (in 10 slots)	8	8	8	8	Type II DM-RS pattern is assumed for NR
	Number of MU layers	8	8		12	
	Number of DM-RS REs per RB	12	16	16	24	
	DM-RS REs in 10 slots	9600	6528	13,568	52,416	
	DM-RS OH	6.79%	9.24%	9.60%	14.83%	
CSI-RS	Number of slots that transmit CSI-RS (in 10 slots)	2	2	2	2	CSI-RS period is assumed to be every 5 slots or TTIs
	Number of CSI-RS REs per RB	32	32	32	32	Cell-specific CSI-RS is assumed (e.g., non-beamformed CSI-RS)
	CSI-RS REs in 10 slots	6400	3264	6784	17,472	
	CSI-RS OH	4.53%	4.62%	4.80%	4.94%	

(continued)

Table 5.12 (continued)

Item		LTE 32 Tx, MU layer = 4–8, with 4 MBSFN subframes	NR 32 Tx, MU layer = 8, SCS = 30 kHz 20 MHz BW	NR 32 Tx, MU layer = 8, SCS = 30 kHz 40 MHz BW	NR 32 Tx, MU layer = 12, SCS = 30 kHz 100 MHz BW	Remarks on Assumptions in overhead evaluation
CSI-IM	Number of CSI-IM slot (in 10 slots)	2	2	2	2	
	Number of CSI-IM REs per RB	4	4	4	4	
	CSI-IM REs in 10 slots	800	408	848	2184	
	CSI-IM OH	0.57%	0.58%	0.60%	0.62%	
CRS	Number of MBSFN subframe (TTI) in 10 TTIs	4	–	–	–	In S subframe, only 8 REs/PRB for 2 CRS ports
	CRS port number	2	–	–	–	
	Number of CRS REs per PRB	12	–	–	–	
	Number of TTIs that transmits CRS (in 10 TTIs)	4	–	–	–	
	CRS REs in 10 slots	4000.000	–	–	–	
	CRS OH	2.83%	–	–	–	

GP				
Number of GP symbols in 10 slots	2	2	2	2
GP REs in 10 slots	2400	1224	2544	6552
GP OH	1.70%	1.73%	1.80%	1.85%
Total number of OH REs in 10 slots and excluding guard band	42,928	21,609	34,316	89,503
Total number of OH REs in 10 slots (including guard band REs)	31461.3	28813.3	63949.3	34846.7
Total overhead, OH	33.7%	33.3%	25.5%	25.8%
Guard band ratio, $\bar{\eta}$	10.00%	8.20%	4.60%	1.72%
Overall overhead, $\Gamma = 1 - (1 - OH)(1 - \bar{\eta})$	40.37%	38.78%	28.88%	27.05%
Gain introduced by reduction of overall overhead	Baseline	2.7%	19.3%	22.3%

Table 5.13 Average length of one OFDM symbol, 2/4/7 OS non-slot, and one slot (ms)

Number of OSs	SCS			
	15 kHz	30 kHz	60 kHz	120 kHz
M = 1 OS	0.0714	0.0357	0.0179	0.0089
M = 2 OS (2 OS non-slot)	0.1429	0.0714	0.0357	0.0179
M = 4 OS (4 OS non-slot)	0.2857	0.1429	0.0714	0.0357
M = 7 OS (7 OS non-slot)	0.5	0.25	0.125	0.0625
M = 14 OS (14 OS slot)	1	0.5	0.25	0.125

Table 5.14 Valid S and L combinations for NR resource mapping

(a) For PDSCH resource mapping

PDSCH mapping type	Normal cyclic prefix			Extended cyclic prefix		
	S	L	$S + L$	S	L	$S + L$
Type A	{0,1,2,3}[a]	{3,…,14}	{3,…,14}	{0,1,2,3}[a]	{3,…,12}	{3,…,12}
Type B	{0,…,12}	{2,4,7}	{2,…,14}	{0,…,10}	{2,4,6}	{2,…,12}

(b) For PUSCH resource mapping

PUSCH mapping type	Normal cyclic prefix			Extended cyclic prefix		
	S	L	$S + L$	S	L	$S + L$
Type A	0	{4,…,14}	{4,…,14}	0	{4,…,12}	{4,…,12}
Type B	{0,…,13}	{1,…,14}	{1,…,14}	{0,…,12}	{1,…,12}	{1,…,12}

$S = 3$ is applicable only if DM-RS-TypeA-Position = 3

transmission can be started anywhere (except the last OFDM symbol) in the slot, with duration of 2, 4, or 7 OFDM symbols, given that the restriction of data transmission not exceeding the boundary of the slot is met. The available starting OFDM symbol index for PDSCH and PUSCH resource mapping type is shown in Table 5.14. In the table, S is the valid starting OFDM symbol, and L is the length of the data transmission for PDSCH or PUSCH. Given the value of S, L should be selected such that $S + L \le 14$ for normal cyclic prefix, and $S + L \le 12$ for extended cyclic prefix.

Therefore, using resource mapping type B, the frame alignment time might be reduced compared to resource mapping type A. This is illustrated in Fig. 5.6. PDSCH transmission is used as example in this plot. The left-hand case and the right-hand case differ from the non-slot length. The left-hand case assumes a 4 OS non-slot (e.g., the data amount is small), while the right-hand case assumes a 7 OS non-slot. In both cases, the ending time of the previous step is assumed to be at the end of symbol#8 (the symbol index number is from 0). For resource mapping type A, since it only allows PDSCH transmission from symbol#0~2, therefore despite the non-slot length, the transmission should wait until the beginning of the next slot, so that the frame alignment time would be 5 OSs. For resource mapping type B, however, it is possible to transmit PDSCH at any of the OFDM symbol within the slot (except the last OFDM symbol), given that the transmission does not exceed the boundary of the slot. Therefore, for 4 OS non-slot, the frame alignment time is 0 because the transmission can happen immediately. In this case, the frame alignment time is reduced.

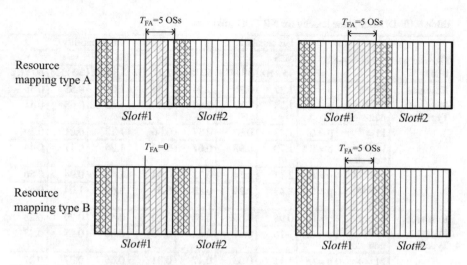

Fig. 5.6 Comparison of resource mapping type A and B

However, for 7 OS non-slot, since the transmission will exceed the boundary of the slot, the transmission should wait until the beginning of the next slot. In this case the frame alignment time is the same as in resource mapping type A.

Based on the above analysis, it can be seen that by the use of larger subcarrier spacing, non-slot transmission, and resource mapping type B, the user plane latency can be greatly reduced.

The DL user plane latency evaluation results for NR FDD are shown in Table 5.15 (see also 3GPP TR37.910). The values are derived following the evaluation method as described in Sect. 5.3.1.6. The value of p indicates the probability of one-time retransmission: "$p = 0$" indicates that there is no retransmission, and "$p = 0.1$" indicates that the one-time retransmission happens with the probability of 10%. Also note that the two UE capabilities as defined in 3GPP are evaluated.

For NR FDD, DL and UL resource always exists on the DL and UL part of the paired spectrum, and therefore the waiting time due to the DL/UL pattern does not exist. It therefore demonstrates the maximum capability of NR on latency performance. On the other hand, since FDD is not impacted by the selection of DL/UL patterns, hence it also provides an insightful assessment on the contribution of frame structure and physical channel structure on latency reduction.

According to Table 5.15, it can be seen that 15 kHz SCS can achieve 0.49 ms user plane latency with 2 OS non-slot and resource mapping type B. And 30 kHz SCS can reach as low as 0.29 ms, which is 41% reduction compared to 15 kHz SCS case, while 60 kHz SCS can further reduce the latency by 20% to 0.23 ms.

On the other hand, for 15 kHz SCS, the latency using 7 OS non-slot is 1.49 ms under resource mapping type B with $p = 0.1$, and the latency is reduced by 13% to 1.30 ms by using 4 OS non-slot, and for 2 OS non-slot, the reduction is 22% to 1.16 ms.

Finally, when comparing resource mapping type A and B, it is seen that around 10% latency reduction can be derived by resource mapping type B in general.

Table 5.15 DL user plane latency for NR FDD (ms)

DL user plane latency—NR FDD			UE capability 1				UE capability 2		
			SCS				SCS		
			15 kHz	30 kHz	60 kHz	120 kHz	15 kHz	30 kHz	60 kHz
Resource mapping Type A	M = 4 (4OS non-slot)	p = 0	1.37	0.76	0.54	0.34	1.00	0.55	0.36
		p = 0.1	1.58	0.87	0.64	0.40	1.12	0.65	0.41
	M = 7 (7OS non-slot)	p = 0	1.49	0.82	0.57	0.36	1.12	0.61	0.39
		p = 0.1	1.70	0.93	0.67	0.42	1.25	0.71	0.44
	M = 14 (14OS slot)	p = 0	2.13	1.14	0.72	0.44	1.80	0.94	0.56
		p = 0.1	2.43	1.29	0.82	0.51	2.00	1.04	0.63
Resource mapping Type B	M = 2 (2OS non-slot)	p = 0	0.98	0.56	0.44	0.29	0.49	0.29	0.23
		p = 0.1	1.16	0.67	0.52	0.35	0.60	0.35	0.28
	M = 4 (4OS non-slot)	p = 0	1.11	0.63	0.47	0.31	0.66	0.37	0.27
		p = 0.1	1.30	0.74	0.56	0.36	0.78	0.45	0.32
	M = 7 (7OS non-slot)	p = 0	1.30	0.72	0.52	0.33	0.93	0.51	0.34
		p = 0.1	1.49	0.83	0.61	0.39	1.08	0.59	0.40

It is also noted that, the eMBB latency requirement of 4 ms is achieved in all configurations, and URLLC latency requirement (1 ms) is achieved under larger subcarrier spacings for DL.

For UL user plane latency, the evaluation results for NR FDD with grant-free transmission are presented in Table 5.16 (see also 3GPP TR37.910). Similar trend are observed. Again, the eMBB latency requirement of 4 ms is achieved in all configurations, and URLLC latency requirement (1 ms) is achieved under larger subcarrier spacings for UL.

5.4.1.3   Contribution to Reliability

As mentioned in the previous sections, the larger subcarrier spacing, e.g., 30 kHz or 60 kHz SCS, can introduce reduced OFDM symbol latency and in turn the reduced one-time transmission latency. According to Sect. 5.3.3.3, reliability is defined as the overall transmission successful probability within the given latency budget, e.g., 1 ms. Therefore, if the one time transmission latency can be reduced, there can be opportunity to have more repetitions within the latency budget. The reliability can be therefore improved.

For 5G NR, slot aggregation is one way for repetition. The latency of different slot aggregation levels can be approximately estimated following the way of user plane latency evaluation. Take downlink slot aggregation for example. By using the

Table 5.16 UL user plane latency for NR FDD with grant-free transmission (ms)

UL user plane latency (Grant free)—NR FDD			UE capability 1				UE capability 2		
			SCS				SCS		
			15 kHz	30 kHz	60 kHz	120 kHz	15 kHz	30 kHz	60 kHz
Resource mapping Type A	M = 4 (4OS non-slot)	p = 0	1.57	0.86	0.59	0.37	1.20	0.65	0.41
		p = 0.1	1.78	1.01	0.69	0.43	1.39	0.75	0.47
	M = 7 (7OS non-slot)	p = 0	1.68	0.91	0.61	0.38	1.30	0.70	0.43
		p = 0.1	1.89	1.06	0.71	0.44	1.50	0.80	0.49
	M = 14 (14OS slot)	p = 0	2.15	1.15	0.73	0.44	1.80	0.94	0.56
		p = 0.1	2.45	1.30	0.84	0.51	2.00	1.06	0.63
Resource mapping Type B	M = 2 (2OS non-slot)	p = 0	0.96	0.55	0.44	0.28	0.52	0.30	0.24
		p = 0.1	1.14	0.65	0.52	0.34	0.62	0.36	0.28
	M = 4 (4OS non-slot)	p = 0	1.31	0.72	0.52	0.33	0.79	0.43	0.30
		p = 0.1	1.50	0.84	0.61	0.39	0.96	0.55	0.37
	M = 7 (7OS non-slot)	p = 0	1.40	0.77	0.55	0.34	1.02	0.55	0.36
		p = 0.1	1.60	0.89	0.63	0.40	1.19	0.64	0.42
	M = 14 (14OS slot)	p = 0	2.14	1.14	0.74	0.44	1.81	0.93	0.56
		p = 0.1	2.44	1.30	0.84	0.51	2.01	1.03	0.63

notations in Table 5.5 (see also in Table 5.17 below), the latency of the slot aggregation is approximately given by

$$T_{SA}(n) = t_{BS\_tx} + t_{FA\_DL} + n \times t_{DL\_duration} + t_{UE\_rx}$$

where n is the number of slots in the slot aggregation.

The calculation results for different subcarrier spacings are shown in Table 5.18. It can be seen that if the reliability is defined within a latency budget of 2 ms, 15 kHz SCS can only support one slot transmission; 30 kHz SCS can support 3-slot aggregation; and 60 kHz SCS can support more than 5-slot aggregation.

If one slot transmission can reach 99% reliability (i.e., 1% error probability), then 3-slot aggregation can achieve $1 - (1 - 1\%)3 = 99.9999\%$ reliability, and 5-slot aggregation can achieve near 100% reliability. The cost is the increased bandwidth request with larger subcarrier spacing for the same packet size of URLLC data packet. Therefore an appropriate subcarrier spacing needs to be selected to make trade-off between the latency and the bandwidth.

Table 5.17 Notations for slot aggregation latency analysis

Notations	Remarks	Values
$t_{BS,tx}$	BS processing delay: The time interval between the data is arrived, and packet is generated	The same calculation method as in Table 5.5
$t_{FA,DL}$	DL Frame alignment (transmission alignment): It includes frame alignment time, and the waiting time for next available DL slot	
$t_{DL\_duration}$	TTI for DL data packet transmission	
$t_{UE,rx}$	UE processing delay: The time interval between the PDSCH is received and the data is decoded	

Table 5.18 Downlink slot aggregation latency for NR FDD (ms)

Latency of DL slot aggregation—NR FDD			UE capability 2		
			SCS		
			15 kHz	30 kHz	60 kHz
Resource mapping Type A	$M = 14$ (14OS slot)	n = 1	1.8	0.94	0.56
		n = 2	2.8	1.44	0.81
		n = 3	3.8	1.94	1.06
		n = 4	4.8	2.44	1.31
		n = 5	5.8	2.94	1.56

5.4.2 NR MIMO, Multiple Access, and Waveform

As introduced in Sect. 2.5, NR supports new physical layer designs including NR massive MIMO, UL CP-OFDM waveform, and UL OFDMA. These new features can bring benefit for spectral efficiency improvement for DL and UL, respectively, and in turn improve the area traffic capacity and user experienced data rate.

5.4.2.1 Contribution to Spectral Efficiency Improvement

NR massive MIMO design is characterized by advanced codebook design, the reference signal design, and the fast CSI information acquisition design.

5.4.2.1.1 Downlink Evaluation

For downlink FDD, Type II codebook is used. As introduced in Sect. 2.5.3, NR type II codebook is a combination of multiple beams that is to improve the accuracy of CSI feedback, at a cost of higher feedback overhead. Type II codebook includes single panel codebook and port selection codebook. NR Type II codebook supports finer quantization granularity, including subband phase reporting, subband amplitude reporting with un-even bit allocation, etc. By this means, the precoder information obtained by NR Type II codebook is usually more precise compared to NR Type I codebook as well as LTE advanced codebook (which is a similar mechanism to NR type II codebook), especially in the case of multi-path propagation scenarios. With the finer granularity and more precise precoder information, the MU-MIMO performance is boosted especially when the number of paired users are relatively high.

Another aspect is the design of DM-RS in NR MIMO application as introduced in Sect. 2.2.4.2. The number of OFDM symbols of DM-RS can be flexibly configured. In the case of low mobility (e.g., for the pedestrian and indoor users), it is sufficient to configure DM-RS on one or two OFDM symbols. This can reduce the DM-RS overhead when possible. Further, DM-RS Type-2 Configuration can support up to 12 orthogonal ports. By this means, up to 12 multi-user layers can be well supported. This is helpful in dense user scenarios. On the other hand, if the number of multi-user layers is less than 12, e.g., eight layers are paired, the overhead of Type-2 configuration can be further reduced compared to Type-1 configuration, at the cost of possible reduced channel estimation accuracy. However, if the channel frequency selectivity is not so severe, such accuracy loss would be very limited.

NR also supports fast CSI feedback by the use of high capability user terminals with very low processing delay. For NR, the CSI feedback can be made right after the slots when the reference signal is received. This can bring some benefit when the channel is quickly varying in time domain.

For downlink TDD, NR can rely on uplink sounding reference signal (SRS) to obtain the DL precoder information by channel reciprocity as in LTE. In addition, NR can use more resources to transmit SRS in the UL slot such that the whole bandwidth can be swept within a shorter time consumption. Such fast sweeping can bring more accurate precoder information for downlink TDD transmission.

To evaluate the benefit brought by the above mechanisms, the system-level simulation is conducted. The spectral efficiency is evaluated by the system-level simulation. The dense urban scenario as defined in Report ITU-R M.2412 is employed for the evaluation, and the system configuration parameters are shown in Table 5.19. This scenario is characterized by dense deployment with 10 users per transmission receive point (TRxP), and 80% of the users are located indoor with low mobility (3 km/h) while 20% of the users are located outdoor with median mobility (30 km/h). The indoor users are located in different floors which implies that the 3D MIMO will be needed. In such scenario, massive MIMO can benefit the spectral efficiency and user data rate improvement. However, the precise precoder information is needed to maximize the massive MIMO capability. It is therefore expected that the Type II codebook and SRS fast sweeping can contribute to the precoder information estimation. On the other hand, the fast CSI feedback will be helpful for the outdoor median mobility users.

The technical configuration parameters for NR are provided in Table 5.20. Besides, LTE is evaluated. For LTE, the following parameters are used as compared to NR: the Turbo code is used for PDSCH coding scheme, 15 kHz SCS is used for

Table 5.19 System configuration assumption for dense urban scenario

System configuration parameters	Value
Carrier frequency for evaluation	1 layer (Macro) with 4 GHz
BS antenna height	25 m
Total transmit power per TRxP	44 dBm for 20 MHz bandwidth 41 dBm for 10 MHz bandwidth
UE power class	23 dBm
Percentage of high loss and low loss building type	20% high loss, 80% low loss
Inter-site distance	200 m
Device deployment	80% indoor, 20% outdoor (in-car) Randomly and uniformly distributed over the area under Macro layer
UE density	10 UEs per TRxP Randomly and uniformly distributed over the area under Macro layer
UE antenna height	Outdoor UEs: 1.5 m Indoor UTs: $3(n_{fl} - 1) + 1.5$; $n_{fl} \sim$ uniform $(1, N_{fl})$ where $N_{fl} \sim$ uniform (4,8)
UE speeds of interest	Indoor users: 3 km/h Outdoor users (in-car): 30 km/h
BS noise figure	5 dB
UE noise figure	7 dB
BS antenna element gain	8 dBi
UE antenna element gain	0 dBi
Thermal noise level	−174 dBm/Hz
UE mobility model	Fixed and identical speed |v| of all UEs of the same mobility class, randomly and uniformly distributed direction
Mechanic tilt	90° in GCS (pointing to horizontal direction)
Electronic tilt	105°
Handover margin (dB)	1
UT attachment	Based on RSRP (see formula (8.1–1) in 3GPP TR36.873) from port 0
Wrapping around method	Geographical distance based wrapping
Traffic	Full buffer

both FDD 10 MHz and TDD 20 MHz, 10% guard band ratio is employed, LTE advanced codebook is used for FDD, two codewords are used in codeword-to-layer mapping, LTE DM-RS configuration for 4 orthogonal ports is used, and for SRS two OFDM symbols are configured in special subframe for SRS transmission. Other parameters are the same as in NR.

The evaluation results are shown in Table 5.21. It is seen that, with the same antenna configuration and bandwidth configuration, NR can introduce additional 18–25% average spectral efficiency gain brought by massive MIMO mechanisms

Table 5.20 Technical parameters for spectral efficiency evaluation for NR downlink MIMO

Technical configuration Parameters	NR FDD	NR TDD
Multiple access	OFDMA	OFDMA
Duplexing	FDD	TDD
Network synchronization	Synchronized	Synchronized
Modulation	Up to 256 QAM	Up to 256 QAM
Coding on PDSCH	LDPC Max code-block size = 8448 bit with BP decoding	LDPC Max code-block size = 8448 bit with BP decoding
Numerology	15 kHz SCS, 14 OFDM symbol slot	30 kHz SCS, 14 OFDM symbol slot
Guard band ratio on simulation bandwidth	FDD: 6.4% (for 10 MHz)	TDD: 8.2% (51 RB for 30 kHz SCS and 20 MHz bandwidth)
Simulation bandwidth	10 MHz	20 MHz
Frame structure	Full downlink	DDDSU
Transmission scheme	Closed SU/MU-MIMO adaptation	Closed SU/MU-MIMO adaptation
DL CSI measurement	Non-precoded CSI-RS based	For 32T: Non-precoded CSI-RS based
DL codebook	Type II codebook; 4 beams, WB + SB amplitude quantization, 8 PSK phase quantization	For DL CSI measurement: Type II codebook; 4 beams, WB + SB amplitude quantization, 8 PSK phase quantization For PDSCH: SRS-based precoder
PRB bundling	4 PRBs	4 PRBs
MU dimension	Up to 12 layers	Up to 12 layers
SU dimension	For 4Rx: Up to 4 layers	For 4Rx: Up to 4 layers
Codeword (CW)-to-layer mapping	For 1~4 layers, CW1; For 5 layers or more, two CWs	For 1~4 layers, CW1; For 5 layers or more, two CWs
SRS transmission	N/A	For UE 4 Tx ports: Non-precoded SRS, 4 SRS ports (with 4 SRS resources), 4 symbols per 5 slots for 30 kHz SCS; 8 RBs per OFDM symbol
DM-RS configuration	Type-2 configuration	Type-2 configuration
CSI feedback	PMI, CQI: every 5 slot; RI: every 5 slot; Subband based	CQI: every 5 slot; RI: every 5 slot, CRI: every 5 slot Subband based
Interference measurement	SU-CQI; CSI-IM for inter-cell interference measurement	SU-CQI; CSI-IM for inter-cell interference measurement
Max CBG number	1	1
ACK/NACK delay	The next available UL slot	The next available UL slot

(continued)

Table 5.20 (continued)

Technical configuration Parameters	NR FDD	NR TDD
Retransmission delay	The next available DL slot after receiving NACK	The next available DL slot after receiving NACK
Antenna configuration at TRxP	32 T: (M,N,P,Mg,Ng; Mp,Np) = (8,8,2,1,1;2,8) (dH, dV) = (0.5, 0.8)λ	32 T: (M,N,P,Mg,Ng; Mp,Np) = (8,8,2,1,1;2,8) (dH, dV) = (0.5, 0.8)λ;
Antenna configuration at UE	4R: (M,N,P,Mg,Ng; Mp,Np) = (1,2,2,1,1; 1,2) (dH, dV) = (0.5, N/A)λ	4R: (M,N,P,Mg,Ng; Mp,Np) = (1,2,2,1,1; 1,2) (dH, dV) = (0.5, N/A)λ
Scheduling	PF	PF
Receiver	MMSE-IRC	MMSE-IRC
Channel estimation	Non-ideal	Non-ideal

Note: In antenna configuration, the notation (M, N, P, Mg, Ng; Mp, Np) means the following:
– M: Number of vertical antenna elements within a panel, on one polarization
– N: Number of horizontal antenna elements within a panel, on one polarization
– P: Number of polarizations
– Mg: Number of vertical panels
– Ng: Number of horizontal panels
– Mp: Number of vertical TXRUs within a panel, on one polarization
– Np: Number of horizontal TXRUs within a panel, on one polarization

with appropriate reference signal design, codebook design, and CSI acquisition design. Besides that, NR demonstrates significant gain on the wideband channel due to the wideband frame structure and physical channel structure as analyzed in Sect. 5.4.1. As well, it can be observed that, the NR spectral efficiency is more than 3 times as compared to IMT-Advanced requirements (see Report ITU-R M.2134), which achieves the 5G vision on spectral efficiency capability.

5.4.2.1.2 Uplink Evaluation

For uplink, the NR MIMO design together with multiple access enhancement could improve the UL spectral efficiency. For NR, OFDMA is supported for UL in addition to SC-FDMA. Using OFDMA, the UL scheduling can be more flexible, and the uplink DM-RS can share the same design as downlink DM-RS (with OFDMA). For small number of transmit antennas, the OFDMA-based DM-RS pattern can reduce the overhead as compared to the SC-FDMA-based DM-RS pattern, where in the latter case, a dedicated OFDM symbol has to be reserved for DM-RS transmission due to the single carrier characteristics need to be preserved. And the SRS fast sweeping will help NR FDD and TDD to get UL CSI in a timely manner. With these features, NR can bring performance enhancement as compared to LTE.

The UL spectral efficiency is evaluated by system-level simulation using the dense urban scenario parameters as shown in Table 5.19. The technical parameters for NR

Table 5.21 Evaluation results for NR downlink Massive MIMO (32T4R)

32T4R DL MU-MIMO	FDD 10 MHz				TDD 20 MHz			
	LTE Release-15 (15 kHz SCS, 6 MBSFN subframes in every 10 subframes)	NR (15 kHz SCS)			LTE Release-15 (15 kHz SCS, DSUDD, 4 MBSFN subframes in every 10 subframes)	NR (30 kHz SCS, DDDSU)		
Bandwidth	10 MHz	10 MHz	20 MHz	40 MHz	20 MHz	20 MHz	40 MHz	100 MHz
Average spectral efficiency (bit/s/Hz/TRxP)	9.138 (baseline)	11.45 (+25%)	12.94 (+42%)	13.83 (+51%)	10.64 (baseline)	13.04 (+22%)	15.17 (+43%)	16.65 (+56%)
5th percentile user spectral efficiency (bit/s/Hz)	0.29 (baseline)	0.376 (+30%)	0.425 (+47%)	0.454 (+57%)	0.30 (baseline)	0.382 (+18%)	0.445 (+51%)	0.488 (+65%)

uplink are shown in Table 5.22. For LTE, the following parameters are used as compared to NR: DFT-S-OFDM is used for UL waveform and SC-FDMA is employed in multiple access, the Turbo code is used for PUSCH coding scheme, 15 kHz SCS is used for both FDD 10 MHz and TDD 20 MHz, 10% guard band ratio is employed, two codewords are used in codeword-to-layer mapping, and for FDD, one OFDM symbol is configured for SRS and for TDD, two OFDM symbols are configured in special subframe for SRS transmission. Other parameters are the same as in NR. For both NR and LTE, the power back off model is considered. For NR, due to the use of OFDMA in uplink, the power back off will be larger than LTE. However, it is noted that for dense urban, this back off power loss is limited for the case of inter-site distance of 200 m.

The evaluation results are shown in Table 5.23. It is seen that, with the same antenna configuration and bandwidth configuration, NR can introduce additional 15–21% average spectral efficiency gain through fast SRS sweeping and UL MIMO with OFDMA multiple access scheme and OFDMA-based DM-RS design.

5.4.2.2 Contribution to Area Traffic Capacity

According to Sect. 5.3.1.5, the area traffic capacity is related to average spectral efficiency and the aggregated bandwidth. Since 5G NR can improve the average spectral efficiency by more than three times through the use of massive MIMO and the appropriate multiple access schemes and waveform, the area traffic capacity can be boosted accordingly. On the other hand, 5G NR supports significantly larger aggregated bandwidth. According to Table 5.24, it can be seen that the maximum aggregated bandwidth ranges from 800 MHz to 1.6 GHz for frequency range 1 (i.e., below 6 GHz). This is 8 to 16 times the capability of LTE Release-10, which is the first version of IMT-Advanced developed in 2010. For downlink, the data rate can be assumed to be proportionally increased by the aggregated bandwidth. Therefore, it could be expected that, even without the introduction of denser network deployment, the area traffic capacity could already increase by 24–48 times by the use of massive MIMO and large aggregated bandwidth.

5.4.3 LTE/NR Coexistence (DL/UL Decoupling)

As introduced in Sect. 2.4, LTE/NR coexistence is supported by NR to enable early 5G deployment on, e.g., C-band. By decoupling of DL and UL band, NR downlink transmission can be carried out on the higher frequency band (e.g., C-band or even higher band) which is usually configured as TDD band with large bandwidth (e.g., more than 100 MHz bandwidth) but suffers from larger propagation loss. The NR uplink transmission can be carried out on the lower frequency band (e.g., 1.8 GHz or 900 MHz) which is usually the uplink part of FDD band with moderate bandwidth (e.g., 10 MHz bandwidth) but has significantly smaller propagation loss. The

Table 5.22 Technical parameters for spectral efficiency evaluation for NR uplink MIMO

Technical configuration parameters	NR FDD	NR TDD
Multiple access	OFDMA	OFDMA
Duplexing	FDD	TDD
Network synchronization	Synchronized	Synchronized
Modulation	Up to 256QAM	Up to 256QAM
Coding on PUSCH	LDPC Max code-block size = 8448 bit [with BP decoding]	LDPC Max code-block size = 8448 bit [with BP decoding]
Numerology	15 kHz SCS, 14 OFDM symbol slot	30 kHz SCS, 14 OFDM symbol slot
Guard band ratio on simulation bandwidth	FDD: 6.4% (for 10 MHz)	TDD: 8.2% (51 RB for 30 kHz 20 MHz)
Simulation bandwidth	10 MHz	20 MHz
Frame structure	Full uplink	DDDSU
Transmission scheme	UL SU-MIMO with rank adaptation	UL SU-MIMO with rank adaptation
UL codebook	For 2Tx: NR 2Tx codebook	For 2Tx: NR 2Tx codebook
SU dimension	Up to 2 layers	Up to 2 layers
Codeword (CW)-to-layer mapping	For 1–4 layers, CW1; For 5 layers or more, two CWs	For 1–4 layers, CW1; For 5 layers or more, two CWs
SRS transmission	For UE 2 Tx ports: Non-precoded SRS, 2 SRS ports (with 2 SRS resources); 2 symbols for SRS in every 5 slots, 8 PRBs per symbol	For UE 2 Tx ports: Non-precoded SRS, 2 SRS ports (with 2 SRS resources); 2 symbols for SRS in every 5 slots, 8 PRBs per symbol
Antenna configuration at TRxP	32R: (M,N,P,Mg,Ng; Mp,Np) = (8,8,2,1,1; 2,8) (dH, dV) = (0.5, 0.8)λ;	32R: (M,N,P,Mg,Ng; Mp,Np) = (8,8,2,1,1; 2,8) (dH, dV) = (0.5, 0.8)λ;
Antenna configuration at UE	2 T: (M,N,P,Mg,Ng; Mp,Np) = (1,1,2,1,1; 1,1); (dH, dV) = (N/A, N/A)λ	2 T: (M,N,P,Mg,Ng; Mp,Np) = (1,1,2,1,1; 1,1); (dH, dV) = (N/A, N/A)λ
Max CBG number	1	1
UL retransmission delay	Next available UL slot after receiving retransmission indication	Next available UL slot after receiving retransmission indication
Scheduling	PF	PF
Receiver	MMSE-IRC	MMSE-IRC
Channel estimation	Non-ideal	Non-ideal
Power control parameter	P0 = -86, alpha = 0.9	P0 = -86, alpha = 0.9
Power backoff model	Continuous RB allocation: follow TS 38.101 for FR1; Non-continuous RB allocation: additional 2 dB reduction	Continuous RB allocation: follow TS 38.101 for FR1; Non-continuous RB allocation: additional 2 dB reduction

(continued)

Table 5.22 (continued)

Note: (M, N, P, Mg, Ng; Mp, Np)
- M: Number of vertical antenna elements within a panel, on one polarization
- N: Number of horizontal antenna elements within a panel, on one polarization
- P: Number of polarizations
- Mg: Number of panels;
- Ng: default: 1
- Mp: Number of vertical TXRUs within a panel, on one polarization
- Np: Number of horizontal TXRUs within a panel, on one polarization

Table 5.23 Evaluation results for NR uplink MIMO (2T32R)

2T32R UL SU-MIMO	FDD		TDD	
	LTE Release-15 (15 kHz SCS)	NR (15 kHz SCS)	LTE Release-15 (15 kHz SCS, DSUDD)	NR (30 kHz SCS, DDDSU)
Bandwidth	10 MHz	10 MHz	20 MHz	20 MHz
Average spectral efficiency (bit/s/Hz/TRxP)	6.738 (baseline)	8.12 (+21%)	5.318 (baseline)	6.136 (+15.4%)
5th percentile user spectral efficiency (bit/s/Hz)	0.238 (baseline)	0.388 (63.0%)	0.244 (baseline)	0.276 (13.1%)

Table 5.24 NR capability on bandwidth

	SCS [kHz]	Maximum bandwidth for one component carrier (MHz)	Maximum number of component carriers for carrier aggregation	Maximum aggregated bandwidth (MHz)
FR1 (below 6 GHz)	15	50	16	800
	30	100	16	1600
	60	100	16	1600
FR2 (above 24 GHz)	60	200	16	3200
	120	400	16	6400

uplink part of the FDD band employed in the DL/UL decoupling is usually referred to as supplementary uplink (SUL) band.

By this means, the higher frequency TDD band can be focused on downlink transmission, such that the downlink dominant DL/UL pattern can be used (e.g., DDDSU), and the lower frequency SUL band can be used to improve the UL coverage such that the UL cell edge user data rate can be improved. Furthermore, SUL band can also reduce the UL CSI and ACK feedback delay for the case of downlink dominant DL/UL pattern.

Otherwise, if only the higher frequency TDD band is used to convey both DL and UL transmission, the DL dominant frame structure will introduce large UL delay, and the UL transmission would be very power limited since the higher frequency band would suffer from relatively large propagation loss. The large UL delay will degrade the DL user data rate because the CSI information could not be reported in time, and the large ACK feedback will increase the RTT time such that the retransmission will experience larger time delay. The UL latency is also a problem. The power limited UL transmission will degrade the UL user data rate.

Therefore, the use of DL/UL decoupling can improve the user experienced data rate for both downlink and uplink, and contribute to reduce UL user plane latency in case of DDDSU is used for TDD band. These benefits are evaluated in this section.

5.4.3.1   Contribution to DL User Experienced Data Rate

For the evaluation of downlink user experienced data rate, the burst buffer system-level simulation is conducted. In this case, user-perceived throughput (UPT) is employed to evaluate the user experience data rate. For simplicity, the UPT is evaluated per packet basis. It is calculated as

$$UPT_i = \frac{S_i}{T_i} \approx \frac{S_i}{T_0 + n(T_{ACK} + T_{re-tx})} \tag{5.17}$$

where S_i is the packet size of the ith packet (usually the same packet size is assumed in the simulation), T_i is the time consumption between the arrival of the ith packet and the correct reception of the ith packet, T_0 is the time consumption of the first transmission of the ith packet, T_{ACK} is the ACK feedback delay which is related to the frame structure, T_{re-tx} is the time consumption of retransmission, and n is the number of retransmissions. The second equation is an approximation because it invokes the assumption that every of the retransmission employs the same MCS and frequency-time resource, and has the same scheduling delay. With the above definition, the contribution of DL/UL decoupling to the user experienced data rate is evaluated.

To guarantee the high downlink user experienced data rate, the TDD band needs to be configured to DL dominant frame structure. This is because much more downlink traffic is usually expected than uplink traffic, see, e.g., [5], where it is predicted that more than 6 times DL traffic will happen compared to UL traffic. On the other hand, the low UL delay should be kept for the DL dominant frame structure.

As mentioned above, DL/UL decoupling is the promising technology that enables the DL dominant frame structure configuration on TDD band, while keeping low UL delay by the use of SUL band.

To adapt to the DL/UL traffic pattern as shown in [5], the DL dominant configuration "DDDSU" with 30 kHz SCS is usually employed, where three downlink slots with one special slot and one uplink slot is used. For the special slot, the number of OFDM symbols for downlink, guard period, and uplink are 10:2:2. With this configuration, the downlink resource is approximately 4 times than uplink resource. The use of 30 kHz SCS is to support 100 MHz bandwidth for one component carrier (see Table 5.8).

When NR is deployed on adjacent band to an LTE TDD band, the NR frame structure should be kept the same to avoid the adjacent band interference. Most of the LTE TDD operators adopts "DSUDD" in their network due to the DL/UL traffic pattern. Considering LTE only supports 15 kHz SCS, while NR is likely to use 30 kHz SCS, the frame structure "DDDDDDDSUU" with 30 kHz SCS is likely to be used to keep the same frame structure as "DDDSU" with 15 kHz SCS in LTE, with the special slot OFDM symbol configuration of DL:GP:UL = 6:4:4.

If only TDD band is used, there are proposals to use "compromise" frame structure such as "DSDU" with 30 kHz SCS. The motivation is to reduce UL delay.

To evaluate the benefit of DL dominant frame structure, the three frame structures are employed: DDDSU, DSDU, and DDDDDDDSUU (all with 30 kHz SCS). The evaluation is applied to TDD band only. However, it is noted that the period of one DL/UL pattern for the three frame structures are 2.5 ms, 2 ms, and 5 ms, respectively. It implies that in DDDDDDDSUU, the UL ACK and CSI feedback delay is more than two times as compared to DDDSU and DSDU. Therefore it can be expected that without the DL/UL decoupling, the user-perceived throughput will be poor for DDDDDDDSUU, especially for the retransmission packet (see the calculation equation of UPT). To combat with the potential loss introduced by the large UL delay, the TDD + SUL is additionally applied to DDDDDDDSUU.

In summary, four cases will be evaluated in this subsection:

– TDD only band: this is applied to DDDSU, DSDU, and DDDDDDDSUU (all with 30 kHz SCS and 14 OFDM symbol slot), which is to demonstrate the benefit of DL dominant frame structure DDDSU over DSDU for downlink user-perceived throughput, and the problem for DDDDDDDSUU without an SUL band. The three frame structures are plotted in Fig. 5.7.
– TDD + SUL band: this is applied to DDDDDDDSUU to demonstrate that with SUL band, the UL delay is reduced such that the DL dominant frame structure DDDDDDDSUU can work well to provide high downlink user-perceived throughput.

In the evaluation, burst buffer traffic is used, and urban macro scenario is employed. The system configuration parameters related to urban macro are shown in Table 5.25. Compared to dense urban parameters as shown in Table 5.19, the urban macro scenario is characterized by larger inter-site distance of 500 m as compared to 200 m in dense urban. However, the indoor and outdoor user distribution is kept the same: 80% of indoor users and 20% of outdoor users as in dense urban.

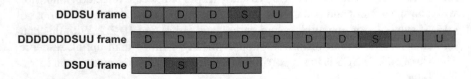

Fig. 5.7 NR frame structures for evaluation

Table 5.25 System configuration parameters for urban macro scenario

System configuration parameters	Values
Test environment	Macro Urban
Carrier frequency	TDD band: 3.5 GHz (1 macro layer) SUL band: 1.8 GHz
BS antenna height	20 m
Total transmit power per TRxP	46 dBm
UE power class	23 dBm
Inter-site distance	500 m
User distribution	Indoor: 80%, Outdoor: 20%
UE speeds of interest	Indoor users: 3 km/h Outdoor users: 30 km/h
Traffic model	Burst buffer: file size 0.5 Mbytes, arrival rate 0.5/1/2/3
UE density	10 UEs per TRxP
UE antenna height	Outdoor UEs: 1.5 m Indoor UTs: $3(n_{fl} - 1) + 1.5$; $n_{fl} \sim$ uniform$(1, N_{fl})$ where $N_{fl} \sim$ uniform$(4,8)$
Mechanic tilt	90° in GCS (pointing to horizontal direction)
Electronic tilt	100°

Therefore, the challenge of this scenario is the requirement of larger coverage (to allow inter-site distance of 500 m with 80% of indoor users).

Before showing the evaluation results, the impact factors of different frame structures on the performance are investigated. The following aspects are found to be important:

- **Guard Period (GP) overhead**:
  GP is introduced at the downlink/uplink switching point. Frequent DL/UL switching will introduce larger GP overhead, which will result in system capacity reduction.

- **UL delay due to the UL slot availability**:
  The UL slot availability affects the Channel State Information (CSI) feedback and ACK/NACK feedback delay. More available UL subframes will reduce the CSI and ACK/NACK feedback delays.

- **DL and UL ratio**:
  The DL and UL ratio associate with a certain frame structure should be well aligned with the DL and UL traffic pattern. Otherwise, the DL or UL system capacity will be degraded.

For the above aspects, it is observed that the DSDU frame structure may benefit from the fast CSI measurement and feedback; however, the frequent downlink and uplink switching in this frame structure will bring extra overhead.

On the other hand, the DDDSU and DDDDDDDSUU frame structure may suffer from a relatively slower CSI feedback, yet they can benefit from the reduced GP overhead.

Taking into account the channel varying nature that depends on the device moving speed distribution (in urban macro scenario one has 80% indoor users with 3 km/h and 20% outdoor users with higher mobility), the trade-off of the CSI/ACK feedback and the overhead introduced by DL/UL switching point are evaluated for different candidate frame structures.

The overhead can be analytically calculated following a similar way as indicated in Sect. 5.4.1.1. Based on the evaluation assumptions listed in Table 5.26, the total overhead for the different frame structures are provided in the same table. It is observed that, among the three frame structures, DSDU has the highest overhead due to the increased CSI-RS and GP overhead for the fast CSI measurement and DL/UL switching. The DDDSU frame structure provides good balance for overhead and CSI acquisition. DDDDDDDSUU has the lowest overhead. However, if the system only operates on the TDD band, the UL feedback (both for CSI and ACK) for DDDDDDDSUU will be very much delayed, and hence the UPT performance of DDDDDDDSUU may be degraded. In this case, the DL/UL decoupling will be helpful since the SUL band will reduce the CSI and ACK feedback delay.

The overall impact of the overhead and the CSI/ACK feedback delay is evaluated by system-level simulation. The technical configuration parameters for the DL user experienced data rate (or DL UPT) evaluation are shown in Table 5.27. It is noted that the UPT in this evaluation is defined on the basis of a data packet. Therefore the

Table 5.26 Overhead assumption for different frame structures in DL

Overhead assumption	DDDSU	DDDDD DDSUU	DSDU
PDCCH	2 complete symbols in the downlink dominant slot	2 complete symbols in the downlink dominant slot	2 complete symbols in the downlink dominant slot; 1 complete symbol in the uplink dominant slot
DM-RS	Type-2 configuration	Type-2 configuration	Type-2 configuration
CSI-RS	4 ports per UE with 5 slots period; 40 REs/PRB for 10 users	4 ports per UE with 10 slots period; 40 REs/PRB for 10 UEs	4 ports per UE with 4 slots period; 40 REs/PRB for 10 UEs
SSB	8 SSBs per 20 ms	8 SSBs per 20 ms	8 SSBs per 20 ms
TRS	2 burst consecutive slots per 20 ms, bandwidth with 51 PRBs	2 burst consecutive slots per 20 ms, bandwidth with 51 PRBs	2 burst consecutive slots per 20 ms, bandwidth with 51 PRBs
GP	2 symbols	4 symbols	2 symbols
Total overhead	0.38	0.35	0.46

Table 5.27 Technical configuration parameters for DL user experienced data rate evaluation

Technical configuration parameters	Values
Multiple access	OFDMA
Numerology	For TDD band: 30 kHz SCS For SUL band: 15 kHz SCS
Simulation bandwidth	20 MHz (51 PRBs)
Frame structure	DDDSU, DDDDDDDSUU, DSDU on TDD band (3.5 GHz) Full uplink on SUL band (1.8 GHz)
Transmission scheme	Closed SU/MU-MIMO adaptation
MU dimension	Up to 12 layers
SU dimension	Up to 4 layers
SRS transmission	Non-precoded SRS, 4 Tx ports, 8 PRBs per symbol, 2 symbols per 10 slots for DDDDDDDSUU, 2 symbols per 5 slots for DDDSU, 2 symbols per 4 slots for DSDU
CSI feedback	CQI/RI feedback every 10 slots for DDDDDDDSUU, CQI/RI feedback every 5 slots for DDDSU, CQI/RI feedback every 4 slots for DSDU, Non-PMI feedback, Sub-band based
Interference measurement	SU-CQI
Max CBG number	1
ACK/NACK delay	N + 1
Retransmission delay	the next available DL slot after receiving NACK
Antenna configuration at TRxP	32T, (M,N,P,Mg,Ng; Mp,Np) = (8,8,2,1,1;2,8), (dH, dV) = (0.5, 0.8)λ, Vertical 1 to 4.
Antenna configuration at UE	4R, (M,N,P,Mg,Ng; Mp,Np) = (1,2,2,1,1; 1,2), (dH, dV) = (0.5, N/A)λ
Scheduling	PF
Receiver	MMSE-IRC
Channel estimation	Non-ideal

UPT performance statistic is conducted per data packet basis, which offers the assessment on data packet throughput distribution.

In Figs. 5.8 and 5.9, the evaluation results on average UPT and the 5th percentile UPT with different arrival rates are illustrated.

It is observed that, with the good balance of overhead and feedback delay, DDDSU has the best performance in most cases, and the performance gain can reach more than 10% when compared to DSDU.

For DDDDDDDSUU, however, since the UL delay is large, the performance is degraded. This is observed despite its low GP overhead. Specifically, when retransmission becomes dominant, e.g., for the worst 5th percentile of packet, the performance degradation is significant. This explains why the 5th percentile UPT of DDDDDDDSUU is worse than DSDU, even the overhead of DDDDDDDSUU is less than DSDU.

In the above case, the DL/UL decoupling helped DDDDDDDSUU recover its performance by the use of SUL. The CSI and ACK feedback could be reported in a

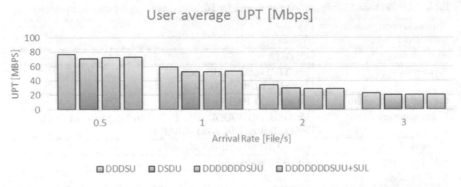

Fig. 5.8 DL average UPT of different frame structures with and without DL/UL decoupling

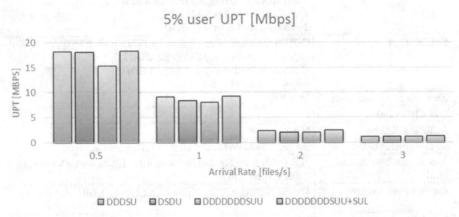

Fig. 5.9 DL 5th percentile UPT of different frame structures with and without DL/UL decoupling

timely manner from SUL band, and the UPT is therefore increased to the similar level of DDDSU.

Based on the above evaluation, it demonstrates that the frame structures of DDDSU and DDDDDDDSUU can introduce high DL user-perceived throughput, and by the use of SUL together, DDDDDDDSUU could boost the 5th percentile user-perceived throughput.

5.4.3.2   Contribution to UL User Experienced Data Rate

The DL/UL decoupling can boost the uplink user experienced data rate. This is because the DL/UL decoupling employs the SUL band which is usually on lower frequency band (e.g., 1.8 GHz) for UL transmission. Compared to TDD only case, where the TDD band is usually on higher frequency band (e.g., 3.5 GHz), the propagation loss of SUL band is much reduced. This can benefit the UL transmission from power limited case on 3.5 GHz to non-power limited case on 1.8 GHz. The power limited case indicates that the UE reaches its maximum transmission power, and therefore the transmit power density would be proportionally decreased with increased bandwidth. It implies that no matter how large bandwidth is assigned to the UE, the data rate cannot be increased. By the use of SUL, the low propagation loss removes the power limitation.

On the other hand, the UL resource in SUL band is continuous on time domain. In contrast, in DDDSU case, the UL resource is limited. It implies SUL can offer more UL transmission opportunities and therefore the UL data rate can be increased.

In summary, the lower frequency SUL band helps to increase the UL data rate from two aspects: one is to remove power limitation by its nature of lower propagation loss compared to higher frequency TDD band, and the other is to provide more available UL slots which is desired by UL transmissions.

In this subsection, the system-level simulation is conducted to evaluate the UL average throughput and the 5th percentile user throughput (cell edge user throughput) for the following cases: TDD band only case for the three frame structures DSDU, DDDSU, DDDDDDDSUU, and TDD band plus SUL band case. The high load network traffic is assumed, and the urban macro scenario as given in Table 5.25 is employed.

For frequency band f_1 and f_2, the carrier frequency 3.5 GHz (TDD band) and 1.8 GHz (SUL band) are used, respectively. By appropriately selecting the RSRP threshold, it is observed that 30% of the users (with low RSRP on 3.5 GHz) select the SUL band and the other 70% of users select the TDD band. Other technical parameters for the UL evaluation are listed in Table 5.28.

The evaluation results are provided in Figs. 5.10 and 5.11. It is observed that, for cell average throughput, SUL band can introduce about 40% gain. This is because, for the 70% of users on the TDD band, more spectrum resources can be allocated after the 30% of users are offloaded to the SUL band. On the other hand, benefiting

Table 5.28 Technical configuration parameters for UL user experienced data rate evaluation

Technical configuration parameters	Values
Multiple access	OFDMA
Numerology	TDD band: 30 kHz SCS; SUL band: 15 Hz SCS
Simulation bandwidth	TDD band: 100 MHz SUL band: 20 MHz
Frame structure	DDDSU, DDDDDDDSUU, DSDU on TDD band (3.5 GHz) Full uplink on SUL band (1.8 GHz)
Transmission scheme	SU adaptation
UL CSI measurement	Non-precoded SRS and wideband PMI
UL codebook	Codebook based
SU dimension	Up to 2 layers
SRS transmission	Non-precoded SRS, 2 Tx ports, 8 PRBs per symbol, For TDD band: 2 symbols per 10 slots for DDDDDDDSUU, 2 symbols per 5 slots for DDDSU, 2 symbols per 4 slots for DSDU, for SUL band: 2 symbols per 5 slots

Table 5.28 (continued)

Technical configuration parameters	Values
Max CBG number	1
Antenna configuration at TRxP	32R, (M,N,P,Mg,Ng; Mp,Np) = (8,8,2,1,1;2,8), (dH, dV) = (0.5, 0.8)λ, Vertical 1 to 4.
Antenna configuration at UE	2T, (M,N,P,Mg,Ng; Mp,Np) = (1,1,2,1,1; 1,2), (dH, dV) = (0.5, N/A)λ
Power control parameters	P0 = −60 dBm, alpha = 0.6
Power backoff	Continuous PRB allocation model: follow 3GPP TS 38.101; Non-continuous PRB allocation: additional 2 dB reduction
Scheduling	PF
Receiver	MMSE-IRC
Channel estimation	Non-ideal

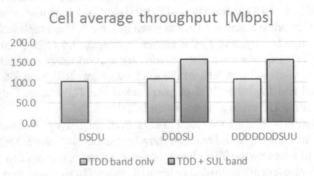

Fig. 5.10 UL average throughput of different frame structures with and without DL/UL decoupling (High load)

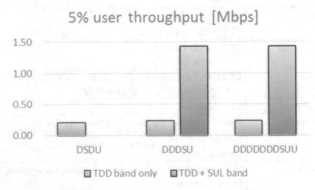

Fig. 5.11 UL cell edge user throughput of different frame structures with and without DL/UL de coupling (High load)

from the lower propagation loss and the sufficient bandwidth, the gain for cell-edge users on SUL band can achieve about 5 times.

5.4.3.3   Contribution to Uplink User Plane Latency

The benefit of DL/UL decoupling on reducing uplink user plane latency is obvious: in DL dominant frame structure, the UL slots are limited. After the use of SUL band, the UL resource always exists. Therefore, the UL latency can be reduced.

For UL user plane latency evaluation, the approach as mentioned in Sects. 5.3.1.6 and 5.4.1.2 is used. Grant-free transmission is assumed. The evaluation results can be found in Table 5.29. Note that in this evaluation, it is assumed that for TDD band with 30 kHz SCS plus SUL band with 15 kHz SCS, the UE and BS processing delay is calculated based on 15 kHz SCS. That is, the worst case is assumed for the UE and BS processing capability. If higher capability BS is used, the latency of TDD + SUL can be further reduced.

It is observed that, with the introduction of SUL band, the UL latency can be reduced by more than 15%. In some cases, the reduction achieves 60%.

5.4.4   NB-IoT

NB-IoT is capable of providing large connection density required by IMT-2020 vision.

In [4], the connection density of NB-IoT is evaluated using system-level simulation.

According to Report ITU-R M.2412, connection density is said to be C (the number of devices per km^2), if, under the number of devices, $N = C \times A$ (A is the simulation area in terms of km^2), that the packet outage rate is less than or equal to 1%, where the packet outage rate is defined as the ratio of

– The number of packets that failed to be delivered to the destination receiver within a transmission delay of less than or equal to 10s

   to

– The total number of packets generated by the ($N = C \times A$) devices within the time T.

The *transmission delay* of a packet is understood to be the delay from the time when uplink packet arrives at the device to the time when the packet is correctly received at the destination (BS) receiver.

For NB-IoT, the transmission delay is related to the delay of each step in the access procedure. In [4], the early data transmission procedure is employed. There are six steps in this procedure:

Step 1: Sync + MIB transmission from base station to device

Table 5.29 UL user plane latency for NR TDD + SUL with grant-free transmission (ms) (Frame structure for TDD carrier: DDDSU)

UL user plane latency—NR TDD (DDDSU) + SUL			UE capability 1 SCS			UE capability 2 SCS		
			30 kHz TDD only	30 kHz (TDD) + 15 kHz (SUL)	30 kHz (TDD) + 30 kHz (SUL)	30 kHz TDD only	30 kHz (TDD) + 15 kHz (SUL)	30 kHz (TDD) + 30 kHz (SUL)
Resource mapping Type A	M = 4 (4OS) non-slot	p = 0	1.86	1.57	0.86	1.65	1.18	0.65
		p = 0.1	2.11	1.79	1.01	1.90	1.38	0.76
	M = 7 (7OS) non-slot	p = 0	1.91	1.68	0.91	1.71	1.29	0.71
		p = 0.1	2.16	1.90	1.06	1.96	1.49	0.82
	M = 14 (14OS slot)	p = 0	2.16	2.18	1.16	1.96	1.79	0.96
		p = 0.1	2.41	2.48	1.32	2.21	2.01	1.12
Resource mapping Type B	M = 2 (2OS) non-slot	p = 0	1.36	1.04	0.59	1.10	0.54	0.33
		p = 0.1	1.60	1.22	0.70	1.35	0.63	0.39
	M = 4 (4OS) non-slot	p = 0	1.63	1.32	0.73	1.39	0.86	0.49
		p = 0.1	1.88	1.52	0.85	1.64	0.97	0.56
	M = 7 (7OS) non-slot	p = 0	1.69	1.43	0.79	1.48	1.04	0.58
		p = 0.1	1.93	1.63	0.91	1.73	1.17	0.66

Step 2: PRACH Message 1 transmission from device to base station

Step 3: NPDCCH + RAR (including UL grant) transmission from base station to device

Step 4: UL data transmission from device to base station

Step 5: RRCEarlyDataComplete transmission from base station to device

Step 6: HARQ ACK from device to base station

In conventional system-level simulations, the device is assumed to be in active mode. In that case, the system-level simulation focuses on Step 4. For connection density evaluation, all the 6 steps needs to be modeled in the dynamic system-level simulator. A simplified SINR-to-delay model is proposed in [5] to enable the evaluation of delays for Step 1 to Step 4 as well as Step 6. Based on this model, the evaluation results are presented in Table 5.30. It is observed that the connection density requirement of 1,000,000 devices per km² can be met by using 2 RBs by NB-IoT.

In addition, the connection efficiency evaluation is provided in Table 5.30, where the connection efficiency is given by

$$CE = \frac{C}{M} \cdot \frac{A}{W} \quad \left(\text{number of device} / \text{Hz} / \text{TRxP}\right)$$

where C is the connection density (number of devices per km²), A is the simulation area in terms of km², M is the number of TRxP in the simulation area A, and W is the UL bandwidth (for FDD). It is seen that NB-IoT demonstrates a good capability of connection efficiency, which implies that the spectrum resources are efficiently used when connecting massive number of devices.

Table 5.30 NB-IoT evaluation results for connection density

		ISD = 1732m, channel mode A as defined in Report ITU-R M.2412	ISD = 1732m, channel mode B as defined in Report ITU-R M.2412
NB-IoT	Devices supported per km² per 180 kHz	599,000	601,940
	Required bandwidth to support 1,000,000 devices	360 kHz	360 kHz
	Connection efficiency (# of devices/Hz/TRxP)	2.88	2.896
eMTC	Devices supported per km² per 180 kHz	369,000	380,000
	Required bandwidth to support 1,000,000 devices	540 kHz	540 kHz
	Connection efficiency (# of devices/Hz/TRxP)	1777	1828

5.4.5   Field Test of LTE/NR Spectrum Sharing

The forgoing discusses results that are mainly for the IMT 2020 evaluation. The next set of result are from field trails there were done to test the LTE/NR spectrum sharing capabilities. The field test of LTE/NR spectrum sharing (a.k.a. UL/DL decoupling) was performed in the 5G technology test campaign organized by IMT-2020 promotion group. The information of IMT-2020 promotion group and the related campaigns of 5G technology test can be found in Sect. 1.2.2.2 and Sect. 6.3.1 respectively. The UL/DL decoupling was tested in both non-standalone (NSA) and standalone (SA) deployment scenario to verify the advantage of uplink coverage, which demonstrate about 2~5x and 2~6x performance gain over uplink of 3.5GHz, respectively. The outdoor test results also show the significant performance gain of UL/DL decoupling especially in the coverage limited scenario.

5.4.5.1   Indoor Test in NSA Deployment

The field test scenario of UL/DL decoupling in NSA deployment is depicted in Fig. 5.12, where the distance between base station and the test UEs is about 340 m. The test UEs are placed inside a building, and it can be observed that there is no barrier between the base station and the building. The locations of the test UEs are shown in Fig. 5.13.

The test results in the case of NSA are summarized in Table 5.31, where the uplink throughput of the secondary cell group (SCG), i.e., NR cell group, is chosen

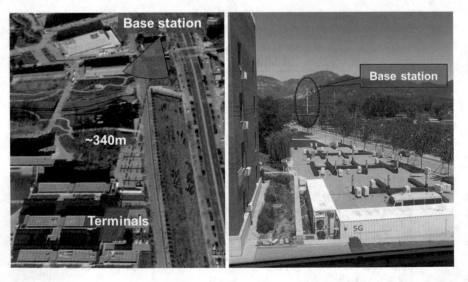

Fig. 5.12 Field test scenario of UL/DL decoupling in the cease of NSA deployment. (**a**) Fourth floor of the building; (**b**) fifth floor of the building

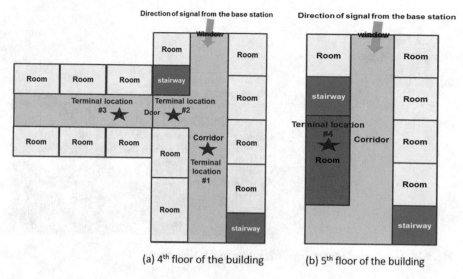

(a) 4th floor of the building   (b) 5th floor of the building

Fig. 5.13 Locations of test UEs in NSA deployment

Table 5.31 Test results of UL/DL decoupling in the case of NSA deployment

Test UE location	Test group	RSRP (dBm)	UL throughput (Mbps)		Gain (%)
			Without UL/DL decoupling	With UL/DL decoupling	
#1	1	−102	16.8	38.7	130
#2	2	−113	2.7	7.1	160
	3	−114	2.85	10.76	270
#3	4	−118	1.3	6.97	430
#4	5	−116	2.6	6.8	160
	6	−115	2.5	7.74	210

to be evaluated. For the case without UL/DL decoupling, only the 3.5 GHz uplink carrier is used for uplink transmission. In the case employing UL/DL decoupling, an additional 2.1 GHz supplementary uplink (SUL) carrier can be also used. It can be observed that the uplink throughput is significantly increased when UL/DL decoupling is utilized compared to without UL/DL decoupling. In particular, the uplink performance gain provided by UL/DL decoupling is more than 100%, and can be up to 430% when the corresponding DL reference signal received power (RSRP) is low. Such a high performance gain is due to the lower propagation loss of the 2.1 GHz SUL carrier.

5.4.5.2   Indoor Test in SA Deployment

The field test scenario is shown in Figs. 5.14 and 5.15, where the UEs used for testing are placed inside a building and the distance between base station and the building is about 300 m. For the case without UL/DL decoupling, only the 3.5 GHz

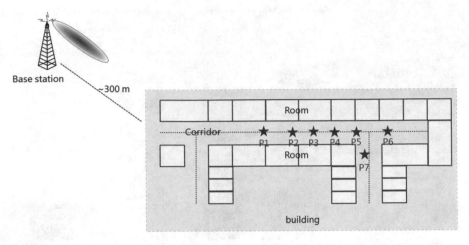

Fig. 5.14 Locations of UE for single UE testing in SA deployment

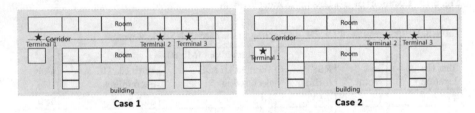

Fig. 5.15 Locations of UE for multiple UE testing in SA deployment

uplink carrier is used for uplink transmission. In the case with UL/DL decoupling, an additional 2.1 GHz SUL carrier can be used.

In Fig. 5.14, it is for single UE test and the locations of the UE are at the points marked from P1 to P7. The test results in the case of single UE testing are summarized in Table 5.32, where the RSRP of all the seven test points is from −117 dBm to −106 dBm. For the test points with RSRP less than −109 dBm, the corresponding UL throughput gain obtained by UL/DL decoupling is higher than 100% and can even reach to 600% where the RSRP of the UE is −117 dBm. As for test point with higher RSRP, i.e., −106 dBm, the throughput gain can be larger than 50%.

The multiple UEs test is shown in Fig. 5.15. In particular, three UEs are scheduled for uplink transmission simultaneously. There are two cases considered:

- **Case 1**: One UE has high RSRP regarded as a cell-center UE, and another two UEs with low RSRP are treated as cell edge UEs.
- **Case 2**: All the three UEs are with low RSRP.

The test results in the case of multiple UE testing are summarized in Table 5.33. When UL/DL decoupling is enabled, the UEs with low RSRP will be switched from 3.5 GHz uplink to 2.1 GHz SUL for uplink transmission. Due to the lower propagation loss of 2.1 GHz SUL compared to 3.5 GHz uplink, the uplink throughput for the UEs

Table 5.32 Test results in the case of single UE testing in SA deployment

Test position	CSI RSRP (dBm)	Without UL/DL decoupling		With UL/DL decoupling		Gain (%)
		UL throughput (Mbps)	UE transmit power (dBm)	UL throughput (Mbps)	UE transmit power (dBm)	
P1	−106	9.75	23	14.75	20	51
P2	−109	6.62	23	16.73	20	153
P3	−111	4.02	23	16.05	20	299
P4	−112	3.13	23	12.08	20	286
P5	−114	2.16	23	10.32	20	378
P6	−115	1.42	23	8.36	20	489
P7	−117	0.57	23	3.98	20	602

Table 5.33 Results of multi-terminal testing for UL/DL decoupling in SA deployment

	Test UEs	RSRP (dBm)	Without UL/DL decoupling		With UL/DL decoupling		Gain (%)
			UL throughput (Mbps)	UE transmit power (dBm)	UL throughput (Mbps)	UE transmit power (dBm)	
Case 1	UE1	−84	34.78	20	84.77	20	144
	P2	−114	1.55	23	4.89	20	216
	UE3	−110	2.08	23	6.48	20	211
Case 2	UE1	−108	3.15	23	7.04	20	123
	P2	−112	1.74	23	6.12	20	251
	UE3	−110	2.33	23	6.41	20	175

with low RSRP can be significantly increased and the performance gain can be higher than 200%. It is worth noting that more than 100% gain in the uplink can be obtained for UE with high RSRP, i.e., UE 1 in Case 1, though it is still on the 3.5 GHz UL carrier after UL/DL decoupling is enabled. The reason is that UE 1 can monopolize all the uplink resource of the 3.5 GHz uplink after the other two UEs are switched to the 2.1 GHz SUL. Thus, it is demonstrated that when UL/DL decoupling is applied, both UEs with high RSRP and low RSRP can obtain significant performance gain in the uplink.

5.4.5.3   Outdoor Test

The scenario of outdoor testing for UL/DL decoupling is shown in Fig. 5.16. The test UE is placed on the car which moves in an anti-clockwise direction at the speed of 20–30 km/h for three rounds.

The first Round: UL/DL decoupling is disabled, and the 3.5 GHz TDD carrier is used for uplink transmission.

The second Round: UL/DL decoupling is enabled. Both 3.5 GHz TDD carrier and 1.8 GHz SUL carrier can be used for uplink transmission. The RSRP threshold for UL/DL decoupling is set to be −105 dBm.

The third Round: UL/DL decoupling is enabled. Both 3.5 GHz TDD carrier and 1.8 GHz SUL carrier can be used for uplink transmission. The RSRP threshold for UL/DL decoupling is set to be −50 dBm.

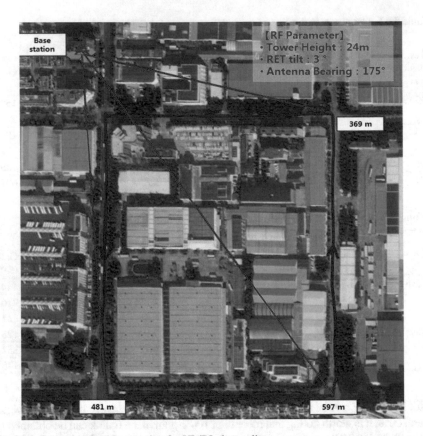

Fig. 5.16 Scenario of outdoor testing for UL/DL decoupling

The configuration parameters including bandwidth, antenna configuration, and transmit power are listed in Table 5.34.

The DL RSRP measured by CSI-RS at different locations during the whole round trip and the frequency of UL carrier selected for UL transmission are illustrated for all the three rounds in Fig. 5.17. Obviously, only the 3.5 GHz TDD carrier can be used for UL transmission in the first round. In the second round, the UE can select either 3.5 GHz TDD carrier or 1.8 GHz SUL for UL transmission in a dynamic manner based on its measured RSRP. In the third round, the UE will always select the SUL since the RSRP threshold is set to be extremely high.

The UL throughput during the three rounds are demonstrated in Fig. 5.18. It can be observed that for the areas that RSRP is less than −105 dBm, the average UL throughput without UL/DL decoupling is about 1.4 Mbps, and the average UL throughput with UL/DL decoupling can reach 10.2 Mbps and 10.8 Mbps with −105 and − 50 dBm RSRP threshold, respectively. The performance gain provided by UL/DL decoupling is more than 600%.

Table 5.34 Parameters of outdoor experiment testing

Parameter	NR TDD 3.5 GHz	NR SUL 1.8 GHz	LTE 1.8 GHz (with SUL)
Bandwidth	100 MHz	10M	15 MHz
Number of RB	273RB	52RB	23RB
Base station Antenna configuration	64T64R	4R	4T4R
Terminal Antenna Configuration	1T4R	1T	1T2R
Transmit power of Base station	200w	\	60w

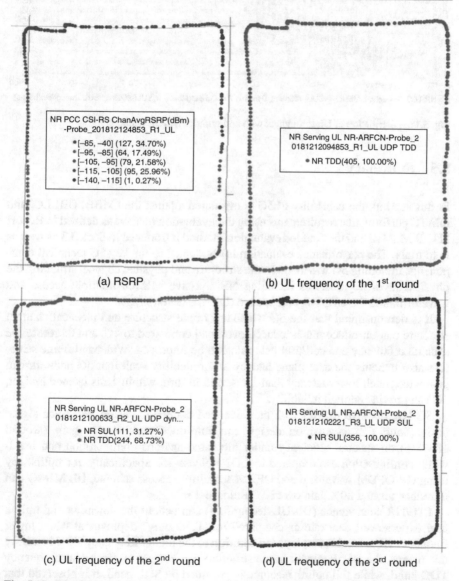

(a) RSRP

(b) UL frequency of the 1st round

(c) UL frequency of the 2nd round

(d) UL frequency of the 3rd round

Fig. 5.17 DL RSRP and frequency of UL carrier on different locations during the whole round trip. (**a**) RSRP, (**b**) UL frequency of the first round, (**c**) UL frequency of the second round, (**d**) UL frequency of the third round

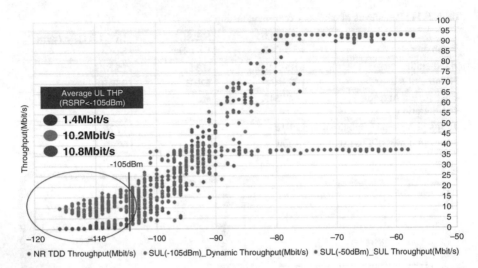

Fig. 5.18 Distribution of UL throughput with and without UL/DL decoupling

5.5   Summary

In this section, the capability of 5G is evaluated against the eMBB, URLLC, and mMTC performance requirements using the evaluation method as defined in Report ITU-R M.2412 and the detailed evaluation method introduced in Sect. 5.3 as well as field trials. The performance evaluation is conducted to the key 5G technical components, including 5G wideband frame structure and physical channel structure, the physical layer key features including NR massive MIMO, multiple access and waveform, and the LTE/NR coexistence (DL/UL decoupling).

It is demonstrated that the 5G wideband frame structure and physical channel structure can introduce much reduced overhead compared to 4G, and therefore the spectral efficiency and cell/user data rate can be improved. Wideband frame structure also benefits the user plane latency and reliability such that 5G network can guarantee much lower latency than 1 ms, and in turn, within 1 ms latency budget, 5G can provide enough reliability.

5G NR massive MIMO with the advanced codebook design, reference signal design, and CSI acquisition design can introduce considerable gain (around 18–25%) on spectral efficiency under the same antenna configuration and bandwidth configuration as compared to LTE Release-15. Specifically, for uplink, by using CP-OFDM waveform and OFDMA multiple access scheme, UL MIMO can introduce around 20% gain over LTE Release-15.

LTE/NR coexistence (DL/UL decoupling) can benefit the downlink and uplink user experienced data rate as compared to "TDD only" deployment when lower frequency SUL band is employed in addition to the higher frequency TDD band. By this means, the DL dominant frame structure can be kept on the higher frequency TDD band, while full uplink resource is provided by SUL band. It is observed that DL user experienced data rate can be improved by more than 10% over "compromise" frame structure, and UL user experienced data rate is improved several times over TDD only configuration. The UL user plane latency can be reduced by more than 15% to as much as 60%.

In this section, not all the technical requirements defined in Report ITU-R M.2412 are evaluated. For example, the evaluation of peak spectral efficiency, peak data rate, energy efficiency, mobility, mobility interruption time, etc. are not provided. The 5G key technical components will contribute to these capabilities as well. The readers interested in the evaluation of those technical performance requirements are referred to 3GPP's self-evaluation report TR37.910 for the detailed results.

However, it is no doubt that 3GPP's 5G technology shows strong capability on fulfilling all the IMT-2020 requirements as defined by ITU-R. Such capability is the basis for further progressing of 3GPP 5G to more industry applications and bridging the gaps of future service requirements and the realistic grounds. The next chapter will discuss the market situation for the 5G wireless systems. It will argue that the industry is in a transition from a more or less homogeneous market with eMBB to that of a heterogeneous market that expand the services to that for a number of vertical industries.

References

1. 3GPP TR38.913, "Study on Scenarios and Requirements for Next Generation Access Technologies", June 2017
2. Report ITU-R M. 2410, "Minimum requirements related to technical performance for IMT-2020 radio interface(s)", November 2017
3. 3GPP TR36.814, "Further advancements for E-UTRA physical layer aspects", March 2017
4. R1-1808080, "Considerations and evaluation results for IMT-2020 for mMTC connection density", Huawei, August, 2018
5. Report ITU-R M. 2370, "IMT Traffic estimates for the years 2020 to 2030", July 2015

Chapter 6
5G Market and Industry

Coauthored with Amanda Xiang

This chapter discusses the current global 5G market. It shows that although the eMBB service remains a significant segment of the commercial wireless market, the capabilities of 5G allow it to move beyond the eMBB service and support vertical industries. This convergence of human-to-human communication with human-to-machine and machine-to-machine communications will be a key enabler of the fourth industrial revolution. How the industry plan to move forward, with a globally unified 5G standards and ecosystem from the different verticals, will be presented. Early 5G field trials and deployment plans will then be elucidated. The chapter and the book will conclude with a look forward on possible 5G developments.

6.1 5G Market

The available ubiquitous commercial wireless communications has, forever, changed the interaction of humans. The 4G revolution was mainly all about human-to-human interactions, the so-called social media revolution. The fundamental technological paradigm shift was the development of the mobile phone from a device for telephony to a smartphone. The applications in the smartphone not only are a harbinger of information but it, as well, enabled socialization with others in a unique and individualized fashion. How we socialized with others can be tailored to our individual flair and wishes.

As we now moved onto the 5G era, humanity will not only just communicate with one and other, but our communication will fuse together, the physical, digital, and biological worlds, and usher in the fourth industrial revolution [1]. It offers an unprecedented opportunity to break the current link between economic growth and environmental sustainability [2]. With innovations enhancing the production cycle and connect manufacturers with their supply chains and consumers to enable closed

Huawei Technologies, USA
Plano, TX

© Springer Nature Switzerland AG 2020
Wan Lei et al., *5G System Design*, https://doi.org/10.1007/978-3-030-22236-9_6

feedback loops to improve products, production processes and shorten time-to-markets, it will allow us to simultaneously drive profitability and sustainability. This could not come at a more opportune time, as consumer pressure has now made embracing sustainability and green principal not just a mere marketing tool. No longer are consumer and regulators satisfied when economic growth are at the expense of environmental sustainability. The main societal benefits, coming from 5G, can be said to be the time and cost savings as well as efficiency gains form specific hyper-customized services and applications while enhancing sustainability.

Within this transformation, human-to-human communication will form just one of the pillars of this revolution. The other pillar will be the ubiquitous availability of wide area, ultra-customized, communication for the different vertical industries. Nevertheless, it is clear that the first deployment will be the enhanced mobile broadband service (eMBB).

6.1.1   5G for Enhanced Mobile Broadband Service

The human-to-human communications, the so-called mobile broadband service (MBB), was the economical bedrock of the cellular communication industry in the 4G era. The service, in essence, provided the user with internet connection anywhere at any time. In a little over a decade, the service has grown from an interesting curiosity into an essential part of the human daily lives. Hardly can an hour go by where we do not check our e-mails! Almost any questions that now come into our minds can be answered by just opening an internet browser application and looking for the answer with an internet search engines such as Google. It changed the way we cooperate at work with video conferencing and how we socially interact with Facebook, You Tube, Twitter, etc. By any measure, this service has transformed the human experience, and has now reached nine billion mobile connections globally from 5.1 billion unique subscribers [3].

The first phase of 5G is to provide a massive improvement of the MBB service with eMBB service. There will be significantly improvement in the capacity, coverage, and latency of the network. The end user will find his internet experience faster and in more places. But perhaps the most significant improvement is in the reduction of the latency which allows for the extension of the internet to another human sensory system. Not only is the internet a visual and auditory experience, 5G allows for the extension of the internet to the tactile senses [4]. These new capabilities will now usher in the era of augmented and virtual reality. The implication of which is that the productivity decrease [5] of the internet age can be potentially erased with the technological advancement provided by 5G.

6.1.2   5G for Vertical Applications

Given that the number of mobile connections currently, it can be argued that almost every person on earth that wants a mobile connection already have one. Consequently, increasing revenue for mobile carriers from eMBB service will be difficult.

Therefore, the success of 5G will not be contingent upon the eMBB service alone. The industry will need to expand the 5G footprint.

Fortunately, with very fast data rate of 10 Gbps, ultra-low latency as small as 1 ms, and flexible network architecture for different deployment options, 5G is not only attracting wide spectrum of vertical applications and industries which needs communications capability, but also can inspire new technical, application and business model innovation. That is, in the future, the number of connected things (the so-called internet of things (IoT)) will far exceed the number of humans on earth. Hung [6] concluded that by 2020, the number of connected devices will reach 20 billion. This growth is further expected to be exponential and unbounded. 5G is going to be the driving force that will make mobile technology a General Purpose Technology and enable $12.3 trillion of global economic output [7]. A similar projection of $14 trillion by 2030 was given in [8].

Comparing to 3G and 4G deployment which are mainly used for consumer Mobile Broadband applications, vertical applications and industries will contribute a significant portion of business revenue generated by 5G, as much as 1.2 trillion [9]. The possible new vertical use cases are shown in Fig. 6.2. Currently, the first likely vertical to adopt 5G are smart manufacturing and connected automobile. In terms of potential revenue, however, the major ones are energy utilities and manufacturing at 20% and 19% share of the total possible revenue respectively [9].

It is not straightforward, however, to realize this vertical industrial growth path for the cellular industry. Unlike MBB and eMBB which are relatively monolithic services, Figs. 6.1 and 6.2 show that there are a number of vertical industries, each with its own service requirement. This is the familiar long-tailed services with the eMBB service representing the head service and the vertical services representing the tail [11]. This is also known as the classic selling less to more strategy. Each vertical service in the long tail does not generate significant revenue, but since the

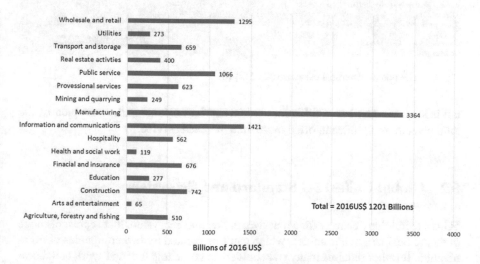

Fig. 6.1 The 5G enabled output of various industries (data from [7])

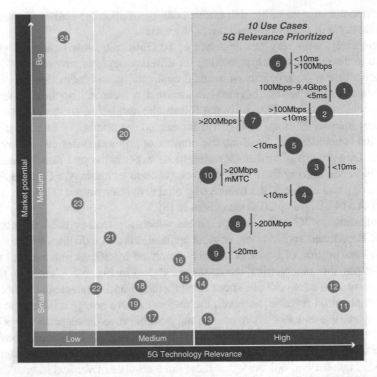

Fig. 6.2 The potential vertical uses cases for 5G [10]

tail is long, the number of different vertical services is large and, consequently, the total revenue is significant, often more than the head service.

6.2   Global Unified 5G Standard and Ecosystem

5G is a technology enabler for a massive market and ecosystem that represents huge potential business opportunities, which can't be created by few companies or organization. It rather requires many stakeholders to come together and work collaboratively with their own expertise and requirements for 5G in order to develop a healthy ecosystem with a unified 5G standards. In this section, we will provide overview of some key standard organizations or industry consortiums which will influence the 5G technologies for various vertical industries.

6.2.1   3GPP

Since 5G is becoming a General Purpose Technology to a much wider ecosystem than the traditional ICT industry, a global unified 5G standard which can accommodate different needs from various vertical industries and applications is critical for the success of 5G. Although the 3G, 4G technologies developed by 3GPP has been used by vertical industries to meet certain communication needs, those technologies are heavily ICT centric, relying on certain deployment model (e.g., heavily reliance on the public network of mobile operators) as well as the communication service performance that maybe good enough for some MBB services but not good enough to meet the stringent requirement of the verticals. Therefore 3GPP as the core standard developing organization for 5G communication technology has been in the center stage of this cross-industry unification effort and has been evolving to be more open and flexible:

1. Besides the traditional ICT players like mobile operators, ICT vendors and players that are continually participating in the 5G standardization effort, more and more vertical industry players are also joining 3GPP and playing important roles in bringing the vertical expertise and knowledge to help 3GPP create a 5G technology that will be suitable for the deployments for their verticals. Currently such verticals as Automobile industry, Smart manufacturing and industrial automation, Railway, Programme Making and Special Events (PMSE) industry and drone industry are actively participating in 3GPP.
2. 3GPP are also closely collaborating with other vertical centric industrial consortiums, such as 5G AA, 5G ACIA, LIPS, and so on to directly bring in those verticals needs, and drive corresponding 3GPP works to fulfill their requirements.

Before 3GPP can start to develop any technical solution, it's important to understand the service requirements from different vertical industrials. The job to collect those service requirement is the responsibility of SA1 (Service Architecture Working Group 1). Upon the publication of this treatise, SA1 will have completed or is working with vertical partners on service requirements for the following industries/applications:

- Cyber-physical control applications (TS 22.104)
- Railway communications (TS22.289)
- V2X (TS 22.186)
- Medical application (TR 22.826)
- Assets tracking (TR 22.836)
- Programme Making and Special Events (PMSE) (TR 22.827)
- Cellular connected Drone and its application and traffic management (UTM) (TS22.125, TR 22.829)

Comparing to normal MBB traffic, some verticals require some specific or even more stringent requirements in order to meet their needs. For example, the smart manufacturing, a key component of *industrie* 4.0, has very stringent URLLC (Ultra-Reliable and Low latency Communication) requirement for 5G, as shown in Table 6.1 defined in 3GPP SA1 TS22.104 [12].

Besides service requirements on QoS key performance indicators (KPI's), 3GPP SA1 are also working on the other service requirements for verticals industries, such as private network, network slicing, QoS monitoring, security, service continuity between public network and private network, and so on.

While the SA1 is developing the service requirement and use cases for the vertical applications, other downstream working group of 3GPP are also developing the 5G technologies to fulfill those requirements, such as:

- RAN WGs (Radio Access Network Working Group) which are the working groups for radio access network and the air-interface between device and base station of 5G, are working on radio technologies to meet the 1 ms latency and high reliability (99,999%) packet deliver rate for verticals.
- SA2 WG (System Architecture WG 2) which is the working group that is identifying the main functions and entities of the network, how these entities are linked to each other and the information they exchange, Basing on the services requirements elaborated by SA WG1. SA2 WG is working on network architecture enhancement and information exchange between network entities for an isolated private network deployment for verticals, as well as other vertical specific needs, such as supporting V2X and UTM.
- SA3 WG (System Architecture WG 3) which is the working group responsible for security and privacy in 3GPP systems, determining the security and privacy requirements, and specifying the security architectures and protocols. They will look into 5G security features to support vertical applications, such as for verticals' private network.

3GPP has spent almost 34 months to successfully complete release-15 3GPP specifications, which is the first 5G release to support standalone 5G New Radio deployment without relying on 4G LTE technology. This can be viewed as the real start of the 5G standard to many vertical players, because this allows vertical industries to be able to test the full benefits of 5G without the need to worry about interworking with 4G technologies. After release 15, 3GPP didn't get much time to celebrate but immediately started to work on the release 16 standard. When completed it will further enhance 5G capability for better commercial deployment, especially starting to focus more on addressing basic requirements from verticals other than eMBB applications, such as those from V2X, and industrial automation. While the Release 15 standard is the first standalone 5G NR standard providing basic capabilities of 5G, and so more eMBB application centric, Release 16 may be considered by many vertical players as the first 5G vertical-ready standard release for initial commercial deployment. Release 16 is expected to be completed by 3GPP around March 2020, but the work of defining Release 17 service

Table 6.1 Requirements for smart manufacturing [12]

	Characteristic parameter				Influence quantity						Remarks
	Communication service availability: target value[a]	Communication service reliability: mean time between failures	End-to-end latency: maximum[b]	Service bit rate: user experienced data rate	Message size [byte]	Transfer interval: target value	Survival time	UE speed	# of UEs	Service area[c]	
	99,999%	Below 1 year but >>1 month	<transfer interval value	≥200 kbit/s	≤200	100 ms	~500 ms	≤160 km/h	<25	50 km × 200 m	Railbound mass transit—Control of automated train (A.3.2)[d]
	99,999%–99,99,999%	~10 years	<transfer interval value	–	50	500 μs	500 μs	≤75 km/h	≤20	50 m × 10 m × 10 m	Motion control (A.2.2.1)
	99,9999%–99,999,999%	~10 years	<transfer interval value	–	40	1 ms	1 ms	≤75 km/h	≤50	50 m × 10 m × 10 m	Motion control (A.2.2.1)
	99,9999%–99,999,999%	~10 years	<transfer interval value	–	20	2 ms	2 ms	≤75 km/h	≤100	50 m × 10 m × 10 m	Motion control (A.2.2.1)
	99,9999%	–	<5 ms	1 kbit/s (steady state)1.5 Mbit/s (fault case)	<1500	<60 s (steady state)≥1 ms (fault case)	TBD	Stationary	20	30 km × 20 km	Electrical distribution—Distributed automated switching for isolation and service restoration (A.4.4)[e]

(continued)

Table 6.1 (continued)

Characteristic parameter			Influence quantity							Remarks
Communication service availability; target value[a]	Communication service reliability: mean time between failures	End-to-end latency: maximum[b]	Service bit rate: user experienced data rate	Message size [byte]	Transfer interval: target value	Survival time	UE speed	# of UEs	Service area[c]	
99.9999%–99,999,999%	~10 years	<transfer interval value		1 k	≤10 ms	10 ms	–	5–10	100 m × 30 m × 10 m	Control-to-control in motion control (A.2.2.2)
>99.9999%	~10 years	<transfer interval value	–	40–250	1–50 ms[f, g]	Transfer interval value	≤50 km/h	≤100	≤1 km²	Mobile robots (A.2.2.3)
99.9999%–99,999,999%	~1 month	<transfer interval value	–	40–250	4–8 ms[g]	Transfer interval value	<8 km/h	TBD	50 m × 10 m × 4 m	Mobile control panels—Remote control of, e.g., assembly robots, milling machines (A.2.4.1)
99.9999%–99,999,999%	~1 year	<transfer interval	–	40–250	<12 ms[g]	12 ms	<8 km/h	TBD	Typically 40 m × 60 m; maximum 200 m × 300 m	Mobile control panels—Remote control of, e.g., mobile cranes, mobile pumps, fixed portal cranes (A.2.4.1)

99,9999%– 99,999,999%	≥1 year	<transfer interval value	—	20	≥10 ms[h]	0	Typically stationary	Typically 10–20	Typically ≤100 m × 100 m × 50 m	Process automation—Closed loop control (A.2.3.1)
99,999%	TBD	~50 ms	—	~100	~50 ms	TBD	Stationary	≤100,000	Several km² up to 100,000 km²	Primary frequency control (A.4.2)[j]
99,999%	TBD	~100 ms	—	~100	~200 ms	TBD	Stationary	≤100,000	Several km² up to 100,000 km²	Distributed voltage control (A.4.3)[i]
>99,9999%	~1 year	<transfer interval value	—	15 k–250 k	10 ms–100 ms[g]	Transfer interval value	≤50 km/h	≤100	≤1 km²	Mobile robots—Video-operated remote control (A.2.2.3)
>99,9999%	~1 year	<transfer interval value	—	40–250	40–500 ms[g]	Transfer interval value	≤50 km/h	≤100	≤1 km²	Mobile robots (A.2.2.3)
99,99%	≥1 week	<transfer interval value	—	20–255	100 ms–60 s[g]	≥3 × transfer interval value	Typically stationary	≤10,000–100,000	≤10 km× 10 km × 50 m	Process monitoring (A.2.3.2), plant asset management (A.2.3.3)

[a]One or more retransmissions of network layer packets may take place in order to satisfy the communication service availability requirement
[b]Unless otherwise specified, all communication includes one wireless link (UE to network node or network node to UE) rather than two wireless links (UE to UE)
[c]Length × width (× height)
[d]Two UEs per train unit
[e]Communication includes two wireless links (UE–UE)
[f]This covers different transfer intervals for different similar use cases with target value by < ±25%
[g]The transfer interval deviates around its target value by < ±25%
[h]The transfer interval deviates around its target value by < ±5%
[i]Communication may include two wireless links (UE–UE)

requirements has started in 3GPP SA1. In Release 17, further 5G enhancements to previously covered vertical applications, as well as the addition of new verticals, such as drones, critical medical applications (remote surgery, wireless operation rooms, so on), PMSE (Programme Making and Special Events), asset tracking, and so on, are expected. Luckily, starting from Release 16, more and more vertical players are actively participating and contributing to the 3GPP 5G standardization effort.

6.2.2 Other Fora

Besides other WG's in 3GPP, a number of fora have developed in the ecosystem to aide the adoption of 5G for various vertical industries. Among the most significant, so far, are 5G ACIA and 5G AA for the smart manufacturing and automotive verticals, respectively.

6.2.2.1 5G ACIA for 5G Development in Manufacturing and Processing Industry

Because of its unprecedented reliability and ultra-low latencies capability, being able to support massive of IoT devices, and flexible and innovative architecture with leveraging new IT technologies as Cloud, MEC, 5G is getting more attraction to factory and processing industry for future smart factories and *industrie* 4.0.

In order to ensure that the specific needs and requirements of manufacturing and processing industry are adequately understood and considered by the telecom industry, especially 3GPP, a close collaboration is desired and required between all relevant players so the capabilities of 5G are fully realized and exploited by the manufacturing and processing industries. With this strong desire within relevant players, the 5G Alliance for Connected Industries and Automation (5G-ACIA) has been established on April 2018, which serves as the central and global forum for addressing, discussing, and evaluating relevant technical, regulatory, and business aspects with respect to 5G for the industrial domain. It reflects the entire ecosystem, encompassing all relevant stakeholder groups from the OT (operational technology) industry, the ICT (information and communication technology) industry, and academia as shown in Fig. 6.3.

While the book is being written, there are 41 member companies, which includes some key players in both ICT and OT industries, such as Siemens, Bosch, Vodafone, T-mobile, Orange, China Mobile, Ericsson, Huawei, Nokia, and Qualcomm.

The activities of the 5G-ACIA are currently structured in five different working groups (WGs) (Fig. 6.4).

- WG1 is responsible to collect and develop 5G requirements and use cases from manufacturing and processing industry and provide those as input to 3GPP SA1.

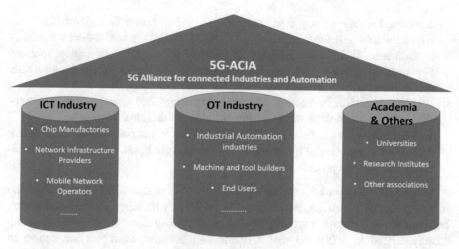

Fig. 6.3 A pictorial representation of the 5G-ACIA ecosystem

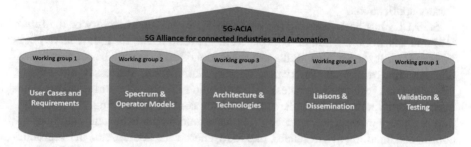

Fig. 6.4 The technical working group structure of 5G-ACIA

It will also educate OT players on existing 3GPP requirements and relevant work
to not only align both sides but also help to identify the 5G gaps in addressing the
needs from those OT partners. For example, the main input of 3GPP SA1 cyber-
physical control application uses and requirements work, such as 3GPP TR22.804,
and TS22.104 came from the joint effort of 5G ACIA members. Furthermore, the
work between the two groups are well distributed because many of the active
members of this 5G ACIA WG1 are also active members of 3GPP.

- WG2 is the working group to identify and articulate the specific spectrum needs
  of industrial 5G networks and explores new operator models, such as for operating private or neutral host 5G networks within a plant or factory, as well as coordinate 5G ACIA participation on relevant regulatory activities.
- WG3 is the main technical working group to shape and evaluate overall architecture of future 5G-enabled industrial connectivity infrastructures, and evaluate
  and integrate some key manufacturing and processing industry concepts with
  5G, such as Industrial Ethernet/TSN. The Working Group will base their work on
  3GPP standard technologies as well as that from other relevant standard organizations, such as IEC and IEEE. While this treatise is being written, WG3 has

completed a radio propagation analysis and evaluation for factory environment, which led to a new work in 3GPP RAN WG to study channel model for factory environment. WG3 is also working on study of private network deployment and Seamless Integration of Industrial Ethernet Technologies in 5G Architecture, whose outcome will be potential input for 3GPP 5G work.

- WG4 takes care of interaction with other initiatives and organizations by establishing liaison activities and initiating suitable promotional measures.
- WG 5 deals with the final validation of 5G for industrial applications, which includes the initiation of interoperability tests, larger trials, and potentially dedicated certification procedures.

Since 3GPP standard 5G technologies and its 5G standardization effort are one of key pillar of 5G ACIA, 3GPP and 5G ACIA are building close collaboration between the two groups. Not only are the works being done in the working group closely related to 3GPP 5G works (some collaboration examples are shown in Fig. 6.5), 5G-ACIA is also approved by 3GPP as a 3GPP MPR (Market Representative Partner) to bring the needed close collaboration to have a unified 5G applicable to factory applications.

5G ACIA has been studying the usage of 5G in the smart factory of the future [13] and has worked with 3GPP to put forth the retirements for industrial applications in [14]. It showed that all the use cases identified so far can be roughly classified into five major characteristics: Factory automation, Processes automation, HMIs (human machine interfaces) and production IT, and monitoring and maintenance. Factory automation deals with automatic control, monitoring, and optimization of factory process and workflows. Process automation is the automation of the control and handling of substances, like food, chemicals, liquids, etc., in the production facility. HMIs and production IT handles the human machine interfaces of pro-

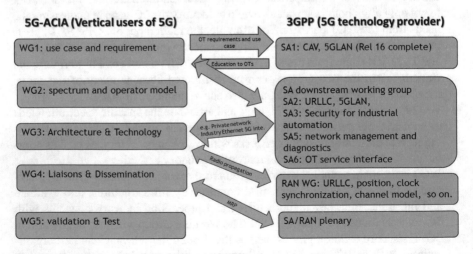

Fig. 6.5 The relationship of 5G-ACIA with 3GPP

duction devices, like panels on a machine, to IT devices, like computers, laptops, printers, etc., as well as IT-based manufacturing applications, like manufacturing execution system (MES) and enterprise resource planning (ERP) systems. Logistic and warehousing is concerned with the storage and flow of materials and products in industrial manufacturing. Monitoring and maintenance covers the storage and flow of materials and products in industrial manufacturing. In addition to the usual 5G performance requirements, these uses cases generally also contain operational and functional requirements. Overall, although 5G needs to be enhanced to cover all the use cases identified, the capabilities of the current release of 5G are sufficient to support about 90% of the use cases.[1]

6.2.2.2   5GAA for 5G Development for Connected Automotive

Since cellular networks are endowed with wide coverage and reliable connectivity, cellular 5G technology can be the ultimate platform to enable C-ITS (Cooperative Intelligent Transportation Systems) and the provision of V2X. In order for Automotive industry to better embrace 5G technologies, the **5G Automotive Association (5GAA)** is founded on September 2016, with over 80 member companies from the automotive and ICT industries, including eight founding members: AUDI AG, BMW Group, Daimler AG, Ericsson, Huawei, Intel, Nokia, and Qualcomm Incorporated. It's a global, cross-industry organization of companies from the automotive, technology, and telecommunications industries (ICT), working together to develop end-to-end solutions for future mobility and transportation services using 5G technology.

There are five working groups in 5G AA to cover various aspects of developing future connected automotive and its applications (Fig. 6.6):

- WG1: This is the "Use Cases and Technical Requirements" group, which defines end-to-end view on use cases and derive technical requirements and performance indicators for the certification of connected mobility solutions (e.g., on communication architecture, radio protocols, radio parameters, frequency spectra and carrier aggregation combinations). It ensures interoperability for V2X and other affected technologies.
- WG2: This is the "System Architecture and Solution Development" group, which defines, develops, and recommends system architectures and interoperable end-to-end solutions to address use cases and Services of Interest. The Working Group also reviews currently available solutions in technical areas such as wireless air interface technologies, wireless network deployment models, radio access networks and networked clouds, connectivity and device management or security, privacy and authentication for applicability to automotive applications.

---

[1] A detailed discussion can be found in Chap. 10 of [11].

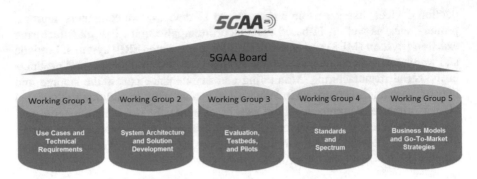

Fig. 6.6 The technical working group structure of 5GAA

- WG3: This is the "Evaluation, Testbeds, and Pilots" group, which evaluates and validates end-to-end solutions through testbeds. Promote commercialization and standardization via pilots and large-scale trials by selecting the use cases in conjunction with go-to-market strategies. This includes multi-phase/multi-year planning with prioritization.
- WG4: This is the "Standards and Spectrum" group, which acts as "Industry Specification Group," to provide recommendations, contributions, and positions to ETSI, 3GPP, and other standard development organizations. It develops spectrum requirements for V2X in ITS, MBB, and unlicensed bands. It also represents the association vis-à-vis other industry organizations.
- WG5: This is the "Business Models and Go-To-Market Strategies" group, which identifies involved organizations and companies, and prioritizes them. It drafts exemplary go-to-market plan as straw man function for agreed use cases under test and business models. It also provides the guideline to best achieve a global approach to certification of the target connected mobility solutions.

As a member of 3GPP MRP, 5GAA is working closely with 3GPP, such as acting as key source of various 3GPP service requirements work on V2X (TS22.186) [15]. The other 3GPP working group will develop the final V2X technical solution based upon those requirements. The V2X feature includes four types of communication: Vehicle-to-Vehicle (V2V), Vehicle-to-Infrastructure (V2I), Vehicle-to-Network (V2N), and Vehicle-to-Pedestrian (V2P). The feature has two complementary communication links: network and sidelink. The network link is the traditional cellular link and it provides long-range capability. The sidelink is the direct link between the different vehicles and it generally has lower latency than the network link and is of shorter range. The sidelink was implemented in LTE as the PC5 link and the first aspect of V2X support in NR is targeted to be completed in Release 16.

5GAA has also done an extensive analysis on the cellular V2X architectural solutions against two main use cases of interest: the intersection movement assist and the vulnerable user discovery [16]. Their conclusion showed that the Release 15 can adequately support both use cases.

6.2.2.3 Other Verticals

The current model is that there will be one industrial forum for each vertical. However, the industry is currently discussing whether such a model can scale when the number of vertical industries are large. The one forum per industry has the distinct advantage of being highly focus and passionate about its particular vertical. The price is that the resources to achieve such coverage are rather high.

Consequently, the industry is currently discussing other models to achieve the development needed for the vertical industries. For example, should there be a single or small number of fora for the vertical industries to discuss their use cases and requirements? Clearly this is more efficient but it has the drawback that some verticals may not be able to find their voice in such diverse forum. At present no clear answer presents itself and there will likely be a mixture of dedicated and general forums as the ecosystem works to find the right answer.

One thing to note is that not only is the discussion of use cases and requirements necessary, the development of proof of concepts (PoCs) for each vertical will also be key to the adoption. Thus these industries not only serves as an upstream entity (feeding use case and requirement) to 3GPP but also as a downstream entity to provide PoC's and reference deployments for the verticals.

6.3 Early Deployments

With the development of 5G NR standard, there is also 5G trial campaign ongoing to verify the candidate technology, test the system design, and promote 5G industry. This will facilitate the maturity of 5G industry and build a good foundation for commercialization in the coming years. Consequently, 5G trial campaign played a very important role in facilitating the earlier deployment of 5G and commercialization. In this section, the 5G trial campaign organized by IMT-2020 (5G) promotion group and the earlier deployment information will be discussed.

6.3.1 5G Trial in IMT-2020 (5G) Promotion Group

5G trial is one of important tasks of IMT-2020 (5G) promotion group, in which operators, infrastructure vendors, chipset vendors, instruments companies, university and research institute come together to help speed up the 5G development. Thanks to the participation of these industry companies, they jointly pushed the progress of 5G industry.

The 5G trial activity includes 5G technologies test (Phase 1: September 2015~December 2018) and 5G product test (Phase 2: 2019~2020). The 5G technologies test is organized by IMT-2020 (5G) promotion group, and the target is to verify 5G technical design, support the standardization, and promote 5G industry. The 5G product test in phase 2 will be driven by the operators to verify the network deployment, improve the industry maturity, and accumulate the experience of commercial deployment. As the test of 5G product just started in 2019, this treatise will mainly cover the phase 1 test.

There are three steps in the test plan of phase 1 including:

– Step 1: verification of key technical components, i.e., the performance of each key technical component is tested by prototype
– Step 2: verification of technical schemes, i.e., the performance of the technical scheme from different companies is tested under the deployment of one base station using the same frequency and specification
– Step 3: verification of system, i.e., 5G system performance test under the network deployment and the demonstration of 5G typical service. There are more test work in this step including the test of indoor, outdoor, device and interoperability in the both NSA and SA deployment scenario.

The trial results of these three steps in phase 1 are summarized from Tables 6.2, 6.3, 6.4, and 6.5 [17].

Table 6.2 Completion status of step 1 trial [17]

	Massive MIMO	New multiple access scheme	New multiple carrier	High frequency	Polar code	Dense network	Full duplex	Spatial modulation
Huawei	●	●	●	●	●		●	
Ericsson				●				
ZTE	●	●	●	●				
Samsung			●	●				●
Nokia Shanghai Bell	●	●		●				
CATT	●	●				●		
Intel	●							

Table 6.3 Completion status of step 2 trials from infrastructure vendors [18]

Vendors	Seamless wide-area coverage	Low-latency high-reliability	Low-power massive connection	High-capacity hot-spot (LF)	High-capacity hot-spot (HF)	High and low frequency mixed scenario	Mixed scenarios	5G high layer protocol	5G core network
Huawei	●	●	●	●	●	●	●	●	●
Ericsson	●	●		●	●	●			
ZTE	●	●	●	●	●	●	●	●	
CATT	●	●	●	●	◖		●		
Nokia Bell	◖	◖		◖	◖				

● Complete   ◖ Partially complete

Table 6.5 Completion status of step 3 trial for SA [19]

Infrastructure vendors	Core network functionalities	Core network performance	Security	Base station functionalities	Base station performance	Field network
Huawei	●	●	●	●	●	●
Ericsson	●	●	●	●		●
CICT	●	●	●	●		◖
Nokia Bell	◗			●		◗
ZTE	●	●	●	●	●	●

Table 6.4 Completion status of step 3 trial for NSA [19]

Infrastructure vendors	NSA core network	3.5 GHz			4.9 GHz	R16
		Base station functionalities	Radio frequency	NSA field network		
Huawei	●	●	●	●	●	●
Ericsson	●	●	●	●	◑	
CICT	●	●	●	●	●	
Nokia Bell	●	●	●	◖	●	
ZTE	●	●	●	●	●	◕

6.3.2 5G Deployment Plan

5G NR standardization for NSA and SA had been frozen in December 2017 and June 2018, respectively, and the 5G industry is becoming mature in the aspects of network, chipset, device, etc. Currently, we are at the stage of deployment, both pre-commercial and commercial. Some countries, moreover, have already announced their 5G commercial plan.

Until February 2019, the global mobile suppliers association (GSA) has identified 201 operators, in 83 countries (see Fig. 6.7) that are actively investing in 5G (i.e., that have demonstrated, are testing or trialing, or have been licensed to conduct field trials of 5G technologies, are deploying 5G networks or have announced service launches) [20].

In the first wave of 5G commercial countries including China, the USA, the UK, Japan, and South Korea, they will officially launch 5G commercial networks from 2019 to 2020 as below [21].

- In Europe, multiple countries have allocated 5G spectrum mainly in C-band and released 5G operational licenses among mobile network operators.
  Finland is one of the first countries in Europe to award licenses for the 3.5 GHz spectrum to allow the construction of 5G networks since 2019. The auction of

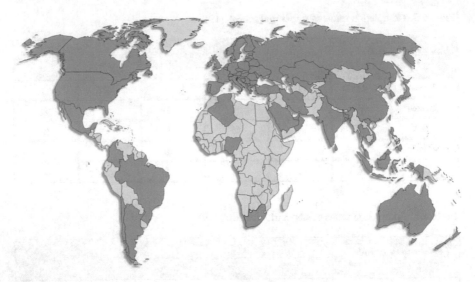

Fig. 6.7 Countries with operators investing in 5G networks (from pre-commitment trials through to pre-commercial and commercial launches) (Data from [20])

spectrum of 3,410–3,800 MHz arranged by the Finnish Communications Regulatory Authority in October 2018, with three major telecommunication operators, Telia Finland, Elisa, and DNA, respectively, won the operating licenses of 3,410–3,540 MHz, 3,540–3,670 MHz, and 3,670–3,800 MHz, with total €77 M paid by the operators.

Germany starts the auction of 2 GHz and 3.6 GHz spectrum in April 2019 with the expectation of over $6.1 Bn, among three existing operators—Deutsche Telekom (80 MHz), Vodafone (90 MHz), and Telefonica Deutschland (70 MHz)—and new entrant 1&1 Drillisch (60 MHz), with all the carriers promised to provide more than 98% high-data rate coverage of German home, highway, and streets.

In Spain, authorities raised about $507 M from selling 3.6–3.8 GHz spectrum among the four mobile carriers, Movistar, Orange, Vodafone, and Masmovil.

Italy's 5G auction in October 2018 includes 75 MHz of spectrum in the 700 MHz band, 200 MHz at 3.6–3.8 GHz, and another 1 GHz in the 26.5–27.5 GHz range, fetching 6.55 billion euros or $7.6 billion, where in C-band Telecom Italia won 80 MHz, Vodafone won 80 MHz, and Iliad and Wind Tre won 20 MHz, respectively.

In Switzerland, the 2019 auction of 700 MHz, 1.4 GHz, 2.6 GHz, and 3.5 GHz spectrum raised 380 million Swiss francs, where in 3.5 GHz band, Swisscom, Sunrise, and Salt got 120 MHz, 100 MHz, and 80 MHz, respectively. Swisscom announced 5G commercial service "inOne Moble" at 3.5 GHz band on April 11, 2019, where the first set of 5G devices include Huawei, Samsung, LG, Oppo.

The latest 5G spectrum auction in March 2019 in Austria fetched €1.88 Bn for 390 MHz of spectrum from 3.4 to 3.8 GHz, Telekom Austria won 100–140 MHz. T-Mobile and 3 Austria won 110 MHz and 100 MHz respectively. In 2018, UK mobilized £1.4 billion from 5G spectrum in 3.4 GHz and 2.3 GHz, among BT-owned EE (40 MHz in 3.4 GHz), Vodafone (50 MHz in 3.4 GHz), Telefonica's O2 (40 MHz in 2.3 GHz), and Hutchison's Three (20 MHz plus previous auctioned 40 MHz in 3.4 GHz). The regulator Ofcom started the auction of low frequency band in 2019.

UK telecom operator BT announced in May 2018 that it would launch 5G commercial services in 2019, actively exploring value-added services such as smart home.

The other countries in Europe are running or plan to run the 5G spectrum auction in 2019 or 2020 as well.

- In China, on December 10, 2018, Ministry of Industry and Information Technology of the People's Republic of China (MIIT) released the 5G trial licenses to China Mobile, China Telecom, and China Unicom, and allocate the 5G trial spectrum of 2.6 GHz, 3.5 GHz, and 4.9 GHz among the three major operators. China Telecom and China Unicom are allocated 100 MHz bandwidth spectrum resources each, i.e., 3.4–3.5 GHz and 3.5–3.6 GHz, respectively. While China Mobile obtains 2,515–2,575 MHz, 2,635–2,675 MHz, and 4,800–4,900 MHz spectrum. If combining with the existing 4G TD-LTE spectrum 2,575–2,635 MHz owned by China mobile, the world biggest carrier has totally 260 MHZ bandwidth available for 5G deployment. In January 2019, MIIT minister announced the plan of releasing the temple 5G commercial licenses in some pilot cities later this year. All the three operators have deployed 5G trials in more than ten cities in 2018 to pilot virtual reality, automatic drive, and other use cases. Furthermore, China Mobile plans to deploy 500k + 5G base stations in 2019–2020; China Telecom announced the eight cities 5G pre-commercialization on 5G joint standalone (SA) and non-standalone (NSA) deployment, and additional nine cities of 5G pilot trial for vertical industries; China Unicom already deployed 5G pre-commercial network in 17 cities in 2018, and plans to provide continuous 5G coverage in seven cities and hotspot 5G coverage in 33 cities in 2019. All three major operators announced their 5G development plans during Mobile World Congress Shanghai in June 2018, and it is clear that they will start 5G pre-commercial deployment in 2019, and release officially scaled commercial 5G deployment in 2020.
- In Japan, operators have also tested typical 5G eMBB use cases, such as virtual reality. According to the 5G plan released by the Ministry of Internal Affairs and Communications (MIC) in 2018, 5G network will be officially commercialized before the Tokyo Olympic Games in 2020. On April 10, 2019, MIC approved the 5G spectrum allocation among four mobile operators NTT Docomo, KDDI, Softbank, and Rakuten, with totally 2.2 GHz 5G spectrum including 600 MHz in C-band and 1.6 GHz in 28 GHz mm-wave spectrum.

The four mobile carriers' plan was approved by Japanese government, which will build 5G wireless networks with investment set to reach JPY 1.6 trillion ($14.4 billion) over the next 5 years, NTT DoCoMo is planning the largest spend with goals to invest at least JPY 795 billion in the deployment of 5G networks during this period. KDDI announced investments of JPY 466 billion, while SoftBank and Rakuten are targeting investments of JPY 206 billion and JPY 194 billion, respectively.

- In South Korea, operators also deployed 5G networks and demonstrated live VR at Pyeongchang Winter Olympics 2018. On June 18, 2018, South Korea has auctioned 280 MHz bandwidth in 3.5 GHz and 2.4 GHz bandwidth in 28 GHz spectrum for 5G network among the three telco carriers, i.e., SKT, KT, and LG-Uplus. SK Telecom and KT each won 100 MHz of the 3.5 GHz spectrum, while LG-Uplus clinched 80 MHz. All three telcos secured 800 MHz of the 2.8 GHz spectrum. In total, the telcos paid 3.6183 trillion won for the spectrum, 340 billion won higher than the starting price of 3.3 trillion won. The Ministry of Science and ICT approved the telcos' usage of the 3.5 GHz spectrum for the next 10 years and 28GHz spectrum for 5 years. In July 2018, the three major operators of South Korea jointly issued a statement that they will launch 5G commercial network in March 2019. On April 3, 2019, SKT released the first 5G network service, which was the first commercial release of 5G service among the world. All of the telcos will likely start service in the capital city of Seoul using 3.5 GHz spectrum, and will then begin 28GHz spectrum use afterwards.

- In the USA, 5G licenses are permitted in the existing 600 MHz, 2.6 GHz and mm-wave frequency bands of 28 GHz and 38 GHz. In addition, Federal Communications Commission (FCC) has done two auctions on 5G spectrum by 2019 Q2. The first millimeter-wave auction is for 28 GHz spectrum since November 2018 to January 2019, with $702 M for 2965 licenses, while the second auction is in April 2019 for 24 GHz with 700 MHz bandwidth, which garnered $1.9 Bn with 2904 licenses. On April 12, 2019, FCC announced the third 5G spectrum auction plan, said as "the largest spectrum auction," in millimeter-wave bands of 37GHz, 39 GHz and 47 GHz with 3.4 GHz bandwidth in total, starting on December 10 this year. In addition, FCC unveiled a plan to provide USD 20.4 billion in the coming decade to connect up to four million rural homes and small businesses to high-speed internet. Verizon, one of America's largest telecommunications companies, announced on April 3 the official operation of its commercial 5G network in two US cities, Chicago and Minneapolis [22], making it the world's second commercial 5G mobile service for customers with 5G-equipped smartphones. The operator Sprint announced it is working with LG Electronics to develop a 5G smartphone compatible with the operator's 2.6 GHz spectrum. The device is scheduled to be available in time for Sprint's launch of 5G services in nine US cities in H1 2019. AT&T has conducted trials to evaluate how 5G can change vertical industries [23] and on April 9, 2019, AT&T mobile 5G service is now live in parts of seven more cities [24].

6.4 Looking Forward

The system that we have elucidated, so far, can be consider the first phase deployment of 5G. If we can learn anything from our experiences in the pass evolution of the commercial cellular system, from 1G to 4G, is that the following:

1. Communications lie at the heart of our human technological development. We are at our core social animals and that communication allows us to move forward as humanity instead of an individual.
2. As we significantly enhance our ability to communicate, the radical shifts in our technological culture are rarely predictable. In the case of 4G, who would have foreseen the development of the smartphone and the impact that it will have on our daily lives.

Although the last few sections have tried to delineate the future use cases for 5G, there is also a significant possibility that we have not even elucidated the use case that will once again completely change our everyday lives. Nevertheless, we may hope that these important use cases will shepherd the industry before the paradigm shift reveals itself. We can, however, surmise from the use cases presented here that although the envelope of the performance of the 5G systems are far greater than that of 4G system, as it should be, not all of its capabilities are needed for every service. This leads us to one of the hallmarks of the future 5G system: hyper-customization. In the future, as we continue to develop 5G beyond the first phase, the network will provide capability to customize the network to the applications. Unlike 4G where the network is one-size-fits-all, the future 5G network will allow for the application to designate a network that is optimized for that particular application. It should also be noted here that the concept of network customization here extends all the way down to the physical layer, the air interface. For example, the 5G IoT air interface may be different than that of the eMBB air interface in future releases of 5G.

Moreover as the services are drastically varied, hyper-customization implies that the future 5G will be a very flexible deployment that can be customized to meet the user's ever-changing needs. One such technology that was discussed extensively was slicing. It not only allows for customization but also the coexistence of customized services in a single system.

This takes us to second hallmark of the future 5G deployment: scalability. Customization and scalability goes hand in hand. As parts of the networks are customized for your service, the network would also need to scale to accommodate the dynamics of the services. Not only should the network scale easily to accommodate the users changing requirements but it should also be able to heal itself. As resources fail, due to, for example, a natural disaster, the network can easily recruit other resources to take over for the fail resources.

The industry is already, in 3GPP, discussing further enhancements to the standard to take a step towards realizing such a future network. The ecosystems are forming and we engineers are already working to change the system again.

References

1. K. Schwab, *The Fourth Industrial Revolution* (World Economic Forum, Geneva, 2016)
2. World Economic Forum, A New Era of Manufacturing in The Fourth Industrial Revolution: $7 Billion Possbibilities Uncovered in Michigan, 2019. [Online], https://www.weforum.org/whitepapers/a-new-era-of-manufacturing-in-the-fourth-industrial-revolution-7-billion-of-possibilities-uncovered-in-michigan
3. Global Data, GSMA Intelligence, [Online], https://www.gsmaintelligence.com/. Accessed 20 Aug 2018
4. M. Simsek, A. Aijaz, M. Dohler, J. Sachs, G. Fettweis, 5G-enabled tactile internet. IEEE Journal on Selected Areas in Communications **34**(3), 460–473 (2016)
5. C. Syverson, *Challenges to mismeasurement explanations for the US productivity slowdown* (Natioal Bureau of Economic Research, Cambridge, 2016)
6. M. Hung, *Leading the IoT* (Gartner Research, Stamford, 2017)
7. K. Campbell, J. Diffley, B. Flanagan, B. Morelli, B. O'Neil, F. Sideco, The 5G Economy: How 5G Technology will Contribute to the Global Economy, IHS Economics and IHS Technologies, 2017. [Online], https://cdn.ihs.com/www/pdf/IHS-Technology-5G-Economic-Impact-Study.pdf
8. Word Economic Forum, Digital Transformation Initiative: Telecommunication, 2017. [Online], http://reports.weforum.org/digital-transformation/wp-content/blogs.dir/94/mp/files/pages/files/dti-telecommunications-industry-white-paper.pdf
9. Ericsson, The 5G Business Potintial, 2017. [Online], http://www.5gamericas.org/files/7114/9971/4226/Ericsson_The_5G_Business_Potential.pdf
10. Huawei, 5G Unlocks a World of Opportunities, 2017. [Online], https://www.huawei.com/us/industry-insights/outlook/mobile-broadband/insights-reports/5g-unlocks-a-world-of-opportunities
11. R. Vannithamby, A.C. Soong, *5G Verticals: Customising Applications, Technologies and Deployment Techniques* (Wiley, New York, in press)
12. 3GPP, TS 22.104: Service Requirements for Cyber-Physical Control Applications in Vertical Domains, [Online], https://portal.3gpp.org/desktopmodules/Specifications/SpecificationDetails.aspx?specificationId=3528
13. 5G-ACIA, 5G for Connected Industries and Automation, Second Edition, 2019. [Online], https://www.5g-acia.org/index.php?id=5125
14. 3GPP, TR 22-804 v16.2.0 Study on Communication for Automation in Vertical Domains (CAV), 2018. [Online], https://portal.3gpp.org/desktopmodules/Specifications/SpecificationDetails.aspx?specificationId=3187
15. 3GPP, TS 22.185 Service Requirements for V2X Services V15.0.0, 2018. [Online], https://portal.3gpp.org/desktopmodules/Specifications/SpecificationDetails.aspx?specificationId=2989
16. 5GAA, Cellular V2X Conclusions based on Evaluation of Available Architectural Opations, 2019. [Online], http://5gaa.org/wp-content/uploads/2019/02/5GAA_White_Paper_on_C-V2X_Conclusions_based_on_Evaluation_of_Available_Architectural_Options.pdf
17. IMT-2020 (5G) Promotion Group, 5G Wireless Technologies Test Progress and Follow Up Plan, 2016. [Online], http://www.imt-2020.cn/zh/documents/1?currentPage=2&content=
18. IMT-2020(5G) Promotion Group, 5G Technologies Step 2 Trial Progress and Follow Up Plan, 2017
19. IMT-2020(5G) Promotion Group, Summary of 5G Technologies Step 3 Trial, 2019
20. Global Mobile Suppliers Association, Global Progress to 5G-Trials, Deployments and Launches, 2019
21. GTI, Sub-6GHz 5G Pre-Commercial Trial White Paper, 2019

22. Verizon, Customers in Chicago and Minneapolis are First in the World to Get 5G-Enabled Smartphones Connected to a 5G Network, [Online], https://www.verizon.com/about/news/customers-chicago-and-minneapolis-are-first-world-get-5g-enabled-smartphones-connected-5g. Accessed 9 Apr 2019

23. AT&T, 5G's Promise, [Online], https://about.att.com/pages/5G. Accessed 9 Apr 2019

24. AT&T, AT&T is the First to Offer Mobile 5G in 7 More U.S. Cities, [Online], https://about.att.com/story/2019/mobile_5g.html. Accessed 9 Apr 2019

Printed in the United States
By Bookmasters